Taste, Olfaction, and the Central Nervous System

CARL PFAFFMANN

Taste, Olfaction, and the Central Nervous System

A Festschrift in Honor of Carl Pfaffmann

Edited by Donald W. Pfaff

The Rockefeller University Press
New York, 1985

Library of Congress Catalogue Card Number 84–043054
ISBN 0–87470–039–6
Printed in the United States of America

Contents

(*continued*)

Neurophysiological Coding of Olfactory Information

Olfactory Determination of Social Behaviors

Preface.
A Career in the Chemical Senses

Born in Brooklyn, Carl Pfaffmann attended New York City public schools, was graduated from Brown University in 1933, and received a second bachelor's degree from Oxford in 1937 as a Rhodes Scholar. Working with Lord Adrian, who won the Nobel Prize for pioneering the application of electrophysiological methods to the nervous system, Carl received his Ph.D. from Cambridge University with a thesis that defined electrophysiological response profiles of primary taste nerves. There he also met Louise, now his wife of more than 40 years.

The main results of Carl's thesis, published in the *Journal of Cellular and Comparative Physiology* in 1941, constituted a classic demonstration of cross-fiber pattern coding in the nervous system. Primary taste nerve axons rarely responded with absolute specificity to an individual gustatory stimulus. Identification of the stimulus in an *n*-dimensional taste space would have to take account of the pattern of electrical discharges across a set of responding axons. As a Professor at Brown University, after military service during World War II, Carl extended these experiments and published the results in the *Journal of Neurophysiology* in 1955. The data that led to the purest statement of the cross-fiber pattern theory were published by Marion Frank and Carl Pfaffmann in *Science* in 1969. Their analysis was geared to show a randomness in the distribution of responses *across* the four primary taste categories. Modern summaries of the requirements for patterns of activity across nerve fibers, as well as the usefulness of more narrowly tuned taste receptive fibers for the direction of particular behavioral responses, are included here in the chapters by Marion Frank, Robert Erickson, and David Smith. In Carl's *Psychological Review* article in 1960, the opportunity of physiological work in the chemical senses became clear: that electrical action potentials could be correlated not only with behavioral reports of sensations but also with the affective states characteristic of these sensory modalities.

During those years Carl was not ignoring olfaction. The earliest such exper-

---

Donald W. Pfaff
The Rockefeller University, New York, New York 10021

iments from his lab at Brown were done by Max Mozell and published in 1954. Carl picked up this interest again when, in experiments done in the 1960s by John Scott, it became clear that a practical way to study input from the chemical senses to the lateral hypothalamus would be to use odor stimuli and electrical stimulation of the olfactory bulb. This led naturally to studies on potential mammalian pheromones, in conjunction with hormonal state (*Brain Research*, 1969), as well as to gas chromatographic separations of hamster sex pheromones (*Science*, 1976). One of Carl's early experiments, on taste nerve electrical activity in adrenalectomized compared with normal rats, opened the subject of possible hormonal effects on primary sensory input, a field of work that remains of considerable interest.

The greater part of Carl Pfaffmann's scientific career has been at Brown University followed by his tenure as Professor and Vice President at The Rockefeller University. Throughout these years, however, he has been a popular guest lecturer noted, in speaking at other universities for example, for his enthusiasm for the experiments of the moment and, at meetings of larger scope, for his keen appreciation of the international nature of science in his field. His role at the Rockefeller was to extend a tradition of excellent neurophysiological research set by the likes of Herbert Gasser, David Lloyd, H. Keffer Hartline, Floyd Ratliff, and Victor Wilson, to include solid physiological approaches to the mechanisms of behavior. On the national scene, through his participation in review groups, presidencies of professional organizations, and his roles in the National Academy of Science, Carl Pfaffmann has been a major figure in shaping the young field of behavioral neuroscience during its "pre-Newtonian" era.

Another index of leadership has been that, in the study of taste and olfaction, no scientist in these decades has been more productive of students who have continued to participate in these fields. Many of them have published with him and some are represented in this volume. All of them would remember Carl's joy in experimenting, based in a solid empiricism, and his fertile intuition, tempered by the sternness of his appearance as he looked over the top of his glasses.

In an age of overwhelming attention by the communications media to the accomplishments and shortcomings of scientists, when even scientific authorities have been questioned, Carl Pfaffmann's career stands as a model of integrity. In as difficult a field of work as that which links neural mechanisms with behavior, it is a rare scientist who can combine a long-standing enthusiasm for primary experimental observations with such a remarkable track record for reliability.

Publications

Pfaffmann, C. 1935. An experimental comparison of the method of single stimuli and the method of constant stimuli in gustation. *Am. J. Psychol.* 48:470–476.

———. 1935. Apparatus and technique for gustatory experimentation. *J. Gen. Psychol.* 12:446–447.

Pfaffmann, C., and H. H. Jasper. 1935. Sensory discharges in cutaneous nerve fibers following chemical stimulation. *Psychol. Bull.* 32:565–566.

Pfaffmann, C. 1936. Differential responses of the new-born cat to gustatory stimuli. *J. Genet. Psychol.* 49:61:–67.

Pfaffmann, C., and H. Schlosberg. 1936. The conditioned knee jerk in psychotic and normal individuals. *J. Psychol.* 1:201–206.

Pfaffmann, C. 1938. Action potentials elicited by gustatory stimulation. *Psychol. Bull.* 35:718 (abstr.).

———. 1939. Afferent impulses from the teeth. *J. Physiol. (Lond.).* 95(19):41–67.

———. 1939. Specific gustatory impulses. *J. Physiol. (Lond.).* 96(8):13–25.

———. 1939. Afferent impulses from the teeth due to pressure and noxious stimulation. *J. Physiol. (Lond.).* 97:207–219.

———. 1939. Afferent impulses from the teeth resulting from a vibratory stimulus. *J. Physiol. (Lond.).* 97:220–232.

———. 1940. Potentials in the isolated medullated axon. *J. Cell. Comp. Physiol.* 16:1–4.

———. 1941. Gustatory afferent impulses. *J. Cell. Comp. Physiol.* 17:243–258.

———. 1942. Review of *Experimental Foundations of General Psychology* by W. L. Valentine, *Am. J. Psychol.* 55:140–141.

———. 1947. A multiple cubicle-bench for the student laboratory. *Am. Psychol.* 2:559–560.

Pfaffmann, C., and R. B. Aird. 1947. Pressure stimulation of peripheral nerves. *Proc. Soc. Exp. Biol. Med.* 66:130–132.

Pfaffmann, C. 1948. Aircraft landings without binocular cues: a study based upon observations made in flight. *Am. J. Psychol.* 61:323–334.

———. 1948. Salt preferences following denervation of the tongue. *Am. Psychol.* 3:254 (abstr.).

———. 1948. Studying the senses of taste and smell. In *Methods of Psychology.* T. G. Andrews, ed. New York: John Wiley & Sons, Inc., pp. 268–288.

———. 1948. Taste and smell. In *Foundations of Psychology.* E. G. Boring, H. S. Langfeld, and H. P. Weld, eds. New York: John Wiley & Sons, Inc., pp. 351–359.

———. 1949. The effect of thirst on salt preference. *Am. Psychol.* 4:224 (abstr.).

———. 1950. The afferent neural determinants of the specific hunger for salt. *Am. Psychol.* 5:272 (abstr.).

Pfaffmann, C., and J. K. Bare. 1950. Gustatory nerve discharges in normal and adrenalectomized rats. *J. Comp. Physiol. Psychol.* 43:320–324.

Pfaffmann, C. 1951. A further study of taste preferences following lingual denervation. *Am. Psychol.* 6:273 (abstr.).
———. 1951. Taste: a monitor of diet? *Res. Rev.* 12:16–21.
———. 1951. Gustatory intensity, its peripheral, behavioral and psychological aspects. In *Proceedings of the XIIIth International Congress of Psychology*, pp. 196–197.
———. 1951. Somesthesis and the chemical senses. *Annu. Rev. Psychol.* 2:79–94.
———. 1951. Taste and smell. In *Handbook of Experimental Psychology*. S. S. Stevens, ed. New York: John Wiley & Sons, Inc., pp. 1143–1171.
Manly, R. S., C. Pfaffmann, D. D. Lathrop, and J. Keyser. 1952. Oral sensory thresholds of persons with natural and artificial dentitions. *J. Dent. Res.* 31:305–312.
Pfaffmann, C. 1952. Review of *Beginning of Experimental Psychology* by S. H. Bartley. *Psychol. Bull.* 49:81–82.
———. 1952. Taste preference and aversion following lingual denervation. *J. Comp. Physiol. Psychol.* 45:393–400.
Pfaffmann, C., and R. M. Benjamin. 1952. Cerebral mechanisms in taste discriminations. *Am. Psychol.* 7:251 (abstr.).
Pfaffmann, C. 1953. Species difference in taste sensitivity. *Science (Wash. DC).* 117:470 (abstr.).
Giddon, D. B., M. E. Dreisbach, C. Pfaffmann, and R. S. Manly. 1954. Relative abilities of natural and artificial dentition patients for judging the sweetness of solid foods. *J. Prosthet. Dent.* 4:263–268.
Mozell, M. M., and C. Pfaffmann. 1954. The afferent neural processes in odor perception. *Ann. NY Acad. Sci.* 58:96–108.
Pfaffmann, C. Sensory mechanisms in taste discriminations. Paper read at the Annual Meeting of American Chemical Society, New York 1954.
———. 1954. Variables affecting difference tests. In *Food Acceptance Testing Methodology*. Washington, DC: National Research Council, Advisory Board on Quartermaster Research, pp. 4–17.
Pfaffmann, C., P. T. Young, V. G. Dethier, C. P. Richter, and E. Stellar. 1954. The preparation of solutions for research in chemoreception and food acceptance. *J. Comp. Physiol. Psychol.* 47:93–96.
Schlosberg, H., C. Pfaffmann, J. Cornsweet, and R. Pierrel. 1954. Selection and training of panels. In *Food Acceptance Testing Methodology*. Washington, DC: National Research Council, Advisory Board on Quartermaster Research, pp. 45–54.
Benjamin, R. M., and C. Pfaffmann. 1955. Cortical localization of taste in the albino rat. *J. Neurophysiol. (Bethesda).* 18:56–64.
Pfaffmann, C. 1955. Gustatory nerve impulses in rat, cat and rabbit. *J. Neurophysiol. (Bethesda).* 18:429–440.
Pfaffmann. C., and E. C. Hagstrom. 1955. Factors influencing taste sensitivity to sugar. *Am. J. Physiol.* 183:651 (abstr.).
Pfaffmann, C. 1956. Taste and smell. *Annu. Rev. Psychol.* 7:391–408.
———. 1956. Training the undergraduate—for science or for life? Review of *Introduction to Psychology* by C. T. Morgan. *Contemp. Psychol.* 1:195–196.

Pfaffmann, C., and H. Schlosberg. 1956. The identification and criticism of ideas: a new approach to the introductory course in psychology. *Am. Psychol.* 11:78–83.
Pfaffmann, C. 1957. Taste mechanisms in preference behavior. *Am. J. Clin. Nutr.* 5:142–147.
Pfaffmann, C., and T. Engen. 1958. Psychology: absolute judgments of odor. *AMA (Am. Med. Assoc.) Arch. Ind. Health.* 17:544–545 (abstr.).
Pfaffmann, C., W. R. Goff, and J. K. Bare. 1958. An olfactometer for the rat. *Science (Wash. DC).* 128:1007–1008.
Warren, R. P., and C. Pfaffmann. 1958. Early experience and taste aversion. *J. Comp. Physiol. Psychol.* 52:263–266.
Church, R. M., and C. Pfaffmann. 1959. A respondent-conditioning apparatus for the student laboratory. *Am. J. Psychol.* 72:267–270.
Engen, T., and C. Pfaffmann. 1959. Absolute judgments of odor intensity. *J. Exp. Psychol.* 58:23–26.
Hagstrom, E. C., and C. Pfaffmann. 1959. The relative taste effectiveness of different sugars for the rat. *J. Comp. Physiol. Psychol.* 52:259–262.
Pfaffmann, C. 1959. The afferent code for sensory quality. *Am. Psychol.* 14:226–232.
———. 1959. The sense of taste. *Handb. Physiol.* I: (sect. 1)507–533.
Pfaffmann, C., B. P. Halpern, and R. P. Erickson. 1959. Gustatory afferent discharges in the medulla. *Fed. Proc.* 18:120 (abstr.).
Warren, R. M., and C. Pfaffmann. 1959. Suppression of sweet sensitivity by potassium gymnemate. *J. Appl. Psychol.* 14:40–42.
Engen, T., and C. Pfaffmann. 1960. Absolute judgments of odor quality. *J. Exp. Psychol.* 59:214–219.
Pfaffmann. C. 1960. Sense, chemical. In *McGraw-Hill Encyclopedia of Science and Technology*, vol. 12. New York: McGraw-Hill, Inc., p. 179.
———. 1960. Sensory and motivating properties of taste stimuli. *Science (Wash. DC).* 131:1321.
———. 1960. The pleasures of sensation. *Psychol. Rev.* 67:253–268.
———. 1960. Smell. In *McGraw-Hill Encyclopedia of Science and Technology*, vol. 12. New York: McGraw-Hill, Inc., pp. 384–386.
———. 1960. Taste. In *McGraw-Hill Encyclopedia of Science and Technology*, vol. 13. New York: McGraw-Hill, Inc., pp. 399–401.
Erickson, R. P., R. L. King, and C. Pfaffmann. 1961. Some characteristics of transmission through spinal trigeminal nucleus of rat. *J. Neurophysiol. (Bethesda).* 24:621–632.
Pfaffmann, C. 1961. Sensory processes, affective response, and reinforcement. *Acta Psychol.* 19:356 (abstr.).
———. 1961. Smell and taste. In *Encyclopaedia Britannica*, vol. 20. Chicago: Encyclopaedia Britannica Inc., pp. 819–823.
———. 1961. The sensory and motivating properties of the sense of taste. In *Nebraska Symposium on Motivation: 1961.* M. R. Jones, ed. Lincoln: University of Nebraska Press, pp. 71–108.

Pfaffmann, C., R. P. Erickson, G. P. Frommer, and B. P. Halpern. 1961. Gustatory discharges in the rat medulla and thalamus. In *Sensory Communication*. W. A. Rosenblith, ed. New York: John Wiley & Sons, Inc., pp. 455–473.

Carr, W. J., B. Solberg, and C. Pfaffmann. 1962. The olfactory threshold for estrous female urine in normal and castrated male rats. *J. Comp. Physiol. Psychol.* 55:415–417.

Miller, N., C. Pfaffmann, and H. Schlosberg. 1962. Aspects of psychology and psychophysiology in the U.S.S.R. In *Some Views on Soviet Psychology*. R. Bauer, ed. Washington, DC: American Psychological Association, pp. 189–252.

Oakley, B., and C. Pfaffmann. 1962. Electrophysiologically monitored lesions in the gustatory thalamic relay of the albino rat. *J. Comp. Physiol. Psychol.* 55:155–160.

Pfaffmann, C. 1962. Psychology self-appraised in Canada. Review of *Training for Research in Psychology*, K. S. Bernhardt, ed. *Contemp. Psychol.* 7:140–142.

———. 1962. Sensory processes and their relation to behavior. In *Psychology: A Study of a Science*, vol. 4. *Biologically Oriented Fields: Their Place in Psychology and in Biological Sciences*. S. Koch, ed. New York: McGraw-Hill, Inc., pp. 380–416.

Pfaffmann, C., and D. H. McBurney. 1962. Gustatory adaptation to saliva and NaCl. *Excerpta Med. Int. Congr. Ser.* 48:1344 (abstr.).

Distinguished Scientific Contribution Awards. 1963. To Roger G. Barker, George A. Miller, and Carl Pfaffmann. *Am. Psychol.* 18:801–811.

Makous, W., S. Nord, B. Oakley, and C. Pfaffmann. 1963. The gustatory relay in the medulla. In *Olfaction and Taste* (Proceedings of the First International Symposium). Y. Zotterman, ed. Oxford: Pergamon Press Ltd., pp. 381–393.

McBurney, D. H., and C. Pfaffman. 1963. Gustatory adaptation to saliva and sodium chloride. *J. Exp. Psychol.* 65:523–529.

Nachman, M., and C. Pfaffmann. 1963. Gustatory nerve discharge in normal and sodium-deficient rats. *J. Comp. Physiol. Psychol.* 56:1007–1011.

Pfaffmann, C. 1963. Taste stimulation and preference behavior. In *Olfaction and Taste* (Proceedings of the First International Symposium). Y. Zotterman, ed. Oxford: Pergamon Press Ltd., pp. 257–273.

Bartoshuk, L. M., D. H. McBurney, and C. Pfaffmann. 1964. Taste of sodium chloride solutions after adaptation to sodium chloride: implications for the "water taste." *Science (Wash. DC)*. 143:967–968.

Pfaffmann, C. 1964. Discussion: on the code for gustatory sensory quality. *Excerpta Med. Int. Congr. Ser.* 49:273–279.

———. 1964. Mechanisms of gustatory adaptation. *Acta Psychol.* 23:117–118 (abstr.).

———. 1964. Taste, its sensory and motivating properties. *Am. Sci.* 52:187–206.

Pfaffmann, C., and J. B. Powers. 1964. Partial adaptation of taste. *Psychon. Sci.* 1:41–42.

Bartoshuk, L. M., and C. Pfaffmann. 1965. Effects of pre-treatment on the water taste response in cat and rat. *Fed. Proc.* 24(2):731–732 (abstr.).

Fisher, G. L., C. Pfaffmann, and E. Brown. 1965. Dulcin and saccharin taste in squirrel monkeys, rats, and men. *Science (Wash. DC).* 150(3695):506–507.
Pfaffmann, C. 1965. Behavioral sciences. In *Basic Research and National Goals, A Report to the Committee on Science and Astronautics* (U.S. House of Representatives). Washington, DC: National Academy of Sciences, pp. 203–236.
———. 1965. De gustibus. *Am. Psychol.* 20:21–33.
———. 1965. Harold Schlosberg (1904–1964). *Am. J. Psychol.* 78:148–152.
———. 1965. L'Adaptation gustative. In *Extrait des Acualités Neurophysiologiques*, sixième série. Masson and Cie, Éditeurs, Paris, pp. 85–97.
———. 1965. The modern study of behavior. In *The Future of Biology* (Symposium, November 26–27, 1965). New York: The Rockefeller University and State University of New York, pp. 42–48.
———. 1966. Taste and smell. In *McGraw-Hill Encyclopedia of Science and Technology*, first ed. New York: McGraw-Hill, Inc., pp. 384–386, 399–401.
Pfaff, D. W., J. Scott, and C. Pfaffmann. 1967. Olfactory input to the medial forebrain bundle of the rat. *Psychon. Bull.* 1:22 (abstr.).
Pfaffmann, C. 1967. Statement before the Subcommittee on Government Research, Committee on Government Operations, United States Senate. *Am. Psychol.* 22(3): 205–210.
Pfaffmann, C., G. Fisher, and M. Frank. 1967. The sensory and behavioral factors in taste preference. In *Olfaction and Taste II* (Proceedings of the Second International Symposium). T. Hayashi, ed. Oxford: Pergamon Press Ltd., pp. 361–381.
Scott, J., and C. Pfaffmann. 1967. Olfactory input to the hypothalamus: electrophysiological evidence. *Science (Wash. DC).* 158:1592–1594.
Frank, M., and C. Pfaffmann. 1969. Taste nerve fibers: a random distribution of sensitivities to four tastes. *Science (Wash. DC).* 164:1183–1185.
Frank, M., and C. Pfaffmann. 1969. The distribution of taste sensitivities among single taste fibers. In *Olfaction and Taste III* (Proceedings of the Third International Symposium). C. Pfaffmann, ed. New York: The Rockefeller University Press, pp. 488–491.
Pfaff, D. W., and C. Pfaffmann. 1969. Behavioral and electrophysiological responses of male rats to female rat urine odors. In *Olfaction and Taste III* (Proceedings of the Third International Symposium). C. Pfaffmann, ed. New York: The Rockefeller University Press, pp. 258–267.
Pfaff, D. W., and C. Pfaffmann. 1969. Olfactory and hormonal influences on the basal forebrain of the male rat. *Brain Res.* 15:137–156.
Pfaffmann, C., ed. 1969. *Olfaction and Taste III* (Proceedings of the Third International Symposium). New York: The Rockefeller University Press.
———. 1969. Summary of olfaction roundtable. In *Olfaction and Taste III* (Proceedings of the Third International Symposium). C. Pfaffmann, ed. New York: The Rockefeller University Press, pp. 226–231.
———. 1969. Summary of taste roundtable. In *Olfaction and Taste III* (Proceedings of

the Third International Symposium). C. Pfaffmann, ed. New York: The Rockefeller University Press, pp. 527–532.

———. 1969. Taste preference and reinforcement. In *Reinforcement and Behavior.* J. Tapp, ed. New York: Academic Press, Inc., pp. 215–240.

———. 1970. Physiological and behavioural processes of the sense of taste. In *CIBA Foundation Symposium on Taste and Smell in Vertebrates.* G. E. Wolstenholme and J. Knight, eds. London: J. & A. Churchill, Ltd., pp. 31–50.

———. 1970. The behavioral science model. *Am. Psychol.* 25(5):437–441.

Bujas, Z., and C. Pfaffmann. 1971. Potassium gymnemate and the sweet and bitter taste provoked electrically. *Percept. Psychophys.* 10(1):28–29.

Edinger, H. M., and C. Pfaffmann. 1971. Single unit activity during drinking. *Fed. Proc.* 30(2):376 (abstr.).

Pfaffmann, C. 1971. Biological basis of oral function. Review of *Second Symposium on Oral Sensation and Perception,* J. F. Bosma, ed. *Contemp. Psychol.* 16(8):515–516.

———. 1971. Sense, chemical. In *McGraw-Hill Encyclopedia of Science and Technology,* 3rd ed. New York: McGraw-Hill, Inc., p. 227.

———. 1971. Sensory reception of olfactory cues. *Biol. Reprod.* 4:327–343.

———. 1971. Smell. In *McGraw-Hill Encyclopedia of Science and Technology,* 3rd ed. New York: McGraw-Hill, Inc., pp. 435–437.

———. 1971. Taste (human). In *McGraw-Hill Encyclopedia of Science and Technology,* 3rd ed. New York: McGraw-Hill, Inc., pp. 422–424.

Pfaffmann, C., L. Bartoshuk, and D. H. McBurney. 1971. Taste psychophysics. In *Handbook of Sensory Physiology,* vol. IV. *Chemical Senses,* pt. 2. L. M. Beidler, ed. Berlin: Springer-Verlag, pp. 75–101.

Smith, D. V., M. Frank, and C. Pfaffmann. 1971. Cross-adaptation between salts in the chorda tympani nerve of the rat. *Psychon. Sci.* 21(5):281 (abstr.). (Paper read at the 11th Annual Meeting of the Psychonomic Society, November 1970.)

Frank, M., D. V. Smith, and C. Pfaffmann. 1972. Cross-adaptation between salts in the rat's chorda tympani response. In *Oral Physiology.* N. Emmelin and Y. Zotterman, eds. Oxford: Pergamon Press, Ltd., pp. 227–237.

Pfaffmann, C. 1972. Clarence Henry Graham (Biographical Memoir). In *American Philosophical Society Year Book 1972.* G. W. Corner, ed. American Philosophy Society, pp. 183–189.

———. 1972. Recent advances in the study of olfaction. *Electroencephalogr. Clin. Neurophysiol. Suppl.* [*Recent Contributions to Neurophysiology* (International Symposium in Neurosciences in honor of Herbert H. Jasper)]. 31:185–203.

Scott, J. W., and C. Pfaffmann. 1972. Characteristics of responses of lateral hypothalamic neurons to stimulation of the olfactory system. *Brain Res.* 48:251–264.

Smith, D. V., M. Frank, and C. Pfaffmann. 1972. Cross adaptation among salts in the rat. In *Olfaction and Taste IV* (Proceedings of the Fourth International Symposium). D. Schneider, ed. Stuttgart: Wissenschaftliche Verlagsgesellschaft MBH, pp. 259–265.

Pfaffmann, C. 1973. Some comments on sensory stimuli, motivation, and hedonism. (NRP Work Session on Neural Control of Motivated Behavior, October 4–6, 1970, E. Stellar, chmn.). *Neurosci. Res. Program Bull.* 11:387–389.

———. 1973. Some current activities in sensory physiology. Review of *Contributions to Sensory Physiology*, vol. 5, W. D. Neff, ed. *Contemp. Psychol.* 18(7):313–314.

———. 1973. Stanley Smith Stevens (1906–1973) (Biographical Memoir). In *American Philosophical Society Year Book 1973*. G. W. Corner, ed. American Philosophy Society, pp. 159–165.

———. 1973. The comparative approach to physiological psychology. *Ann. NY Acad. Sci.* 223:57–64.

———. 1974. A story which has no end. Review of *A Man with a Shattered World: The History of a Brain Wound* by A. R. Luria. *Contemp. Psychol.* 19(4):303–305.

———. 1974. De gustibus (Mark II). In *The Psychologists*, vol. 2. T. S. Krawiec, ed. New York: Oxford University Press, Inc., pp. 403–439.

———. 1974. Sensory reception, human. In *Encyclopaedia Britannica*, 15th ed. Chicago: Encyclopaedia Britannica, Inc., pp. 547–555.

———. 1974. Specificity of the sweet receptors of the squirrel monkey. *Chemical Senses and Flavor*, vol. 1, no. 1. EPA Symposium, Washington, DC, May 1973, pp. 61–67.

———. 1974. The sensory coding of taste quality. *Chemical Senses and Flavor*, vol. 1, no. 1. EPA Symposium, Washington, DC, May 1973, pp. 5–8.

Norgren, R., and C. Pfaffmann. 1975. The pontine taste area in the rat. *Brain Res.* 91:99–117.

Pfaffmann, C. 1975. Introduction. In *Nutrition and Mental Functions*. G. Serban, ed. New York: Plenum Publishing Corp., pp. 9–11.

———. 1975. Introduction (Day II). In *Sweeteners: Issues and Uncertainties* (Academy Forum). Washington, DC: National Academy of Sciences, pp. 121–123; summary by the co-chairmen, pp. 234–237.

———. 1975. Phylogenetic emergence of salt appetite (Discussion and chairman's comments). In *Olfaction and Taste V* (Proceedings of the Fifth International Symposium, University of Melbourne, October 1974). D. A. Denton and J. P. Coghlan, eds. New York: Academic Press, Inc., pp. 277–280.

———. 1975. Phylogenetic origins of sweet sensitivity. In *Olfaction and Taste V* (Proceedings of the Fifth International Symposium, University of Melbourne, October 1974). D. A. Denton and J. P. Coghlan, eds. New York: Academic Press, Inc., pp. 3–10.

Pfaffmann, C., M. Frank, L. M. Bartoshuk, and T. C. Snell. 1976. Coding gustatory information in the squirrel monkey chorda tympani. *Progr. Psychobiol. Physiol. Psychol.* 6:1–27.

Singer, A. G., W. C. Agosta, R. J. O'Connell, C. Pfaffmann, D. V. Bowen, and F. H. Field. 1976. Dimethyl disulfide: an attractant pheromone in hamster vaginal secretion. *Science (Wash. DC)*. 191(4230):948–950.

Norgren, R., H. J. Grill, and C. Pfaffmann. 1977. CNS projections of taste to the dorsal pons and limbic system with correlated studies of behavior. In *Food Intake and Chemical Senses* (Proceedings of the First International Symposium on Food Intake and Chemical Senses, Fukuoka, Japan, September 1–4, 1976). Y. Katsuki, M. Sato, S. F. Takagi, and Y. Oomura, eds. Tokyo: University of Tokyo Press, pp. 233–243.

Pfaffmann, C. 1977. Biological and behavioral substrates of the sweet tooth. In *Taste and Development: The Genesis of Sweet Preference.* J. M. Weiffenbach, ed. Bethesda, MD: National Institute of Dental Research, DHEW Publication no. (NIH) 77–1068, U.S. Department of Health, Education and Welfare, National Institutes of Health, pp. 3–24.

———. 1977. Introduction. In *Drinking Behavior: Oral Stimulation, Reinforcement, and Preference.* J. A. W. M. Weijnen and J. Mendelson, eds. New York: Plenum Publishing Corp., pp. xvii–xxx.

———. 1977. Introduction to the behavioral aspects of flavor measurement. In *Flavor: Its Chemical, Behavioral, and Commercial Aspects* (Proceedings of the Arthur D. Little, Inc., Flavor Symposium, Cambridge, MA, April 28–29, 1977). C. M. Apt, ed. Boulder, CO: Westview Press, pp. 79–87.

———. 1977. Sensory, central, and behavioral mechanisms of the sense of taste (abstract of lecture). In *Proceedings of the International Union of Physiological Sciences*, vol. XII (Paris, July 18–23, 1977). Paris: Actes du Congrès, pp. 88–89.

Pfaffmann, C., V. G. Dethier, and D. M. Hegsted. 1977. Concluding comments. Chapter 24. *The Chemical Senses and Nutrition.* M. Kare and O. Maller, eds. New York: Academic Press, Inc., pp. 463–475.

Pfaffmann, C., R. Norgren, and H. J. Grill. 1977. Sensory affect and motivation. *Ann. NY Acad. Sci.* 290:18–34.

O'Connell, R. J., A. G. Singer, F. Macrides, C. Pfaffmann, and W. C. Agosta. 1978. Responses of the male golden hamster to mixtures of odorants identified from vaginal discharge. *Behav. Biol.* 24:244–255.

Pfaffmann, C. 1978. Biological and behavioral substrates of the sweet tooth. In *Proceedings of Symposium on Diet, Nutrition and Dental Caries* (University of Michigan, November 17–18, 1978). Ann Arbor: The University of Michigan Press, pp. 39–64.

———. 1978. Introduction (Symposium: Sense of Taste and Nutrition). *Am. J. Clin. Nutr.* 31:1057.

———. 1978. Neurophysiological mechanisms of taste. *Am. J. Clin. Nutr.* 31:1058–1067.

———. 1978. The vertebrate phylogeny, neural code, and integrative processes of taste. In *Handbook of Perception*, vol. VIA. *Tasting and Smelling.* E. C. Carterette and M. P. Friedman, eds. New York: Academic Press, Inc., pp. 51–123.

O'Connell, R. J., A. G. Singer, C. Pfaffmann, and W. C. Agosta. 1979. Pheromones of hamster vaginal discharge—attraction to femtogram amounts of dimethyl disulfide and to mixtures of volatile components. *J. Chem. Ecol.* 5:575–585.

Bujas, Z., M. Frank, and C. Pfaffmann. 1979. Neural effects of electrical taste stimuli. *Sensory Processes.* 3:353–365.
Pfaffmann, C. 1979. Leonard Carmichael. In *International Encyclopedia of the Social Sciences*, biographical suppl. 18. New York: The Free Press, pp. 101–104.
Pfaffmann, C., M. Frank, and R. Norgren. 1979. Neural mechanisms and behavioral aspects of taste. *Annu. Rev. Psychol.* 30:283–325.
Nowlis, G., M. E. Frank, and C. Pfaffmann. 1980. Specificity of acquired aversions to taste qualities in hamsters and rats. *J. Comp. Physiol. Psychol.* 94:932–942.
Pfaffmann, C. 1980. Leonard Carmichael. In *National Academy of Sciences Biographical Memoir*, vol. 54. Washington, DC: National Academy of Sciences, pp. 25–47.
———. 1980. The sense of taste and behavior. In *Physiology, Past, Present and Future: A Symposium in Honour of Yngve Zotterman.* D. Anderson, ed. Oxford: Pergamon Press, Ltd., pp. 109–128.
———. 1980. Wundt's schema of sensory affect in the light of research on gustatory preferences. *Psychol. Res.* (Wundt Centennial Issue). 42:165–174.
Pfaffmann, C., and T. Pritchard. 1980. Ion specificity of "electric taste." In *Olfaction and Taste VII* (Proceedings of the Seventh International Symposium). H. van der Starre, ed. London: IRL Press, pp. 175–178.
Pfaffmann, C. 1982. Taste: a model of incentive motivation. In *The Physiological Mechanisms of Motivation.* D. W. Pfaff, ed. New York: Springer-Verlag, pp. 61–97.

I.
De Gustibus: Præteriti, Præsentis, Futuri

Introduction

The chemical senses taste and smell have often been referred to as the lower senses, implying thereby a less rich, less discriminatory cognitive function but a closer and important relationship to the basic biological functions and needs of the organism. One feature of the current scene is the growing concern with the biological role of the chemical senses and their relationship to metabolic, nutritional, behavioral, and, now, clinical considerations.

This is not to say that the biological aspects were neglected in the history of studying these senses. In addition to proposing that the basic taste qualities consisted of sweet, bitter, sour, salty, astringent, pungent, and harsh, Aristotle noted that the function of taste is nutritive (Aristotle, 330 B.C.). Taste directed the choice of foods and beverages: pleasant tasting substances being consumed, unpleasant ones being rejected.

In the first century B.C., Lucretius's didactic poem "De Rerum Natura" (Lucretius, 55 B.C.) noted: "Moreover the liquids honey and milk excite a pleasant sensation of tongue when held in the mouth; but on the other hand the nauseous nature of wormwood and of harsh centaury writhes the mouth with a noisome flavour; so that you may easily see that the things which are able to affect the senses pleasantly, consist of smooth and round elements; while all those on the other hand which are found to be bitter and harsh, are held in connexion by particles that are more hooked and for this reason are wont to tear open passages into our senses." Galen (A.D. 180–200) added little to the question of taste perception per se, citing the same seven taste

---

This text is a composite edited version of the Fourth Givaudan Lecture delivered at the 1982 meeting of the Association for Chemoreception Sciences, Sarasota, Florida, and the Inaugural Lecture delivered at the 1982 dedication of the new Clinical Smell and Taste Research Center at The University of Pennsylvania Hospital, Philadelphia, Pennsylvania.

Carl Pfaffmann
The Rockefeller University, New York, New York 10021

qualities given by Aristotle, but he did note that an abnormal appearance of the tongue as well as certain disorders of taste could be symptoms of pathology.

There is a long history of concern with hedonic properties of the chemical senses, especially the sense of smell, in health and disease. Foul smelling air was thought to cause disease due to the property "miasma," while pleasant odors were used to protect the healthy from disease. During the outbreak of the plague, the use of aromatic herbs to ward off disease reached its zenith, eau de Cologne being a survival of one of these plague waters or essences (Haggard, 1929). Thus the ancients and medieval and early modern physicians and philosophers (some close to the mark, some wide of the mark) noted the connection between health and nutrition and these sense organs, as well as their role as organs of conscious experience.

With the maturing of experimental sciences in the nineteenth century came major advances in the physiology of sensation. The first professorship in physiology (University of Berlin, 1833) went to Johannes Müller, father of the doctrine of "specific energies of nerves"—that the quality of the perceived sensation resides not in the stimulus but in the nerve stimulated (Müller, 1830). Pressure against the skin is felt as touch, pressure against the eyeball is perceived as light. Concerning the sense of taste, Sir Charles Bell wrote: "Of these, the papillae of one kind form the seat of the sense of taste; the other papillae (more numerous and smaller) resemble the extremities of the nerves in the common skin, and are the organs of touch in the tongue. When I take a sharp steel point and touch one of these papillae, I feel the sharpness. The sense of touch informs me of the shape of the instrument. When I touch a papilla of taste, I have no sensation similar to the former. I do not know that a point touched the tongue, but I am sensible of a metallic taste, and the sensation passes backward on the tongue" (Bell, 1811).

Early classification of taste sensations in the eighteenth century had tended toward multiplicity. Linnaeus enumerated eleven categories: sweet, acid, astringent, sharp, viscous, fatty, bitter, insipid, aqueous, saline, and nauseous. Other equally long lists appeared and each nearly always included salt, sour, bitter, and sweet. Ultimately such other terms as sharp, harsh, pungent, and aromatic were attributed to combinations of pain, touch, or odor. Alkaline and metallic remained among the basic tastes but by 1893, Wundt in his *Grundzüge der Physiologische Psychologie* (1893), called alkaline and metallic doubtful. Thus, by fiat, he reduced the basic list to only four: salt, sour, sweet, and bitter (Boring, 1942).

An important finding in support of specificity within the subqualities of taste was the report that masticating the leaves of an Indian herb, *Gymnema sylvestre*, completely abolished the perception of sweetness (Falconer, 1847). Shore (1892) reported this effect to be more intense for sweet, less for bitter, slight for salt, and nil for acid. More recent studies with gymnemic acid have indicated that the effect is largely upon sweetness, the effect on bitterness being due to adaptation of bitterness by other components in the crude extract. The case of gymnema even today remains one of the strongest instances of a differential blockade of a specific taste quality.

Öhrwall and Kiesow painstakingly mapped the sensitivity of individual fungiform papillae, which appear as vascularized bright pink spots on the top of the tongue (Öhrwall, 1891; Kiesow, 1894). They showed that the basic taste qualities might occur in isolation with punctate stimulation. Thus, some taste papillae react to a single taste, others to two or three, and still others to all four basic qualities. Öhrwall argued that not only were there specifically different receptors for the different basic qualities, but these were, in fact, different modalities (Öhrwall, 1901). Furious academic debate raged between Öhrwall and Kiesow on the nature of the basic tastes, Kiesow believing that the basic tastes were different qualities within a single modality (Kiesow, 1896).

Until the advent of the electronic age, with the exception of anatomical studies all senses were studied by the indirect methods of psychophysics, that is, the precise physical and chemical control of different kinds of stimulation to the several organs of sense (eye, ear, skin, tongue, and nose) to which the human observer responded with reports of detection, quality identification, scaling of magnitude of sensation, etc. My first paper, published in 1935 in the *American Journal of Psychology* (Pfaffmann, 1935) while I was still a graduate student, studied a new psychophysical method for determining taste difference thresholds in humans. However, for my Ph.D. thesis, I applied the new electrophysiological method to recording from the chorda tympani, the main sensory nerve for taste on the anterior tongue of mammals (Pfaffmann, 1939). I had the good fortune of being able to do so in the Cambridge laboratory of Lord Adrian, one of the founding fathers of this method.

When I retired as a vice president at The Rockefeller University in 1979 I had more time for research at the bench. One subject with some obvious loose ends, and upon which no one else was working, or so I thought at the time, was "electric taste." To be sure, the coding and other problems have loose ends but enough of my own students and my students' students are actively engaged on such projects, so I thought it best to let them thrash it out among themselves.

"Electric Taste"

The history of electric taste is as old as the study of electricity itself. In 1792 Volta rediscovered Sulzer's early observation (1752) that dissimilar metals when touching each other and the tongue produced a taste sensation. Volta, however, correctly ascribed the phenomenon to the "electric fluid" flowing through the tissues from one metal to the other. Volta also reported that the positive pole tasted sour but the negative pole alkaline and bitter. Breaking the anodal current caused no new tastes, but only a discontinuation of the sourness, whereas the break of cathodal current produced a very brief sourness coupled with sweet taste. In the intervening years before the electrophysiological era, many investigators confirmed and extended these observations on human subjects. Bujas (1971) gives an excellent review of this early work.

In my 1939 recording of taste afferent impulses in the cat, I also applied polarizing electrodes to the tongue's receptive field of taste units responsive to acid. Figure 1 shows the response of such a unit to acetic acid. Figure 2 shows that the response to acid can be duplicated by anodal stimulation but with a shorter latency. I did little more with this for a number of years until Zoran Bujas came to Rockefeller as an Academy Exchange Professor over 10 years ago. He, Marion Frank, and I studied electric taste in rats electrophysiologically, although it took us until 1979 to finally publish that work (Bujas et al., 1979). There are many parallels between human psychophysical results and the rat chorda tympani response. Anodal currents have the lowest threshold, "cathode-off" only slightly higher, and the "cathode-on" considerably higher. We found that units responsive to acid or salt were activated by both the anode-on and the termination of cathodal current. Our small sample of only one or two sugar units responded to cathode-on.

I would like to report on my present experiments, which were ably aided by Thomas Pritchard who, as it happens, is an academic great-grandson with a

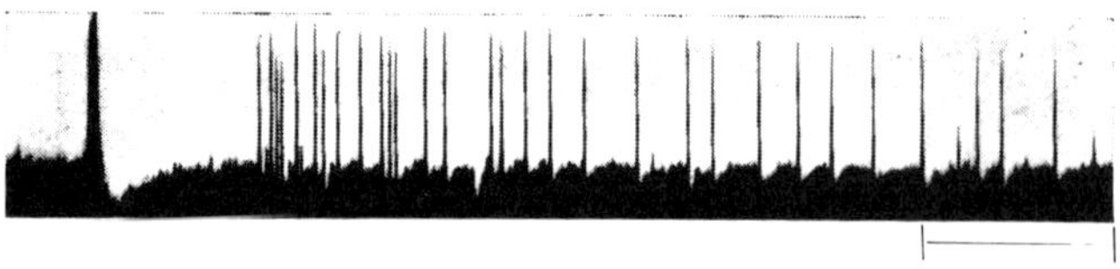

Figure 1. The response of a single taste fiber of the cat to 0.05 N acetic acid. The large baseline deflection at the beginning of the record signals application of the stimulus solution. The spike discharges begin after a latency of about 0.1 s. Time line, 0.1 s. Reproduced with permission from *J. Physiol. (Lond.)*. 1939: 96:41 P.

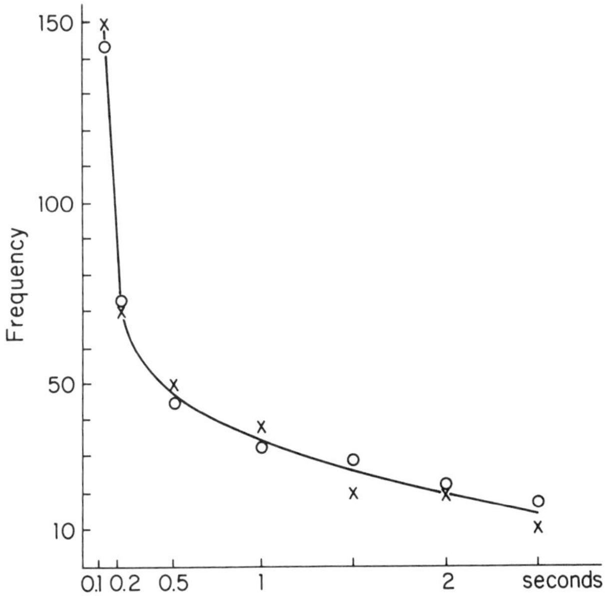

Figure 2. The equivalence of the frequencies of discharge to acid (o) and to 0.2 mA(x) of anodal polarization. Redrawn from Pfaffmann, 1941.

Ph.D. under Thomas Scott, who in turn took his degree with my student Robert Erickson. However, no sooner had I begun studying electric taste than I discovered a number of other people had also become interested. Zoran Bujas had of course continued his psychophysical studies at Zagreb, and in the mid-1960s von Bekesy aroused considerable interest with his studies of sweetness aroused by electrical stimulation (von Békésy, 1964). Najad did his Ph.D. thesis on the subject 20 years ago with Lloyd Beidler. Smith and Bealer (1975) had used anodal currents with various rise-time parameters, with indications that the anode was equivalent to chemical stimulation. The proceedings of the 1980 Japanese Taste and Smell Symposium included a number of papers on electric taste. In 1981 Kashiwayanagi and colleagues at Hokkaido University published a study entitled "Similar effects of various modifications of gustatory receptors on neural responses to chemical and electrical stimulation in the frog" (Kashiwayanagi et al., 1981). Ninomiya, Funakoshi, and colleagues at the Seventh International Symposium in Olfaction and Taste (ISOT VII, 1980) and in recent issues of the *Japanese Journal of Physiology* (Ninomiya and Funakoshi, 1981a; Ninomiya and Funakoshi, 1981b) have studied elec-

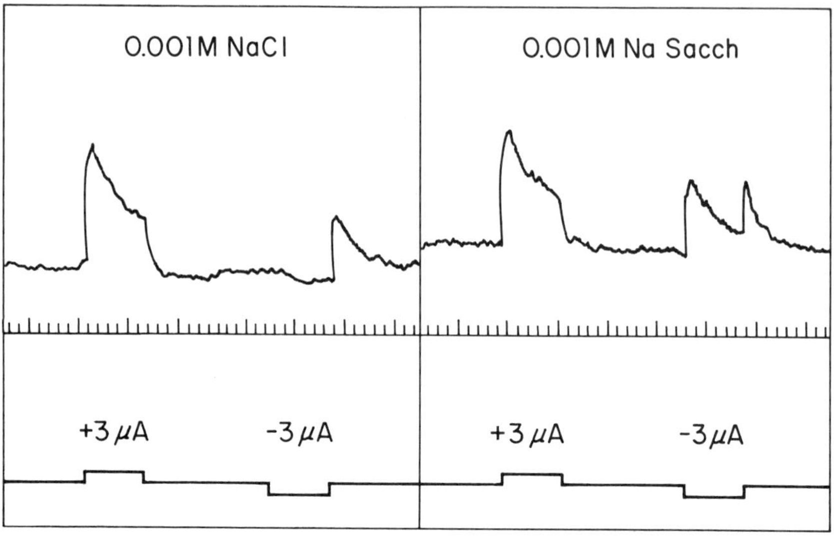

Figure 3. Integrated records of hamster chorda tympani responses to +3 and −3 square wave pulses of 5-s duration. Time marker indicates seconds. In left panel current was passed through a 0.001 M NaCl adapting solution. Note the decrementing excitation to the anode and the drop in baseline (inhibition of neural activity) to cathodal polarization. Current delivered via 0.001 M Na saccharin solution caused excitation at both anode- and cathode-on (right panel). There is a brief transient response at cathode-off in both panels.

trical stimulation in rat chorda tympani and single fibers therein and the effect of afferent ions. Scott Herness has recently completed a Ph.D. thesis on the biophysics of electric taste in Lloyd Beidler's lab. So the plot has thickened!

In our recent experiments, we fitted our tongue chemical flow chamber with a side arm and porous glass window containing agar gel–saline solution plus silver/silver-chlorided wire. An indifferent, nonpolarizable electrode makes contact with a buttonhole incision under the chin. We chose the hamster rather than the rat because of the former's greater sugar sensitivity. By using sodium saccharin rather than sodium chloride as the bridging electrolyte, I thought to optimize stimulation of the sweet receptors. This permits iontophoresis either of the sodium cation toward the tongue surface by anodal polarization or saccharin anion by the cathode, using a Grass electronic stimulator and isolation unit. Figure 3 shows the basic responses of the whole nerve preparation. I should emphasize that such stimulation does not activate tri-

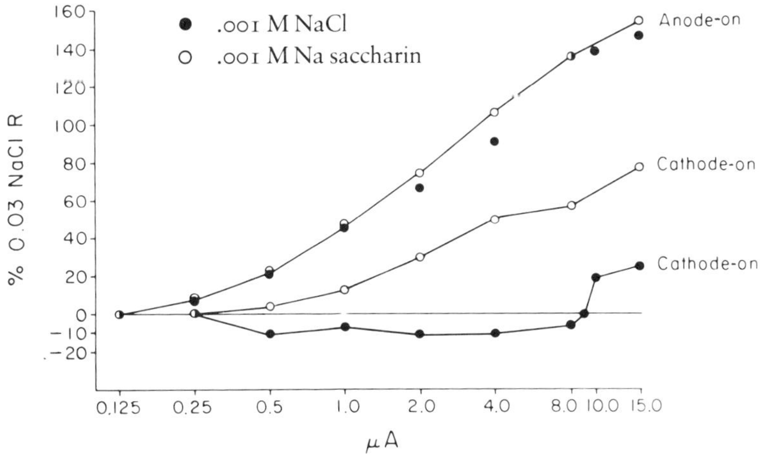

Figure 4. Magnitudes of initial integrated deflection at anode-on and cathode-on via 0.001 M NaCl or 0.001 M Na saccharin. Response magnitudes are shown relative to the neural responses to a solution of 0.03 M NaCl showing that electrical polarization excites responses comparable to those caused by chemical solutions. Reproduced with permission from *Olfaction and Taste VII*, 1980. (London: IRL Press, p. 177).

geminal tongue receptors until much higher current values are used. Numerous control studies including Pritchard's unit studies with single trigeminal lingual afferents and Herness' selective lingual or chordal de-afferentations attest to this point.

Figure 4 shows the quantitative relations between current intensity and response magnitude. I will neglect the cathode-off response for the remainder of this chapter. The anodal response is essentially the same for sodium chloride and sodium saccharin. The cathodal responses, however, are very different. The saccharin anion produces a response at a relatively low current value at which the NaCl cathodal current is inhibitory. The current values shown here can be converted to microamperes per square centimeter for the hamster tongue by dividing all the abscissa values by 3. Note, however, that the cathode-on via NaCl solution does in fact stimulate at a higher threshold value. All the curves reach plateaus, the anode at the highest level, the cathode-on with saccharin next, and soldium chloride close to but not quite up to the saccharin level. When I originally presented these data at ISOT VII at Nordwijkerhout, The Netherlands, I hypothesized that such electrical effects were due to iontophoresis (Pfaffmann and Pritchard, 1980). Therefore, we should expect the character of the cations during anodal polarization to reflect their relative differences in stimulating ability as chemical stimuli.

Solutions of NaCl (with no polarization) are more effective taste stimuli than equimolar solutions of KCl for the hamster's chorda tympani receptors. Equal currents delivered via 0.001 M NaCl similarly produce larger chorda tympani responses than via 0.001 KCl, showing a parallel between anodal and chemical stimulation. It would be interesting to see how K+ compares with Na+ and other cations in DeSimone, Heck, and DeSimone's transepithelial potential measurement (DeSimone et al., 1981).

Hamster sugar receptors were shown by Ogawa et al. (1969) to be relatively specific, the overlap in sensitivity residing primarily among the electrolyte salt, acid, and quinine units. It was my hypothesis that the whole nerve chorda response functions for anodal stimulation could be attributed primarily to salt and/or electrolyte-sensitive units; the cathodal response, to sugar-sensitive units. As a further test that sugar receptors were involved to cathodal stimula-

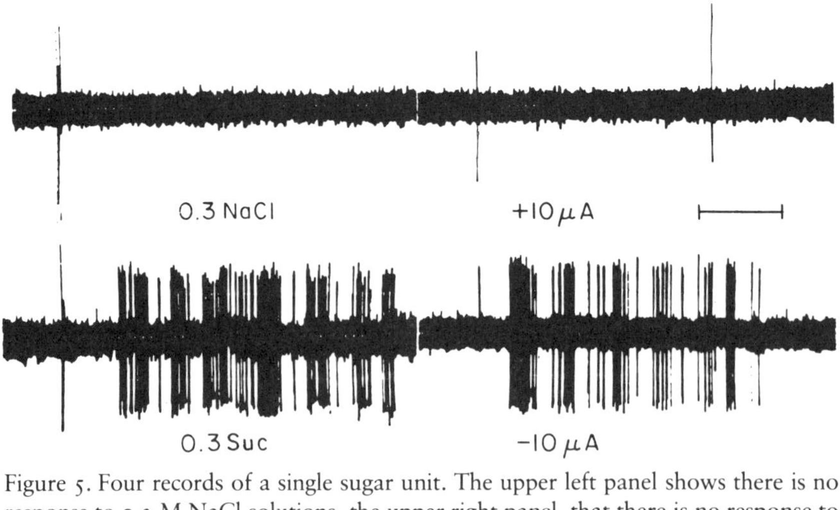

Figure 5. Four records of a single sugar unit. The upper left panel shows there is no response to 0.3 M NaCl solutions, the upper right panel, that there is no response to +10 μA polarization. The lower left record shows a response to a solution of 0.3 M sucrose, the lower right panel, a response of the same unit to −10 μA via 0.001 M sodium saccharin. The latency of response to sucrose is ~ 680 ms, to −10 μA via saccharin solution, ~ 350 ms. This unit did not respond to solutions of NaCl, HCl, or quinine. Bar, 1s. Reproduced with permission from *Olfaction and Taste VII* (London: 1980, IRL Press, p. 178).

tion, I applied a crude decoction of a sweetness-blocking agent, gymnema, via the flow chamber. This diffierentially blocked the cathodal but not the anodal response, thus implicating sweet receptors.

Pritchard, while in our lab, obtained data with regard to the question of ion-afferent nerve specificity. Chorda tympani units that respond to sodium chloride on the tongue also respond to the anode but not to the cathode except at cathode-off (as Adachi previously reported in 1969). The single units that respond to sugar on the tongue but not to salt, respond to cathodal polarization of the tongue via Na saccharin, but do not respond to the anode as shown in Figure 5. Some units respond to NaCl and to sucrose on the tongue, and to both polarities of current.

Latency measurements of salt-sensitive unit activity to the anode fall within the range of 20–40 ms, as compared with a much longer latency, > 300 ms for the cathode-on of a sugar-sensitive unit with saccharin. A similar difference in latency occurs with chemical solutions, sodium chloride (fast) vs. sugar solutions (slow) for salt-sensitive and sugar/saccharin-sensitive units, respectively. Dual-sensitive units that respond to both NaCl and to sucrose show these same latency differences, fast for NaCl, slow for sucrose or saccharin. Thus the latency of response depends not on the afferent nerve fiber being driven thereby but on the receptor site per se being stimulated either by the chemical alone or by polarization. Kelling and Halpern (1983) recently described measurement of human reaction time for the sweetness of 2 mM sodium saccharin, ranging from 800 to 1,150 ms compared with a reaction time of 475–500 ms for saltiness of a 500 mM NaCl solution. Such profound latency differences both in electrophysiology and in psychophysics, previously noted by others but not especially remarked upon, suggest that the transducer mechanisms for salt vs. sweetness receptor sites are fundamentally different.

There is much more to the electric taste story than can be included in this chapter; rather than being a quaint historical oddity, electric taste used analytically may be an important probe of the mechanisms of taste transduction. In addition, electric taste (not nerve but taste-bud stimulation) with a small agar/Na-saccharin-filled glass miniprobe is useful for stimulating a restricted number of the taste receptors in the oral cavity or in determining the receptive field of a gustatory neuron. By a "flip of the switch," we can test a unit's receptor specificity (e.g., salt-acid vs. sugar-saccharin sensitivity). Other cations and anions in the conducting electrolyte might tap other sensitivities. In addition, "electric taste" provides a kind of "gustatory click" which Ralph Norgren, Susan Travers, and I are using in studying evoked CNS responses. (Yamamoto et al. reported doing this in 1980 but only with the anode.)

Finally, electric taste is used clinically in "electrogustometry" to measure the area and intensity of reduced or aberrant taste sensitivity of the patient's tongue. The devices most commonly used consist of metallic (often stainless steel) electrodes of one polarity with a complex sour-metallic taste. We expect that fluid electrodes of the appropriate polarity with selected cations and anions could elicit sweet, bitter, sour, or salty qualities.

Behavior and the Maltose Paradox

Work as early as 1859 by Kussmaul and 1882 by Genzmer and others permitted Luciani's 1917 *Textbook of Physiology* to affirm: "Agreeable and disagreeable substances excite different expressional movements of the facial muscles; indifferent and insipid substances produce no facial movements, or at most arouse an expression of indifference or slight disgust . . . by means of these expressional reactions it is possible even in babies of a few months old and in many animals to distinguish clearly between the sensations aroused by different tastes in the mouth. A sweet taste always gives them a pleasurable sensation, even when it is in excess. Other substances, on the contrary, give a disagreeable sensation in concentrated solutions, or are indifferent if dilute. In the first case the reaction is a movement of sucking or licking; in the second there are efforts at repulsion and evidences of displeasure or disgust." Jacob Steiner's photographs of 1972 convincingly demonstrate such gustofacial responses in normal as well as anencephalic infants (Steiner, 1973).

Sherrington (1917) reported that decerebrate cats reject food soaked either in quinine, acetic acid, or sodium chloride, whereas dextrose-soaked food was not rejected; Macht in Bard's lab (1951) showed the same for chronic decerebrates. Well known by now is the work of Harvey Grill and Ralph Norgren (1978a and 1978b) in our lab at Rockefeller, on the orofacial responses in normal and chronic decerebrate rats when taste solutions are injected directly into the oral cavity. Quinine produces the most dramatic aversive response, including whole body movements (see chapter 10). However, the distinction among salt, sugar, and acid paradoxically is not as obvious as one might have expected from the preference aversion data but subtler measurements of longer duration do discriminate among these three.

Since, speaking loosely, the hindbrain can tell good from bad tastes, one might ask what the more cephalad parts of the brain have to do with processing taste information. The inability of the hindbrain to acquire a conditioned taste aversion indicates one such head-end capacity. In conditioned taste aversions, the cortex may also participate in recognizing the novel from the familiar. Some cortical electrophysiology hints at different spatial loci for different tastes in the exclusively gustatory cortex, but cortical unit recording on a par with that for vision, audition, or somesthesis, in the primate taste areas outlined by Burton and Benjamin (1971), is sorely needed. Does the cortical taste area have a columnar arrangement? Does taste have another type of organization better suited to its status as a special visceral sense and its CNS relation to the general visceral afferent and limbic confluences?

Let us turn from speculations about the CNS to certain behavioral consequences of taste stimulation. Taste provides a good model of incentive motivation. Preferred substances like sugar for most animals (except the cat) and salt for rats and certain other species (but not hamsters) in a Richter two-bottle 24-hr preference test show a preference-aversion function that starts at some threshold level, increases to a peak intake value, and then turns down with reduced preference and ultimately aversion. In some cases these biphasic functions can be attributed to purely oral factors, but in many cases there is also a

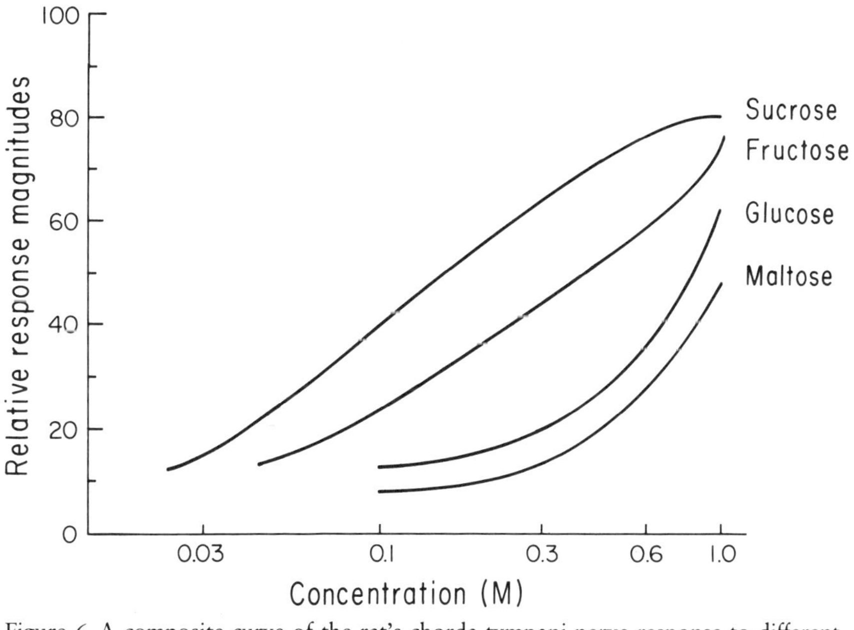

Figure 6. A composite curve of the rat's chorda tympani nerve response to different sugars. Reproduced with permission from *Physiological Mechanisms of Motivation*, 1982. New York: Springer-Verlag, p. 79.

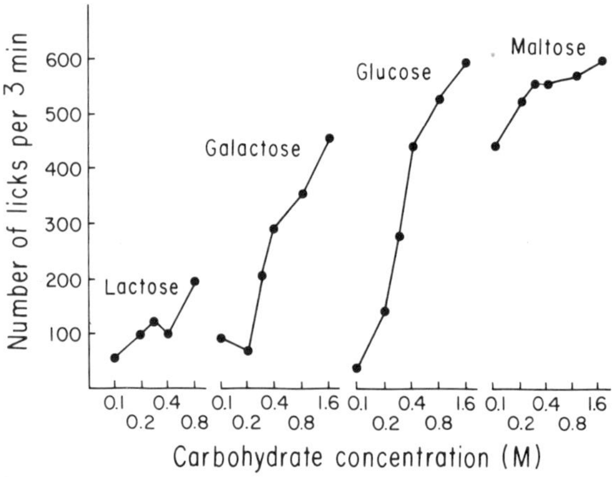

Figure 7. The mean number of lick response elicited by four different sugars at different concentrations. Rats were presented with daily 3-min tests to each of the sugars in ascending order of concentrations. Reproduced with permission from *Physiological Mechanisms of Motivation*, 1982. New York: Springer-Verlag, p. 81.

gastric and/or nutritional consequence of ingestion. There are various stratagems to eliminate this postingestive effect in order to reveal gustatory influences, per se. One is to use a brief exposure method to limit intake, as P. T. Young did in some of his early experiments (Young, 1955) and Cagan and Maller did later (1974). To humans, sucrose, fructose, glucose, lactose, and maltose differ, in that order, in degree of sweetness, sucrose being the sweetest. Even in the rat, with its lesser sucrose response, the efficacy of sugars can be seriated by the relative mangitudes of the whole nerve chorda responses, i.e., sucrose > fructose > glucose > maltose (see Figure 6 and Pfaffmann, 1982). Data by Jack Davis and his colleagues measuring lick rate at very brief exposures is shown in Figure 7 (Davis, 1973; Davis et al., 1975). Different sugars show an increasing effectiveness with concentrations but maltose is almost as effective as sucrose and surpasses other sugars in eliciting licking. Why is maltose so effective behaviorally when it is relatively weak as a chorda tympani taste stimulus for the rat?

So far I have dealt only with the quantitative degree of sensory efficacy as the major variable, but the taste of different sugars may vary as well in quality. Sucrose is largely sweet to humans, but glucose is reported to be sweet as well as mildly sour or tart, lactose sweet as well as salty, etc. (Cameron, 1947). I am also reminded that in the electrophysiological study of the dog, Andersson and colleagues (1950) found an order of effectiveness similar to what I have described, but they also remarked on the occurrence of a few single units that were especially reactive to maltose, more so than to the other sugars. In conditioned taste aversion tests in both hamsters and rats, these common sugars, including the D-amino acids, generalize from each other to sucrose. However, we did not specifically include maltose until quite recently. Nowlis and Frank (1981) found that although an aversion to sucrose produces some aversion to maltose, an aversion conditioned to maltose *does not* generalize to sucrose. Maltose may share some sensory qualities with sucrose but it may also have some other distinctive qualities as well. In the gerbil also, Jakinovich (1982a) found greater acceptance of maltose than sucrose or other sugars.

Jakinovich (1982b) has also shown electrophysiological evidence for separate or at least dissociable sites for sucrose and for saccharin in the gerbil. Shimada and colleagues (1974) clearly demonstated two independent glucose (pyranose) and fructose (furanose) molecular sites in chemoreceptors of the flesh fly. Psychophysical work on humans (Faurion et al., 1980a) has shown evidence suggesting different molecular receptor sites for sweetness. There is a clustering of sensitivities within which thresholds and sensory magnitude judgments tend to co-vary, but there are also other groups among the stimuli for which there is a lesser co-variation. Faurion et al. (1980b) have in fact shown a similar set of relations in their method of antidromic recording from microelectrodes inserted into one taste papilla when a nearby circumscribed area of the tongue is stimulated with sugars, amino acids, artificial sweeteners, etc. Quite a bit earlier McBurney (1969) showed evidence for two different bitter sites in cross-adaptation studies. Phenylthiocarbamide (PTC) sensitivity in relation to other bitter sensitivities could be interpreted this way (Hall et al., 1975). So it is entirely possible that a subquality as well as intensity of sweetness may be coded in the afferent nerve input. Although there may be multimolecular sites within one qualitative domain, they do share the final common path of sweetness, which unites the submolecular components within a common or closely related qualitative domain.

Maltose, as such, is not abundant in nature, occurring only occasionally (White et al., 1964). In apple, for example, 69% of the total solids are made up of glucose, fructose, and sucrose, but there is only a trace of maltose (Shallenberger, 1974). Maltose is a disaccharide of two glucose molecules joined with a glycosidic linkage. It can be readily hydrolyzed to glucose by the enzyme maltase distributed throughout the small intestine's mucosa. Glucose, in a sense, is the major physiological form of sugar in the internal systemic environment, where it is most readily absorbed and metabolized. The greater taste effectiveness of sucrose, the disaccharide of fructose and glucose, may appear to be a biological anomaly. Taste, however, does provide the chemical contact with the external world where sucrose occurs widely in a variety of fruits, seeds, leaves, flowers, and roots, especially sugar cane and beets, the main commercial sources. Thus there appears to be functional specialization of taste receptors for those forms of sugar most likely to be encountered in the environment. The external world and internal world, with their different chemical orders of merit, come together in the oral cavity. Does maltose composed of two glucose units with a ready dissociation by enzymes provide a unique bridge between these two orders of merit?

Quality Coding

The single taste units from the chorda tympani of the cat recorded at the Cambridge Physiology Lab showed three main types of fiber (Pfaffmann, 1941). One responded only to HCl or acetic acid, a second to sodium chloride and to acid, and a third to quinine and acid. I found sugar to be quite ineffective, as other investigators have also since reported. Because sodium chloride and quinine each stimulate a different group of receptors, I suggested that discrimination between these two substances could be coded by impulses in two quite different sets of nerve fibers reaching the CNS in a classic labeled-line manner. But whereas impulses in one set of fibers might indicate saltiness, discharges in the same fibers combined with activity in all the remaining gustatory fibers, it might be presumed, could indicate a sour taste. Sensory quality then would not depend simply on the "all or nothing" activation of a particular fiber group alone but on the pattern of other fibers concurrently active. This was the initial statement and origin of the pattern theory of taste quality coding. Apparent support for the pattern idea appeared in our subsequent work on rats and other species. Erickson (1963) was the first to develop the cross-fiber cross-correlation measure as a quantitative statement of degree of overlapping sensitivities. He and his students subsequently went on to develop more sophisticated concepts of cross-fiber coding.

Let me just suppose that I had not started with the cat preparation nor the chorda tympani preparation but had recorded single units in the glossopharyngeal nerve of the rat. Figure 8 is a summarization of just this preparation used by Marion Frank while still a graduate student in my lab at Brown University. These data (published in 1981) document the responsivities of units in vallate and foliate papillae separately in the order they were obtained, arranged only by their best stimulus classes to the basic four taste stimuli

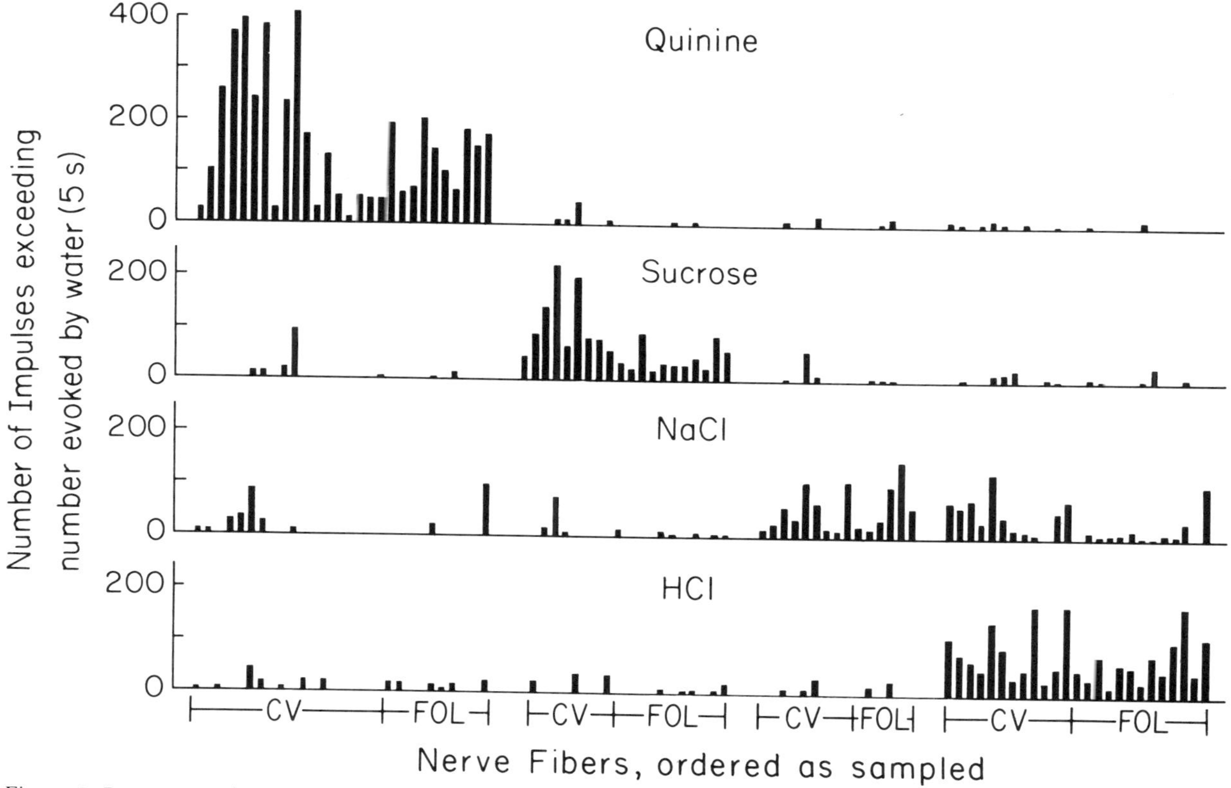

Figure 8. Responses of single nerve fibers of the rat to quinine, sucrose, NaCl, and HCl. Responses of a single neuron to the four stimuli are plotted in vertical columns, groups of quinine-best, sucrose-best, NaCl-best, and HCl-best units are plotted progressively from left to right divided into innervating circumvallate (CV) or foliate (FOL) papillae. Within each category, the fibers are ordered as they were sampled. Reproduced with permission from Nowlis and Frank, *Perception of Behavioral Chemicals*, 1981. New York: Elsevier-North Holland, Inc., pp. 58–80).

(Nowlis and Frank, 1981). In this species and in this nerve with the concentrations used, not only do we see a response to sucrose but also a significant response to quinine at a relatively lower concentration (0.001 M) than that required to elicit a chorda tympani response. There is only slight overlap of sucrose and quinine units with the other units. The N-best units do not respond much to the other three stimuli but the H-best show both acid and salt sensitivity (my acid salt units of the cat?). What might I have concluded if this had been the first taste single unit data to surface? This looks rather more like a degenerate basic taste formulation. With the accumulation of many more units and many more stimuli in more species, the multiple sensitivity of taste units has been established. The unit data from the macaque chorda tympani of Sato et al. (1975) indicate that all four basic tastes are well represented with clear overlap in acid and salt sensitivity but not with sugar and quinine. The question of multiple sensitivity of many taste units as recorded is not at issue so much as the characterization of that multiple sensitivity in clusters or modes of sensitivity for the basic stimuli in the chorda tympani nerve.

Where do I stand on pattern coding vs. clusters of sensitivity and even labeled-line concepts? I see evidence of peripheral specificity and also grouping in clusters thereof. Sophisticated statistical analysis cast doubts in some as to the validity of clusters and how discrimination of a large variety of different taste compounds can be discriminated thereby. It has happened that in theories concerning other senses previously divergent positions on one coding system versus another have been settled. Is it possible that the final resolution in taste may also turn out to be an amalgam of positions once thought to be irreconcilable? Are specificity and cross-fiber patterning necessarily antithetical?

I would like to introduce some concepts for which I am indebted to my Rockefeller colleague, Gerald Edelman, a molecular biologist who has become intrigued by neuroscience. In his studies of the molecular biology of the immune system, for which he was awarded the Nobel Prize in 1972, Edelman described how the immune system can recognize and distinguish positively among different objects by means of a molecular complementarity between the shapes of antigens and combining sites of antibodies in a lock and key relationship, as proposed by Paul Ehrlich. Subsequent to Ehrlich, the foreign body was thought to give instructions to the immune system in order to form the matching lock and key combination. Edelman, on the contrary, characterizes the immune system as a selective one. All the information required to recognize foreign antigens as well as self-antigens is present before the first encounter with these antigens. An encounter with a particular antigen serves to elicit or select responses by those antibody-producing cells that happen to have antibodies with sites complementary to the shape of the antigen (Edelman, 1975). The lymphocyte bearing an antibody on its surface is stimulated to mature and divide, producing a clone of progeny cells, each capable of synthesizing more antibodies of the same kind (Figure 9). The immune system has a repertoire provided by a large collection of antibody molecules each with a different combining site, and sufficient diversity above a certain threshold amount to deal with molecules that could have been encountered in evolution or have never been encountered, as in the case of new man-made molecules. A heterogeneous collection of antibodies exists to even a single well-defined antigen, so that there are several more or less effective ways in which recognition between antigen and antibody can take place. Edelman (1978) refers to this as degeneracy, i.e., there must be more than one way of satisfactorily recognizing a given input signal. The need for degeneracy is perhaps most easily seen by assuming extreme cases; one with no degeneracy, the other with complete degeneracy, as seen in Figure 10. In the upper figure, there is only one

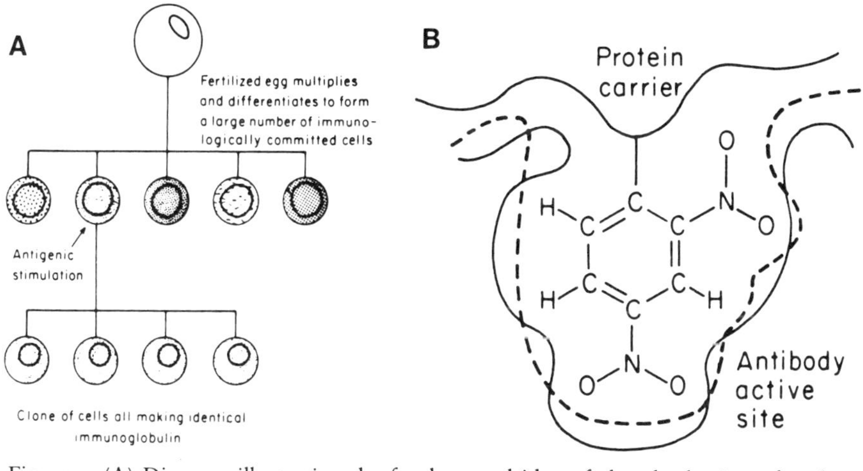

Figure 9. (A) Diagram illustrating the fundamental idea of clonal selection after development of a repertoire of antigen-binding cells carrying different antibody molecules on their surfaces. Encounter by an antigen with the appropriate complementary surface antibody molecule results in cell maturation, mitosis, and increased production of antibodies of the same type by the daughter cells. (B) Diagram illustrating the principle of molecular complementarity by means of which an antibody molecule can bind an antigen of a given shape. Reproduced with permission from Edelman, *Neurosciences: Paths of Discovery*, 1975. Cambridge, MA: MIT Press, p. 66.

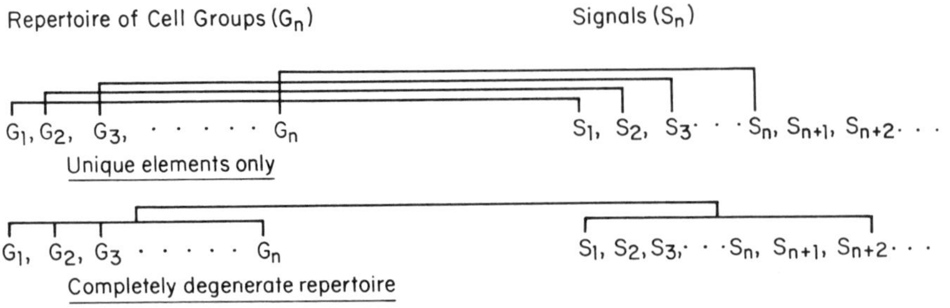

Figure 10. Two extreme cases of repertoires having unique (nondegenerate) and completely degenerate elements. In the upper example, extension of the range of signals to be recognized (for example, beyond S_n) leads to a frequent failure of recognition. In the lower example, there is a loss of specificity and a frequent failure to distinguish different signals since each G can respond to all signals. Reproduced with permission from Edelman, *The Mindful Brain*, 1978. Cambridge, MA: MIT Press, p. 58.

cell group capable of recognizing the signal. With this mode, the system that is capable of recognizing previously unencountered signals would fail in range, and many inputs would go unrecognized. In the other extreme, every element in the repertoire will match any input signal. In this case the range requirement would be satisfied but there would be severe loss of specificity and the capacity to distinguish between two different but closely related signal patterns. Rather, the system must be so constituted as to fall between these two extremes so that there will be several and possibly many different cell groups capable of distinguishing a given input relatively well. Degeneracy is a property fundamental to reconciling specificity of recognition with range of recognition. Figure 11 illustrates the distinction between redundancy and degeneracy. Degenerate groups are isofunctional but not nonisomorphic. Redundant groups, in the strict sense, are isofunctional and isomorphic.

Although Edelman has a far more ambitious and elaborate application of these concepts to the structure and function of the nervous system, including cortical functions, memory, and perception, my aim is much more modest—to suggest that these ideas are pertinent to our conceptions of the chemoreceptors of taste and smell. The chemical senses can be regarded as molecular detectors, indeed recognizers that function not only in relation to well-known or familiar stimuli but also in a manner somewhat like the immune system, able to respond meaningfully to some novel or unknown molecule yet to be discovered or synthesized. Saccharin is an example of a man-made molecule never encountered in nature before it was accidentally found to be sweet.

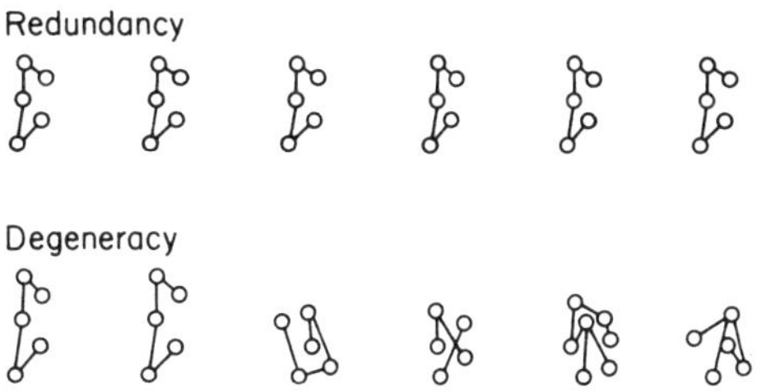

Figure 11. Diagrams illustrating the distinction between redundancy and degeneracy. The structures of degenerate groups carrying out more or less the same functions can differ in many respects, sharing only certain common features among their elements. Redundant groups have the same form and functions. Reproduced with permission from Edelman, *The Mindful Brain*, 1978. Cambridge, MA: MIT Press, p. 59.

Molecular distinctiveness as between sugars and saccharin recently shown by Jakinovich (1982b) may not necessarily indicate a molecular site per se for saccharin, i.e., a saccharin site as compared with a sucrose site. How could one envisage its emergence in evolution? The concept of degeneracy seems adequate to cause nearly isofunctionality among the many diverse other sweet-tasting molecules as well as saccharin and sugar. The taste system as studied electrophysiologically seems to me to display many properties of degeneracy, degeneracy in the strict sense, not in the metaphorical sense, with both specificity and breadth of sensitivity. Specificity and cross-fiber patterning are not necessarily irreconcilable.

Futuri

One good index to the future is given in the series of papers that make up the major portion of this book. In April 1982, at the time of the Givaudan lecture, some of these developments were under way; others, for example Beidler's analysis of the receptor mechanisms for sugars and other sweeteners, was not yet known. The nature of the very first step in the interaction of the chemical molecule or ion with the taste receptor is, of course, a crucial issue for understanding chemosensory transduction. Beidler's seminal formulation and quantitative theory of an adsorption mechanism in taste stimulation still stands as a major theoretical foundation that most other theories and formulations will have to accommodate (Beidler, 1954). But we need a deeper, more mechanistic analysis of this process. Beidler's extension of the multi-site nature of receptor mechanisms beyond each of the different basic stimuli to heterologous molecular sites within each of the basic taste categories is now convincingly demonstrated for sweetening agents. That such might be the case for sweeteners, both natural and artificial, had already begun to surface in psychophysical and more recent electrophysiological studies, but the strictly biophysical evidence had not been brought together quite as Beidler has done. In a similar vein, Jakinovich has now explicated this in considerable detail in his chapter. The final evidence on this problem and also the resolution of certain issues he raises remain to be settled. It may be that other conceptual formulations that I suggested, such as those deriving from immunology, may apply. Nonetheless, the chapters by Beidler and Jakinovich clearly define one important path of future development. They and a number of other workers are currently pursuing such issues with newer techniques including those based on use of monoclonal antibodies and other methods of receptology.

Bruce Oakley has updated his earlier classic work on the regeneration of taste buds with different neural cross innervations in an analysis of the factors determining both sensitivity patterns as well as growth of taste buds per se. It can be expected that the factors determining specific sensitivities as well as breadth of receptor cell tuning will be revealed by further studies of the developmental biology of the taste receptor, especially with in vitro preparations.

Erickson has well defined in his chapter the age-old problem and debate of sensory coding in the gustatory system. This question has set a pattern for a good deal of psychophysical and physiological study in past decades; indeed, it guided my early research when my paradoxical 1941 findings only seemed to complicate the issue of the specific energies of the different taste qualities. The ensuing debate implicit in Marion Frank's antithesis to the implications of Erickson's philosophy, I think, approaches an interesting resolution in David Smith's formulations. There is no denying the multiple sensitivity of individual receptors at the periphery, although the existence of clear-cut, best-stimulus clusters seems well documented as presented by Frank. But the added degradation of specific information through the CNS, at least to the pontine taste areas via the nucleus of tractus solitarius (NTS) which Erickson and his students first indicated and which Smith has now quantified with his concept of entropy, leaves little doubt of the degeneration of the input when categorized by the stimuli that activate them. There is no evidence as yet of sharpening in the gustatory CNS. This may not be surprising if one notes that the gustatory system is not a topographical sense but is a qualitative discriminator without regard to place in the world or on the tongue except for local differences in sensitivity. However, Smith's demonstration that without the clusters of sugar-best or salt-best units, for example, the cross-fiber patterning of differences becomes diluted and is weakened, if not lost. His is one of the best analyses that cross-fiber patterning and receptor specificity are not irreconcilable. Both phenomena demonstrate that taste receptors respond to different but overlapping ranges of the chemical spectrum. It seems to me that Edelman's concept of degeneracy is an apt characterization of the nature of the chemosensory specificity. I would even venture an analogy with the famous paradox regarding the nature of light. For years the evidence on one side argued for the particle nature of light. On the other hand, much evidence argued for its wave nature. As we now know both aspects of light coexist happily in the modern quantum theory of radiation.

Linda Bartoshuk in her chapter also reviews the history of the early studies and conflicts of taste psychophysics in the spirit of ascertaining the nature of its underlying physiology. What comes through clearly from Bartoshuk's presentation is that even with all the debate and disagreement, psychophysics of taste as such is a subject to be pursued on its own merits without excessive concern for basic physiology. She discusses whether in mixtures there may be predictable consequences depending on whether or not there is a compression, synergy, or linearity in scales of subjective intensity of the components. There are other interesting contrast effects and interactions indicating not simply peripheral but also central interactions. Psychophysics has its own domain, necessary for appreciation of taste sensations as we experience them, as well as for the applied field of flavor evaluation.

Bruce Halpern in his chapter on temporal aspects has covered a wide domain reflecting both aspects of peripheral mechanisms to which I have alluded in my discussion of latencies of electric and chemical taste stimulation. With "taste flashes" he explains how we are able to detect and identify transient stimuli before we are able, during or within a few hundred milliseconds of these transients, to respond. Increased stimulus duration actually slows the reaction time as if there is more quantitative information to be analyzed and more CNS input to be processed. Usually in natural situations the temporal properties of gustatory stimulation are not of this order, but, by appropriate instrumentation, further analysis of the temporal properties may be used analytically to study information processing in the chemosensory domain.

Ralph Norgren's studies of the CNS mechanisms of taste showed the existence of previously unsuspected mammalian brainstem way stations such as the pontine parabrachial nucleus and its related ventral pathway to the hypothalamus and central nucleus of the amygdala. Taste has long been known from behavior studies to be an incentive motivator. The degree to which the brainstem can mediate many appetitive functions in relation to visceral mechanisms and internal chemoreceptors of the vagal and sympathetic afferents promises to widen the influences of chemoreception upon biological as well as appetitive behavior functions. The future still holds the methodological developments that will permit study of single cortical neurons in the normal behaving animal in relation to the coding problem, preferred or aversive stimuli, and appetitive states. Such studies of CNS physiology will open new vistas on the neuropsychology of taste.

Taste and olfaction often work in collaboration. We speak of the taste of food even when there is a major olfactory component in sensory experience. Olfaction as a distance chemoreceptor thus has certain different biological functions than the more contactual chemoreception of taste. Concerning molecules dispersed in the vapor phase, Max Mozell discusses how direct physical processes in the nasal mucosa resembling gas chromatography could function cooperatively with selective sensitivity at different locations of the olfactory mucosa. Gordon Shepherd, long a recognized expert in the electrophysiology of the olfactory system, has shown by use of the 2-deoxyglucose method, the differential excitation of glomeruli in the olfactory bulb to different odor stimuli. Further, his concern for behaviorally relevant stimuli has given his physiological studies a new dimension. The existence of a modified glomerular complex, which seems particularly important in the young suckling's life and behavior, provides a unique tool for the further analysis of pheromonal stimuli as well as the question of labeled lines and olfactory transduction. Further, John Scott has shown subclasses of granular and mitral cells with special connectivities and functions necessary for the ultimate analysis of olfactory neurophysiology of the myriad connections and relationships of olfactory CNS anatomy.

Finally, Robert Johnston well systematizes the comparative analysis of olfactory vomeronasal communication among animal species. That olfactory stimuli serve as communicative devices in animals is now well established, with strong hormonal as well as behavioral consequences. The identification of dimethyl disulfide as the major component in the estrous female's vaginal secretion by O'Connell and Singer and their colleagues (1978 and 1979) followed directly from explication of the secretion's arousal function in the male. Johnston's interesting and provocative systemization of the several functions subserved by the main olfactory bulb and its receptors as compared with the accessory olfactory bulb and its vomeronasal receptors suggests a division of labor between them. Differences in the mechanism of stimulation and in their neural structures and pathways with consequent differences in behavioral outcome provides a framework not only for the behaviorist and physiologist but ultimately for the chemist looking for regularities among the many natural odorants in the animal kingdom.

Acknowledgment

The work discussed here was supported in part by a grant from the National Science Foundation (BNS 8111816), particularly our research in the section on "electric taste." I also wish to express my appreciation to Drs. Linda Bartoushuk and William Cain for help with certain of the historical material and references.

References

Adachi, A. 1969. Responses of the chorda tympani to electrotonic, electrolytic, and nonelectrolytic stimulations on the tongue. In (Proceedings of the Third International Symposium). *Olfaction and Taste III*. C. Pfaffmann, ed. New York: The Rockefeller University Press, p. 611 (abstr.)

Andersson, B., S. Landgren, and Y. Zotterman. 1950. The sweet taste fibres of the dog. *Acta Physiol. Scand.* 21:105–119.

Aristotle. 1906 (originally published ca. 330 B.C.). In *Aristotle: De Sensu and de Memoria*. G. R. T. Ross, ed. and trans. Cambridge: Cambridge University Press, pp. 41–99.

Beidler, L. M. 1954. A theory of taste stimulation. *J. Gen. Physiol.* 38:133–139.

Bell, C. 1811. Idea of a new anatomy of the brain. (Facsimile of original article). In *The Way In and the Way Out*. P. F. Cranefield, ed., Mt. Kisco, N.Y.: Futura Publishing Co., Inc., 1974, pp. 9–10.

Boring, E. G. 1942. *Sensation and Perception in the History of Experimental Psychology*. New York: Appleton-Century.

Bujas, Z. 1971. Electrical taste. In *Handbook of Sensory Physiology*, vol. IV. *Chemical Senses*, sect. 2, *Taste*. L. M. Beidler, ed. Berlin, Heidelberg, New York: Springer-Verlag, pp. 180–199.

Bujas, Z., M. Frank, and C. Pfaffmann. 1979. Neural effects of electrical taste stimuli. *Sensory Processes*. 3:353–365.

Burton, H., and R. M. Benjamin. 1971. Central projections of the gustatory system. In *Handbook of Sensory Physiology*, vol. IV. *Chemical Senses*, sect. 2, *Taste*. L. M. Beidler, ed. Berlin, Heidelberg, New York: Springer-Verlag, pp. 148–164.

Cagan, R. H., and O. Maller. 1974. Taste of sugars: brief exposure single-stimulus behavioral method. *J. Comp. Physiol. Psychol.* 87:47–55.

Cameron, A. T. 1947. The taste sense and the relative sweetness of sugars and other sweet substances. *Scientific Report Series no. 9*. New York: Sugar Research Foundation, pp. 11–43.

Davis, J. D. 1973. The effectiveness of some sugars in stimulating licking behavior in the rat. *Physiol. Behav.* 11:39–45.

Davis, J. D., B. J. Collins, and M. W. Levine. 1975. Peripheral control of meal size: interaction of gustatory stimulation and postingestional feedback. In *Hunger: Basic Mechanisms and Clinical Implications*. D. Novin et al., eds. New York: Raven Press, pp. 395–408.

DeSimone, J. A., G. L. Heck, and S. K. DeSimone. 1981. Active ion transport in dog tongue: a possible role in taste. *Science (Wash. DC)*. 214:1039–1041.

Edelman, G. M. 1975. Molecular recognition in the immune and nervous systems. In *The Neurosciences: Paths of Discovery*. F. G. Warden et al., eds. Cambridge, MA: MIT Press, pp. 65–74.

———. 1978. Group selection and phasic reentrant signaling: a theory of higher brain function. In *The Mindful Brain*. G. M. Edelman et al., eds. Cambridge, MA: MIT Press, pp. 51–100.

Erickson, R. P. 1963. Sensory neural patterns and gustation. In *Olfaction and Taste* (Proceedings of the First International Symposium). Y. Zotterman, ed. Oxford: Pergamon Press Ltd., pp. 205–213.

Falconer, B. 1847/48. Über eine merkwürdige Eigenschaft einer indischen Pflanze (*Gymnema sylvestre*). *Pharm. J. Trans.* 7:551–552.

Faurion, A., B. B. Bonaventure, and P. MacLeod. 1980a. Multiple approach of the sweet taste sensory continuum: psychophysical and electrophysiological data. In *Olfaction and Taste VII* (Proceedings of the Seventh International Symposium). H. van der Starre, ed. London: IRL Press, p. 86.

Faurion, A., S. Saito, and P. MacLeod. 1980b. Sweet taste involves several distinct receptor mechanisms. *Chem. Senses*. 5:107–121.

Grill, H. J., and R. Norgren. 1978a. The taste reactivity test. I. Mimetic responses to gustatory stimuli in neurologically normal rats. *Brain Res.* 143:263–279.

———. 1978b. The taste reactivity test. II. Mimetic responses to gustatory stimuli in chronic thalamic and chronic decerebrate rats. *Brain Res.* 143:281–297.

Haggard, H. W. 1929. *Devils, Drugs and Doctors: The Story of the Science of Healing from Medicine-Man to Doctor*. New York: Harper and Bros. 405 pp.

Hall, M. J., L. M. Bartoshuk, W. S. Cain, and J. C. Steven. 1975. PTC taste blindness and the taste of caffeine. *Nature (Lond.)*. 253:442–443.

Jakinovich, W. 1982a. Stimulation of the gerbil's gustatory receptors by saccharin. *J. Neurosci.* 2:49–56.

———. 1982. Taste aversion to sugars by the gerbil. *Physiol. Behav.* 28:1065–1071.

Kashiwayanagi, M., K. Yoshi, Y. Kobataki, and K. Kurihara. 1981. Taste transduction mechanism, similar effects of various modifications of gustatory receptors on neural responses to chemical and electrical stimulation in the frog. *J. Gen. Physiol.* 78:259–275.

Kelling, S. T., and B. P. Halpern. 1983. Taste flashes: reaction times, intensity, and quality. *Science (Wash. DC)*. 219:412–414.

Kiesow, F. 1894. Beiträge zur physiologischen Psychologie des geschmacksinnes. *Philos. Stud.* 10:523–561.

———. 1896. Beiträge zur physiologischen Psychologie des geschmacksinnes. *Philos. Stud.* 12:255–278.

Luciani, L. 1917. The sense of taste. In *Human Pysiology*, vol. 4, *The Sense Organs*. G. M. Holmes et al., eds. London: Macmillan and Co. Ltd., pp. 126–159.

Lucretius. 1940 (originally published, 55 B.C.). De rerum natura (On the nature of things). In *The Stoic and Epicurean Philosophers: The Complete Extant Writings of Epicurus, Epictetus, Lucretius, and Marcus Aurelius*. W. J., Oates, ed. New York: Random House, pp. 69–219.

Macht, M. B. 1951. Subcortical localization of certain "taste" responses in the cat. *Fed. Proc.* 10:88 (abstr.).

McBurney, D. H. 1969. Effects of adaptation on human taste functions. In *Olfaction and Taste*

III (Proceedings of the Third International Symposium). C. Pfaffmann, ed. New York: The Rockefeller University Press, pp. 407–419.
Müller, J. 1830. *Handbuch der Physiologie des Menschen für Vorlesungen*. Coblenz: J. Holscher.
Ninomiya, Y., and M. Funakoshi. 1980. Responses of the rat chorda tympani fibers to electric current of varying rate of rise applied to the tongue. In *Olfaction and Taste VII* (Proceedings of the Seventh International Symposium). H. van der Starre, ed. London: IRL Press, p. 217.
———. 1981a. Responses of rat chorda tympani fibers to electrical stimulations of the tongue. *Jpn. J. Physiol.* 31:559–570.
———. 1981b. Role of ions in generation of taste nerve responses to electrical tongue stimulation in rats. *Jpn. J. Physiol.* 31:891–902.
Nowlis, G. H., and M. E. Frank, 1981. Quality coding in gustatory systems of rats and hamsters. In *Perception of Behavioral Chemicals*. D. W. Norris, ed. New York: Elsevier-North Holland Inc., pp. 59–80.
O'Connell, R. J., A. Singer, F. Macrides, C. Pfaffmann, and W. C. Agosta. 1978. Responses of the male golden hamster to mixtures of odorants identified from vaginal discharge. *Behav. Biol.* 24:244–255.
O'Connell, R. J., A. G. Singer, C. Pfaffmann, and W. C. Agosta. 1979. Pheromones of hamster vaginal discharge: attraction to femtogram amounts of dimethyl disulfide and to mixtures of volatile components. *J. Chem. Ecol.* 5:575–585.
Öhrwall, H. 1891. Untersuchungen über der Geschmacksinn. *Skand. Arch. Physiol.* 2:1–69.
Öhrwall, H. 1901. Die Modalitäts- und Qualitätsbegriffe in der Sinnesphysiologie und deren Bedeutung. *Skand. Arch. Physiol.* 11:245–277.
Ogawa, H., M. Sato, and S. Yamashita. 1969. Gustatory impulse discharges in response to saccharin in rats and hamsters. *J. Physiol. (Lond.)*. 204:311–329.
Pfaffmann, C. 1935. An experimental comparison of the method of single stimuli and the method of constant stimuli in gustation. *Am. J. Psychol.* 48:470–476.
———. 1939. Specific gustatory impulses. *J. Physiol. (Lond.)*. 96:41P–42P.
———. 1941. Gustatory afferent impulses. *J. Cell. Comp. Physiol.* 17:243–258.
———. 1982. Taste: a model of incentive motivation. In *The Physiological Mechanisms of Motivation*. D. W. Pfaff, ed. New York: Springer-Verlag, pp. 61–97.
Pfaffmann, C., and T. Pritchard. 1980. Ion specificity of "electric taste." In *Olfaction and Taste VII* (Proceedings of the Seventh International Symposium). H. van der Starre, ed. London: IRL Press, pp. 175–178.
Sato, M., H. Ogawa, and S. Yamashita. 1975. Response properties of macaque monkey chorda tympani fibers. *J. Gen. Physiol.* 66:781–810.
Shallenberger, R. S. 1974. Occurrence of various sugars in foods. In *Sugars in Nutrition*. H. L. Sipple, ed. New York: Academic Press, Inc., pp. 67–80.
Sherrington, C. S. 1917. Reflexes elicitable in the cat from pinna, vibrissae and jaws. *J. Physiol. (Lond.)*. 51:404–431.
Shimada, I., A. Shiraishi, H. Kijima, and H. Morita. 1974. Separation of two receptor sites in a single labellar sugar receptor of the flesh-fly by treatment with *p*-chloromercuribenzoate. *J. Insect Physiol.* 20:605–621.
Shore, L. E. 1892. A contribution to our knowledge of taste sensations. *J. Physiol. (Lond.)*. 13:191–217.
Smith, D., and S. L. Bealer. 1975. Sensitivity of the rat gustatory system to the rate of stimulus onset. *Physiol. Behav.* 15:303–314.

Steiner, J. E. 1973. The gusto-facial response: observation on normal and anencephalic newborn infants. In *Fourth Symposium on Oral Sensation and Perception*. J. F. Bosmas, ed. Washington: U.S. Government Printing Office, pp. 254–278.

Sulzer, M. 1971. Recherches sur l'origine des sentiments agréables et desagréables. Troisième partie: Des plaisirs des sens. Histoire de l'académie des sciences et belle lettres de Berlin (année 1752). Quoted from Z. Bujas. 1971. Electrical taste, Chapt. 10. In *Handbook of Sensory Physiology*, vol. IV, *Chemical Senses*, sect. 2, *Taste*. L. M. Beidler, ed. New York: Springer-Verlag, pp. 180–199.

Volta, A. 1792. Briefe über thierishce Electricität. In Ostwald's *Klassiker der exakten Wissenschaften*. A. J. Oettingen, ed. Leipzig: Engelmann, 1900. Quoted by Z. Bujas. 1971. Electrical taste, Chapt. 10. In *Handbook of Sensory Physiology*, vol. IV, *Chemical Senses*, sect. 2, *Taste*. L. M. Beidler, ed. New York: Springer-Verlag, pp. 180–199.

von Békésy, G. 1964. Sweetness produced electrically on the tongue and its relation to taste theories. *J. Appl. Physiol.* 19:1105–1113.

White, A., P. Handler, and E. Smith. 1964. *Principles of Biochemistry*. 3rd edition. New York: McGraw-Hill Inc.

Wundt, W. 1893. *Grundzüge der Physiologische Psychologie*. 4th edition. Leipzig: Engelmann, pp. 438–441.

Yamamoto T., N. Yuyama, and Y. Kawamura. 1980. Responses of cortical taste cells and chorda tympani fibers to anodal DC stimulation of the tongue in rats. *Exp. Brain Res.* 40:63–70.

Young, P. T. 1955. The role of hedonic processes in motivation. In *Nebraska Symposium on Motivation*. M. R. Jones, ed. Lincoln: University of Nebraska Press, pp. 193–238.

Taste Receptors

2.
Multiple Sweet Receptor Sites and Taste Theory

The Problem

For many years, chemists have tried to correlate molecular structure of taste substances with their sweetness. The majority of such studies used either human taste threshold or equi-sweet concentrations to provide a measure of sweetness. These measures are derived from responses of a large population of receptor cells. The question then arises whether all the components of the population (taste fibers or taste cells) respond similarly to sweet stimuli. In other words, are all sweep receptor sites of taste cells identical or do multiple sites exist? This is the question to be addressed in this paper. A brief review of current knowledge concerning the nature of sweet molecules and the response of taste units will precede a discussion of data concerning the response of single rat and hamster taste nerve fibers to six sugars.

A Brief Review of Background Information

Sweet Stimuli

The most general property of sweet molecules is their ability to hydrogen bond to complementary structures in the receptor site located on the surface of the microvilli of the taste cells. Shallenberger and Acree (1967) studied such properties and concluded that each sweet molecule can form two hydrogen bonds separated by a distance of ~ 3Å. This property is generally regarded as a necessary but not sufficient condition for a sweet molecule. A third point of attachment by hydrophobic forces also occurs at a distance 3.5–5.5Å from the hydrogen bonds (Kier, 1972). Other investigators have also studied the conformation of specific sweet molecules and deduced aspects of the shape of the sweet receptor sites to which they adsorb (Temussi et al., 1978). Further

Lloyd M. Beidler and Keiichi Tonosaki, Dept. of Biological Sciences Florida State University, Tallahassee, Florida 32306*

*Current address: Gifu College of Dentistry, Takano, Hazumi-cho, Motosu-gun, Gifu, Japan.

details of the molecular properties of sweet substances will be discussed by Jakinovich in this volume.

Chemical Spectrum of Single Taste Fibers

Early anatomists searched for morphologically distinct receptor cells specific for various sensory qualities and modalities. The discovery of visual rods and cones provided an anatomical basis for the duplexity theory of vision. On the other hand, anatomically distinct color receptors that correlate with the trichromatic color theory were not found and only in recent years have they been identified electrophysiologically. Similarly, morphologically distinct taste receptors related to specific taste qualities have been searched for but never found. It was often assumed, however, that a single taste bud may be associated with but one taste quality. Surely one would expect that a single taste fiber would respond to stimuli associated with but only one of the primary taste qualities. It was therefore a great surprise when Carl Pfaffmann (1941) found that single taste fibers respond to a variety of taste stimuli.

At the time of his taste experiments, Pfaffmann had considerable previous experience recording from sensory receptors. In particular, his recording of afferent impulses from the teeth was well known (Pfaffmann, 1939). Similar methods of recording neural activity from the chorda tympani and glossopharyngeal taste nerves of the cat were used by him to study taste. He chose acid, salt, sugar, and quinine as taste stimuli and recorded from single fibers. Their responses allowed him to classify the taste fibers as those stimulated by (1) acids, (2) acids and NaCl, and (3) acids and quinine. Since his choice of stimuli was arbitrary, he stated "perhaps some other choice of stimuli would be more in keeping with a strictly functional classification. This might lead to a different designation of fibers." He also concluded that "each fiber type has a chemical spectrum." It is interesting to note that Pfaffmann also discovered "water taste" or fibers of the cat that respond to water on the tongue but not to Ringer's solution.

Pfaffmann's classic paper lead to many other investigations of neural coding of taste qualities. Most of these dealt with coding of the four primary taste qualities—sour, salty, bitter, and sweet.

Single Taste Fibers–Single Taste Cells

The single taste nerve fiber innervates several taste buds and many taste cells within each taste bud. It is approximated that a total of 10–50 taste cells are innervated by one taste nerve fiber (Beidler, 1969; Miller, 1971, 1974).

How specific are the responses of individual cells of the taste bud? By use of microelectrodes, researchers found that each taste cell responds to a variety of taste stimuli and with various sensitivities (Kimura and Beidler, 1961). It appears that single taste cells as well as single taste nerve fibers lack high specificity and each has its own chemical spectrum.

One might ask, how many receptor sites are located on each taste cell? Although the answer is not known, an approximation of the density of anionic and cationic sites may be calculated. Ling (1962) suggested that tissue protein contains ~ 8% free cationic and ~ 10% free anionic groups. There is one ionizable group every 20 Å along a polypeptide chain and each chain is ~ 20Å apart. Let us assume that there are 5 microvilli/taste cell, each being ~ 0.2×5 µm in size. Then there are ~ 16^6 sites where H^+ ions can bind to microvilli of each taste cell. If 25 taste cells are innervated by a single taste fiber and if there are 5 microvilli/taste cell, then the response of this fiber to acid stimulation may involve up to 2×10^8 receptor sites. This may seem like many H^+ ions but remember that there are 10^{17} H^+ ions/cm^3 of acid solution of pH=3.

The above calculation assumes dense packing of anionic receptor sites on the surface of the microvillus membrane. There is some evidence that such sites are also available deeper into the membrane. Thus, the above value may be too low by a factor of 10. On the other hand, not all anionic receptor sites may be involved in taste transduction even though they bind taste cations.

The receptor sites for sugars and many bitter substances are much less numerous than the anionic or cationic sites. This statement is based upon the fact that such sites are larger and thus fewer per unit area. The lipid solubility of many of the bitter substances suggests that interaction is with the membrane lipid component or else the bitter taste stimulus must pass through a lipid layer in order to reach its receptor site.

Species Differences

We have seen that each single taste fiber has its own spectrum of chemicals to which it is most sensitive. However, if one measures the total response of a large part of the entire population of taste fibers, then one observes that the chemical spectrum of the response of the population of one animal is very similar to that of another of the same species. If however, one compares the responses of the receptor population of an animal of one species with that of another species, large differences are observed (Beidler et al., 1955; Pfaffmann, 1955). Such species differences have also been observed for various sugar stimuli, as shown in Figure 1.

Analysis of Single Fiber Responses to Six Selected Sugars

If all the sweet receptor sites are identical, then the response spectrum or profile of the single taste fibers that innervate taste cells should be similar. That is, the sequence and relative magnitude of responses to six chosen sugar stimuli should be identical in all taste fibers although the sensitivities of the fibers may vary. To test this hypothesis, the responses of 32 hamster and 32 rat single chroda tympani fibers were studied. Half-molar concentrations of sucrose and of maltose were used to identify responding fibers in the nerve bundle. Single fibers were then isolated and stimulated with 0.5 M sucrose,

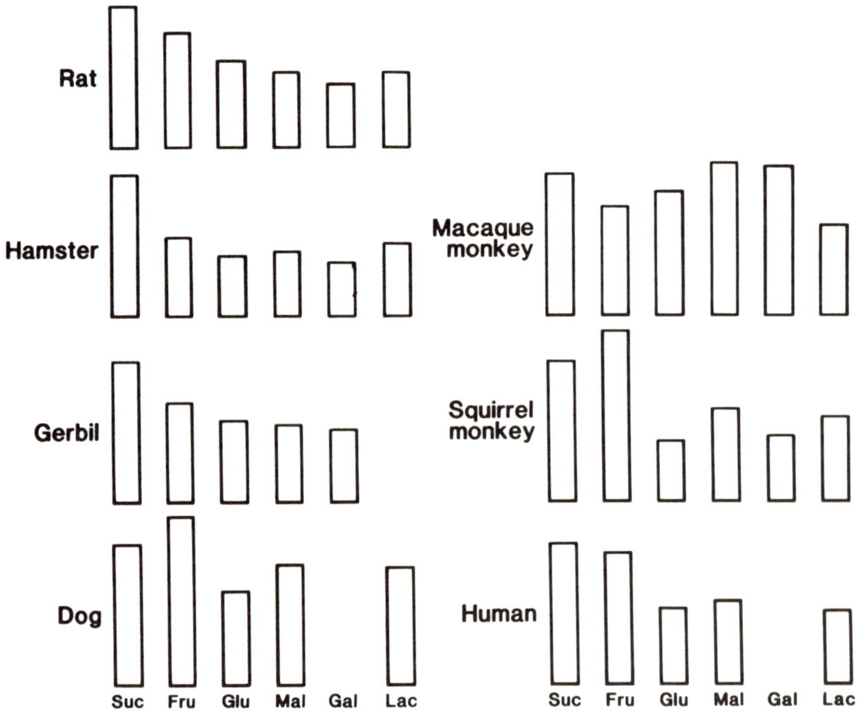

Figure 1. Summated responses to 0.5 M sucrose, fructose, glucose, maltose, galactose, and lactose. Reported values converted to bar graphs for comparison (from Andersson et al., 1962; Diamant et al., 1963; Hardiman, 1964; Noma et al., 1971, 1974; Pfaffmann et al., 1976; Sato et al., 1977).

maltose, galactose, fructose, glucose, and 0.3 M lactose. From the above group, 12 rat and 14 hamster fibers were also stimulated with 0.1, 0.2, 0.5, and 1.0 M concentrations of most of the sugars to obtain response-concentration functions.

Examination of the data shows that each single fiber produces its own characteristic response profile for the six sugars tested. This is true for both rats and hamsters. If all receptor sites were identical, then all fibers would produce the same response profile. Since this is not the case, multiple receptor sites must exist.

Search of the literature revealed that three other studies included a small number of single fiber responses to 0.5 M sugars. The first was by Andersson et al. (1962) who studied dog single fibers. Figure 2 shows the conversion of their data to bar graphs to illustrate the response profiles. Notice that no two fibers show identical profiles. The authors did stress the fact that some fibers responded better to maltose than to sucrose. The second study by Pfaffmann et

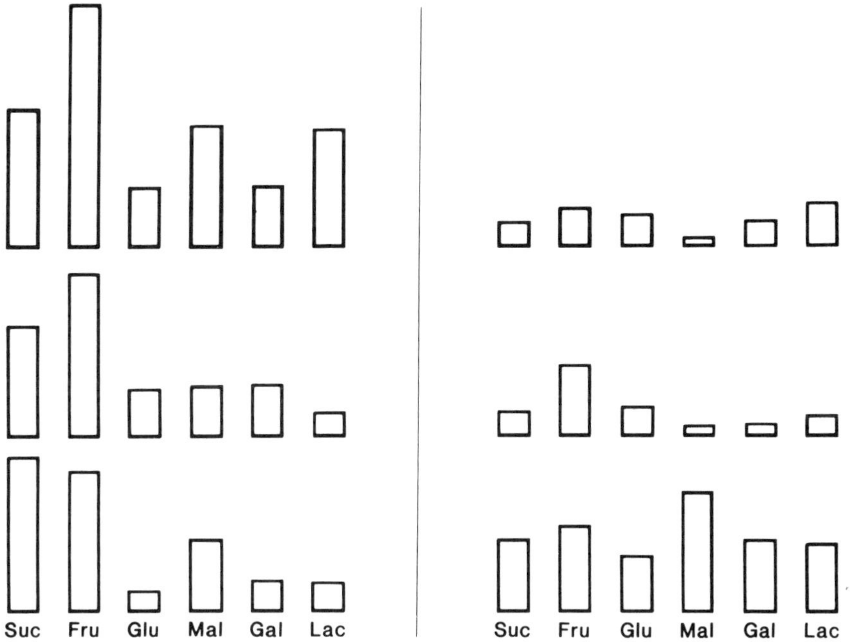

Figure 2. Responses of six dog single chorda tympani fibers to six 0.5 M sugars (from Andersson et al., 1962).

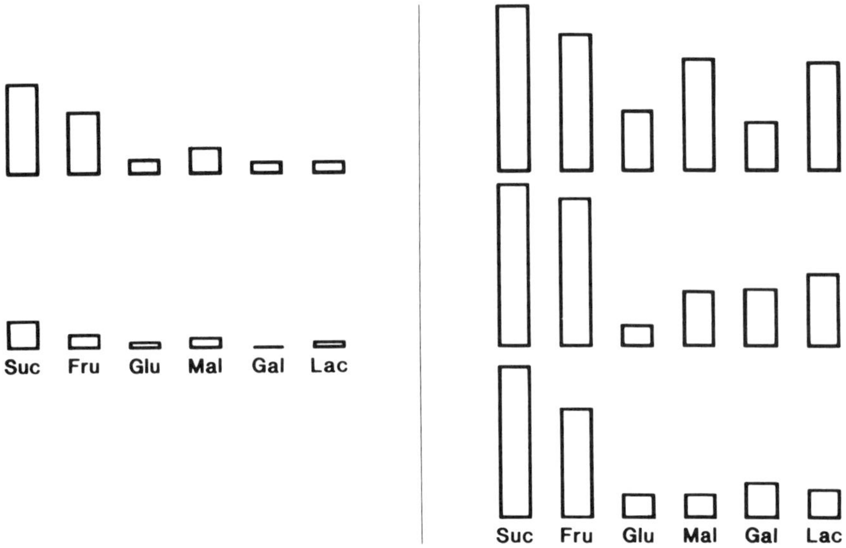

Figure 3. Responses of five squirrel monkey single chorda tympani fibers to six sugars (from Pfaffmann et al., 1976).

al. (1976) used the squirrel monkey. Only the responses of five single fibers could be ascertained from their published figure. Conversion of the data to bar graphs produced Figure 3. Again note the differences in response profiles. The third study was that of Ninomiya et al. (1981) in which they recorded from mouse single fibers. Converted bar graphs of nine fibers are shown in Figure 4.

The magnitude of sucrose response was compared with that of the fructose response for each hamster single fiber in our study. The results are illustrated in Figure 5. Note that there is little correlation between the two. In fact, the absolute magnitude of their differences averages ~ 70%.

Since previous researchers noted some dog fibers that responded well to maltose, we compared our single fiber responses to sucrose with those to maltose. Figure 6 indicates that the rat has many fibers that respond with high frequency of neural firing to maltose. Sucrose, on the other hand, stimulates a larger number of fibers that elicit low firing frequencies. The opposite is true for the hamster. Figure 7 indicates a larger number of fibers that respond with high frequency to sucrose but maltose stimulates a larger number of fibers with low frequency. These results suggest that the ratio of maltose to sucrose magnitudes of responses depends upon the selection of range of firing frequency studied. Since the researcher chooses a minimum value of firing frequency to determine whether the fiber is responding, it is important to know how selective is the choice. For example, Figure 8 shows this ratio for hamster fibers. Note that the ratio value increases as more fibers are included that respond with low frequencies. If the responses of all studied fibers are included, the ratio approaches that attained by conventional electronic summation of the nerve impulses from the entire chorda tympani nerve bundle. This indicates that the 32-hamster single fiber selection is a good predictor of the response of the entire fiber population. A similar treatment of the rat data is shown in Figure 9. Again the maltose-sucrose ratio approaches that of the integrated response of the entire taste nerve bundle, but the agreement is not nearly as good as that of the hamster. The reason for this is unclear.

We have demonstrated above that multiple receptor sites for sugars do

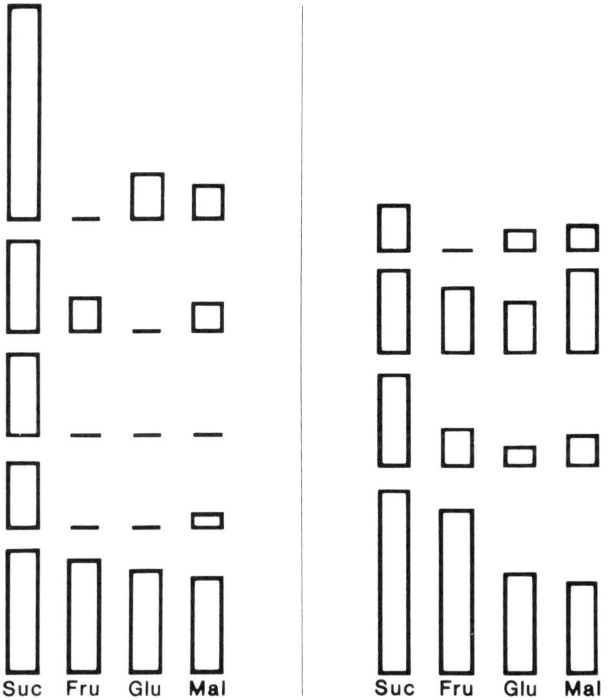

Figure 4. Responses of nine mouse single chorda tympani nerve fibers to four sugars—sucrose, fructose, glucose, and maltose (from Ninomiya et al., 1982).

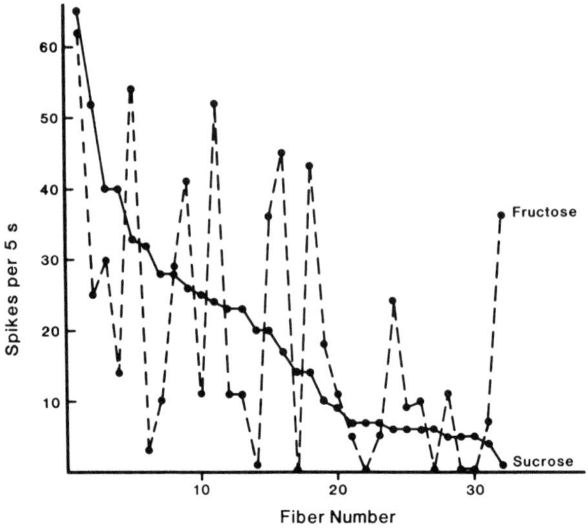

Figure 5. The responses of 32 rat single fibers to 0.5 M sucrose are given in order of decreasing magnitudes (solid line). The responses to 0.5 M fructose are also shown (dashes). The difference in magnitude of responses averages 70%.

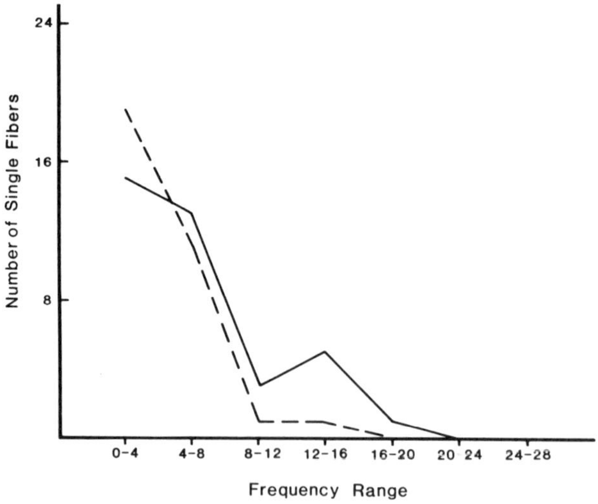

Figure 6. Frequency range of firing during the first 5 s for each of 32 single rat chorda tympani fibers. Note that a larger number of fibers respond to 0.5 M maltose (solid line) with high firing frequencies than to 0.5 M sucrose (dashes).

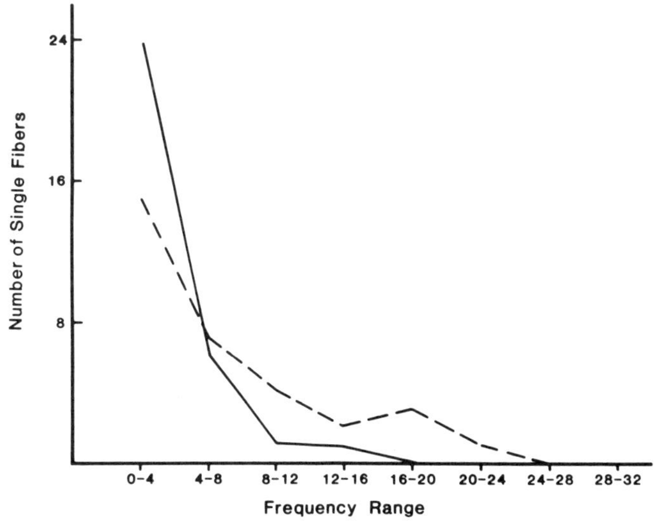

Figure 7. Frequency range of firing during the first 5 s for each of 32 single hamster fibers. Note that a larger number of fibers respond to 0.5 M sucrose (dashes) with high frequencies than to 0.5 M maltose (solid line).

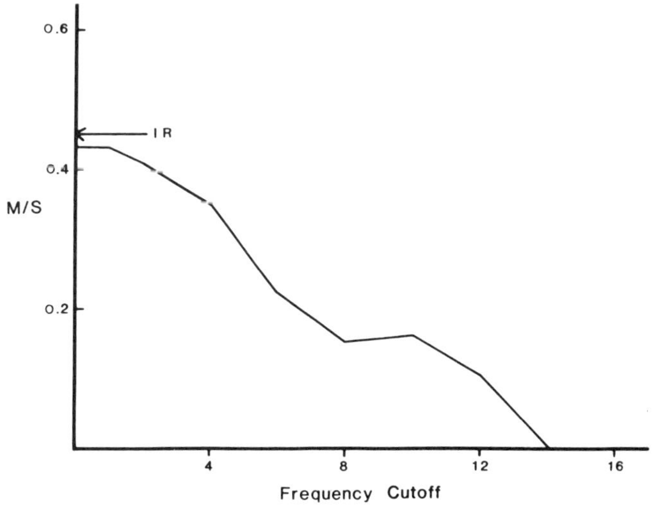

Figure 8. Ratio of hamster response to 0.5 M maltose and to 0.5 M sucrose as a function of frequency of firing of the fibers. The ratio (M/S) is plotted as a function of responses of all fibers whose firing rate is the shown value or higher.

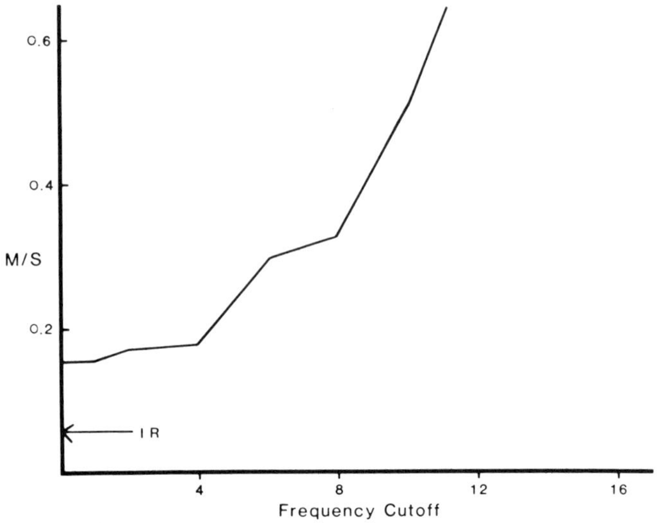

Figure 9. The same type of plot as shown in Figure 8, except the data are for rat fibers.

indeed exist. However, in order to study the site differences in more detail, it is necessary to obtain the entire response-concentration curves. Previous research (Beidler, 1954) suggested that the taste stimulus adsorbs to the receptor site and the magnitude of response, R, is a function of the concentration (C), the maximum response (R_S), and the association constant (K), according to $R=CKR_S / (1 + CK)$. This equation can be rearranged to $C/R = C / R_S + 1/KR_S$. A plot of C/R vs. C for each single fiber reveals the value of K and R_S. Typical values for eight fibers are shown in Table I. Note the range of values for both K and R_S or all single fibers responding to the six sugars.

If it is assumed that the adsorption process alone determines the magnitude of single fiber response, then K reflects the binding affinity of the taste stimulus to the receptor site, and R_S is a function of both the number of total receptor sites available for a given sugar and the efficacy of the bound stimulus-receptor site to produce a transduction. The sensitivity of the single fiber is directly proportional to KR_S.

Table 1. *Hamster Single Fibers*[1]

Sucrose		Fructose		Maltose		Lactose		Glucose		Galactose	
K	R_s	K	R_s	K	R_s	K	R_s	K	R_s	K	R_s
2.8	78	2.5	64	14	48	0.65	170	4.7	65	2.0	106
26.6	96	1.4	154	18	54	4.2	71	10.0	36	3.2	28
3.6	64	5.0	62	19	27	12.0	55	25.0	23	3.0	26
6.0	89	22.0	6.4	6.5	14	17.0	10	3.5	59	1.9	70
7.1	78	3.5	28	15	8	2.3	87	5.4	27	4.3	37
38.0	98	10.0	73	3.4	35	11.0	17	4.7	32	13.0	19
10.0	115	10.0	88	7.6	32	2.1	13	1.5	11	6.0	102
10.0	69	0.61	100	4.8	65	8.9	27	9.5	35	2.2	76

1. The values of K and R_s are shown for eight hamster single fibers in response to the six sugars indicated. The values were obtained by plotting R vs. C (C = concentration, R = response) of data from a concentration series for each sugar.

Single Fiber Spectra and Taste Theories

Transduction Process

Sensory receptors translate stimulus parameters into electrical potentials. This process, called sensory transduction, consists of a series of events, some of which can be experimentally observed. They are separated conceptually as shown in Figure 10.

Stimulus adsorption is considered to occur at the taste cell microvilli membranes. The simplest process is physical adsorption as described by the reaction $S + P \leftrightharpoons SP$ where S is the stimulus concentration, P, the number of unfilled receptor sites, and the SP the number of filled or activated receptor sites. If it is assumed that the remainder of the transduction events are linear, then the magnitude of neural response at equilibrium is mathematically de-

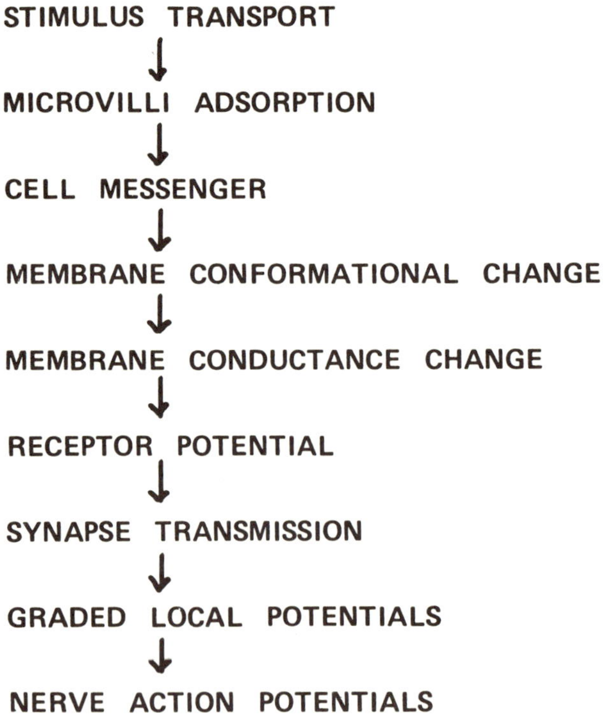

Figure 10. Schematic of events that occur between stimulus application and elicitation of taste nerve action potentials.

fined by $R = CKR_s/1 + CK$, as shown previously. This simple equation has been widely used to describe experimental data from a variety of animal species.

When the above model was proposed in 1954, it was assumed that when the receptor site is filled, a change of state occurs that leads to conformational changes in the molecules of the membrane associated with the receptor site. In 1965, Monod et al. proposed a more complex "regulator" or allosteric model that describes the details of receptor site reactions leading to molecule activation. Morita (1969) integrated the allosteric model with a mathematical description of the changes in membrane transduction to describe electrical events in insect taste cells. Later, Mooser (1980) used the allosteric model to describe taste responses to salt stimulations.

Conformational changes in the microvilli membrane molecules do not lead to electrical events in the microvilli. Rather, the electrical events occur toward the base of the taste cells where nerve synapses are present.

How the occurrence of conformational changes at the microvilli are signaled to other areas of the taste cell is not known. A similar problem is present in visual cells where photons are captured in rod discs and the electrical events occur in the plasma membrane. In this case it is thought that Ca^{++} is released at the discs and diffuses toward the plasma membrane to initiate changes in its electrical polarization. Thus, Ca^{++} may serve as the messenger in visual cells.

Taste cell stimulation is followed by a change in membrane electrical polarization or receptor potentials. These potential changes are associated with changes in membrane conductance due to alterations of membrane permeability to specific ions (possibly Na^+ and Ca^{++}). Microelectrodes inserted into single rat taste cells measure membrane conductance changes as a function of taste stimulus concentration (Ozeki, 1971; Sato and Beidler, 1975). Morita (1969) assumed the equivalent circuit of the membrane to be:

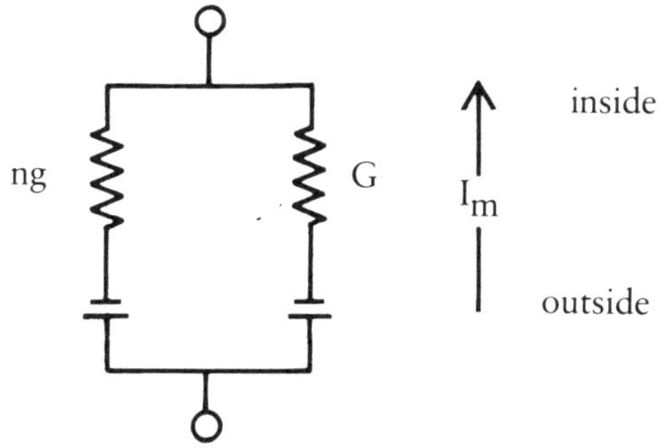

where G = membrane conductance when not stimulated, g = membrane conductance per activated receptor site, and n = number of sites activated by given concentration of taste stimulus. Applying Beidler's taste theory, the relative response is obtained as
$R/R_s = CK^*[(sg + G)/G]/\ 1 + CK^*[(sg + G)/G]$, where s = maximum number of sites, that can be activated at high taste stimulus concentrations, C = taste stimulus concentration, and K^* = association constant.
However, Beidler's taste equation is: $R/R_s = CK/(1 = CK)$.
Comparison of the two equations indicates that $K^* = K[G/(sg + G)]$. How large a discrepancy is inherent when the conductance changes are not considered? Ozeki measured the change in membrane conductance of a single rat taste cell when 0.5 M sucrose was applied to the tongue. The conductance rose ~ 10%. Thus $K^* = K[100/(10 + 100)] = 0.91$ K. The maximum change in membrane conductance was also observed when high NaCl concentrations were used as taste stimuli. This change may be as high as 100%. From these data one would expect a correction of a 10–50% decrease in the value of the association constant K used by Beidler. The correction for the binding energies involved would be reasonably small since they are proportional to the logarithm of K.

The receptor potentials measured with microelectrodes are mirrors of the conductance changes. They have been measured and related to taste stimulus concentration in several animal species (Kimura and Beidler, 1961; Tateda and Beidler, 1964). It is assumed that these receptor potentials initiate local graded potentials in the fibers innervating the taste cell, where the frequency of nerve impulses generated is proportional to the magnitude of the local potential.

Nerve Fiber Innervating Multiple Taste Cells

The single nerve fiber innervates many taste cells. Thus, the firing frequency of the single nerve fiber is proportional to the summation of the generated local potentials of its endings. This fact is of importance in the analysis of the relationship between receptor sites and single fiber responses.

We have shown conclusively that multiple receptor sites occur in rat and hamster taste cells. The number of different receptor sites is unknown, although there must be at least two or more. If two different sites exist that are not mutually exclusive for all six sugars, then it may be possible to explain experimentally all response profiles seen. Suppose activated site I has a profile of maltose = 10, lactose = 1, and sucrose = 5, and another type of activated site, II, has a profile of sucrose = 10, maltose = 3, and lactose = 2. Assume that each taste cell has only one kind of receptor site but all cells have the same number of total sites. Also assume the single fiber response is merely the addition of the responses of the taste cells it innervates. Then the response profiles can be calculated as shown in Table II.

Notice the variety of response profiles generated with these simple assumptions. If the number of taste cells innervated by a single fiber is increased to a realistic value of 10–50 cells, then the variety is increased greatly. Note, however, that with the initial assumption of the specific response profiles of the two receptor sites, there is no way to generate a fiber response profile in which lactose is more effective than sucrose or maltose.

Table II. *Response Profiles*

Taste cells innervated	Response profiles		
	Sucrose	Maltose	Lactose
Taste cell I + taste cell II	15	13	3
Taste cell I + taste cell I	10	20	2
Taste cell II + taste cell II	20	6	4
2 Taste cells I + 1 taste cell II	20	23	4
2 Taste cells II + 1 taste cell I	25	16	5

Is it possible that a given taste cell has many different types of receptor sites and that no two taste cells have similar response profiles for the sugars? The answer is not known for the rat and hamster. However, at least three different sugar receptor sites have been identified in insects. The "pyranose" site interacts with monosaccharides like glucose, the "furanose" site with D-galactose, D-fructose, D-fucose, and some aromatic amino acids, and the "carboxylate" site with carboxylic acids and some aliphatic amino acids. Are these three receptor site types randomly distributed over various taste cells? Fortunately, one fly single fiber innervates only one taste cell so that profile analysis is easy. One concludes that all taste cells of a given type of fly hair have the same response profile (Wieczorek and Kopple, 1982).

Characteristics of Taste Receptor Sites

Man can respond to the tastes of tens of thousands of different molecules. According to the adsorption theory of taste, all of these molecules are able to weakly interact with some of the molecules that compose the plasma membrane of the microvilli of the taste cells, presumably proteins and lipids. The major bindings are due to electrostatic or coulombic bonds, hydrogen bonds, and van der Waals bonds. The bond energies are only a few kilocalories per mole in contrast to the weak thermal energy of 0.6 kcal/mol or the strong convalent bonds of 50–110 kcal/mol.

A single receptor site is only a small part of the total protein or lipid membrane molecule although there may be many such sites per molecule. For example, a protein may contain many carobxyl groups that can bind cations. However, not all such carboxyl groups may be involved in the subsequent transduction process. The cation binding (salts and acids) is fairly weak, not stereospecific, and the binding force drops off slowly with distance ($F \sim \frac{1}{R^2}$). Cation specificity of the receptor site depends upon its electric field strength, which in turn determines the number of water molecules that separate the cation from the receptor site. If the field strength is weak, then the binding of K^+ is greater than that of Na^+. On the other hand, a strong field strength results in a better binding of Na^+ than of K^+.

Sweet and bitter compounds are bound to the taste receptor site with predominantly hydrogen and van der Waal bonds. The strength of attraction of the latter drops rapidly with distance ($F \sim \frac{1}{R^6}$) so that the stimulus molecule must fit snugly with the receptor site in contrast to that of cation binding. The maximum hydrogen bond strength occurs when the three components (donor, acceptor, H atom) are colinear. Thus, stereospecificity is very important for sweet and bitter tastes. Since several hydrogen bonds as well as hydrophobic bonds are involved at specific locations, the stereospecificity is three-dimensional.

The specificity of sweet and bitter molecules may lead one to assume a lock and key concept for the stimulus-receptor site interaction. However, there are many thousands of different bitter and sweet molecules and there is no evidence that a specific receptor site exists for each. One must conclude, therefore, that the lock and key concept is a little too rigid.

Most sweet receptor sites modeled by researchers involve a cleft in the protein spatial arrangement. Stereospecificity is determined by the spatial placement of the hydrogen bonds and the hydrophobic region. However, changes in the microenvironment within the cleft are expected to lead to small changes in the relative strength of bindings of sweet or bitter molecules. Such secondary considerations of the receptor sites may be responsible for the fact that no two taste fiber response profiles are identical. Thus, there may be a limited number of grossly different sweet receptor sites (such as found for flies), but minor variations of each may be present because of their microenvironments. Furthermore, the number and kinds of receptor sites may vary among the taste cells. This would be consistent with the single fiber data presented in this chapter.

Electrophysiological data are often compared with behavioral data. No such attempt is made here. The selection of fibers studied was based upon their ability to respond to 0.5 M sucrose or 0.5 M maltose. How well the fiber population selected on this basis conforms to the total population of sweet fibers is not known. Furthermore, the problem of neural coding of taste qualities was not considered. On the other hand, human psychophysical data have been used to suggest that multiple receptor sites mediate sweetness (Faurion et al., 1980; Schiffman et al., 1981).

References

Andersson, H., M. Funakoshi, and Y. Zotterman. 1962. Electrophysiological investigation of the gustatory effect of various biological sugars. *Acta Physiol. Scand.* 56:362–375.

Beidler, L. M. 1954. A theory of taste stimulation. *J. Gen. Physiol.* 38:133–139.

———. 1969. Innervation of rat fungiform papilla. In *Olfaction and Taste III* (Proceedings of the Third International Symposium). Carl Pfaffmann, ed. New York: The Rockefeller University Press, pp. 352–369.

Beidler, L., I. Fishman, and C. Hardiman. 1955. Species differences in taste response. *Am. J. Physiol.* 181:235–239.

Diamant, H., M. Funakoshi, L. Strom, and Y. Zotterman. 1963. Electrophysiological studies on human taste nerves. In *Olfaction and Taste* (Proceedings of the First International Symposium). Y. Zotterman, ed. Oxford: Pergamon Press Ltd., pp. 193–203.

Faurion, A., S. Saito, and P. MacLeod. 1980. Sweet taste involves distinct receptor mechanisms. *Chem. Senses Flav.* 5:107–121.

Hansen, K., and H. Wieczorek. 1981. Biochemical aspects of sugar reception in insects. In *Biochemistry of Taste and Olfaction.* R. H. Cagan and M. Kare, eds. New York: Academic Press, Inc., pp. 139–162.

Hardiman, C. W. Rat and hamster chemoreceptor responses to a large number of compounds and the formulation of a generalized chemosensory equation. Ph.D. dissertation, Florida State University, 1964.

Kier, L. 1972. A molecular theory of sweet taste. *J. Pharm. Sci.* 61:1394–1397.

Kimura, K., and L. Beidler. 1961. Microelectrode study of taste receptors of rat and hamster. *J. Cell. Comp. Physiol.* 58:131–140.

Ling, G. 1962. *A Physical Theory of the Living State.* New York: Blaisdell Publishing Co.

Miller, I. 1971. Peripheral interactions among single papilla inputs to gustatory nerve fibers. *J. Gen. Physiol.* 57:1–25.
———. 1974. Branched chorda tympani neurons and interactions among taste receptors. *J. Comp. Neurol.* 158:155–166.
Monod, J., J. Wyman, and J. Changeux. 1965. On the nature of allosteric transitions: a plausible model. *J. Mol. Biol.* 12:88–118.
Mooser, G. 1980. Sodium and potassium salt stimulation of taste receptor cells: an allosteric model. *Proc. Natl. Acad. Sci. USA.* 77:1686–1690.
Morita, H. 1969. Electrical signs of taste receptor activity. In *Olfaction and Taste III* (Proceedings of the Third International Symposium). C. Pfaffmann, ed. New York: The Rockefeller University Press, pp. 370–381.
Ninomiya, Y., K. Tonosaki, and M. Funakoshi. 1982. Gustatory neural response in the mouse. *Brain Res.* 244:370–373.
Noma, A., J. Gota, and M. Sato. 1971. The relative taste effectiveness of various sugars and sugar alcohols for the rat. *Kumamoto Med. J.* 24:1–9.
Noma, A., M. Sato, and Y. Tsuzuki. 1974. Taste effectiveness of anomers of sugars and glycosides as revealed from hamster taste responses. *Comp. Biochem. Physiol.* 48A:249–262.
Oakley, B., L. Jones, and J. Kaliszewski. 1979. Taste responses of the gerbil IXth nerve. *Chem. Senses Flav.* 4:79–87.
Ozeki, M. 1971. Conductance change associated with receptor potentials of gustatory cells in rat. *J. Gen. Physiol.* 58:688–699.
Pfaffmann, C. 1939. Afferent impulses from the teeth due to pressure and noxious stimulation. *J. Physiol.* (*Lond.*). 97:207–219.
———. 1941. Gustatory afferent impulses. *J. Cell. Comp. Physiol.* 17:248–258.
———. 1955. Gustatory nerve impulses in rat, cat and rabbits. *J. Neurophysiol.* (*Bethesda*). 18:429–440.
Pfaffmann, C., M. Frank, L. Bartoshuk, and T. Snell. 1976. Coding gustatory information in the squirrel monkey chorda tympani. *Prog. Psychobiol. Physiol. Psychol.* 9:131–150.
Sato, T., and L. Beidler. 1975. Membrane resistance change of the frog taste cells in response to water and NaCl. *J. Gen. Physiol.* 66:735–763.
Sato, M., Y. Hiji, H. Ito, T. Imoto, and C. Saku. 1977. Properties of sweet taste receptors in macaque monkeys. In *Food Intake and Chemical Senses*, Y. Katsuki, M. Sato, S. Takagi, and Y. Oomura, eds. Tokyo: University of Tokyo Press, pp. 187–199.
Schiffman, S., H. Cahn, and M. Lindley. 1981. Multiple receptor sites mediate sweetness: evidence from cross adaptation. *Pharm. Biochem. Behav.* 15:377–388.
Shallenberger, R. and T. Acree. 1967. Molecular theory of sweet taste. *Nature* (*Lond.*). 216: 480–482.
Shimada, I. 1978. The stimulating effect of fatty acids and amino acid derivatives on the labellar sugar receptor of the fleshfly. *J. Gen. Physiol.* 71:19–36.
Shimada, I., and K. Isono. 1978. The specific receptor site for aliphatic carboxylate anion in the labellar sugar receptor of the fleshfly. *J. Insect Physiol.* 24:807–811.
Tateda, H., and L. Beidler. 1964. The receptor potential of the taste cell of the rat. *J. Gen. Physiol.* 47:479–486.
Temussi, P., F. Lelj, and T. Tancredi. 1978. Three-dimensional mapping of the sweet taste receptor site. *J. Med. Chem.* 21:1154–1158.
Wieczorek, H., and R. Kopple. 1982. Reaction spectra of sugar receptors in different taste hairs of the fly. *J. Comp. Physiol. B. Biochem. Syst. Environ. Physiol.* 149:207–213.

3. Sugar Taste Reception in the Gerbil

The current review covers all our gerbil sugar taste experiments, which have led to belief in a sucrose receptor site. Our research, along with other animal taste experiments, is based on available human psychophysical work. In the research presented here, we have tried to compare the electrophysiological responses of our gerbils with those of other mammals and to correlate them with the human behavioral responses, in an attempt to better understand the mechanism of sweet taste response. Along with many similarities observed between animal and human responses, several discrepancies were noted which suggest that further research is needed in this area.

Methodology

Based on the literature (Beidler, 1953), the summated response method, often called the integrated response (representing a running average of neural spike activity from the whole chorda tympani nerve), was determined to be best for our experiments. The "ideal" method, recording from the taste receptor itself was not technically feasible. Another method, recording from the individual neurons that innervate the receptors in a complex manner (Beidler, 1969), provides only a small sample of the responses of a large and varied population of taste receptors. Furthermore, the single neuron recording technique, itself, is only stable for a few moments and is variable in its response. In contrast, the summated response is stable for many hours. In addition, Kimura and Beidler (1961) found that the summated response is correlated with the responses of taste receptor cells. Therefore, it appears that the summated response is a valid measurement of taste receptor activity.

Figure 1 is a typical series of recordings. A response is the upward deflection of the baseline, which can be seen first at 0.003 M. The weakest effective stimulus in the concentration series is defined as the threshold concentration. A response is measured with a ruler from the spontaneous baseline to the top of the pen deflection. In this research, all summated responses are computed as a percentage of the maximum sucrose response. I prefer this to the usual

---

William Jakinovich, Jr., Department of Biological Sciences
CUNY, Herbert H. Lehman College, Bronx, New York 10468

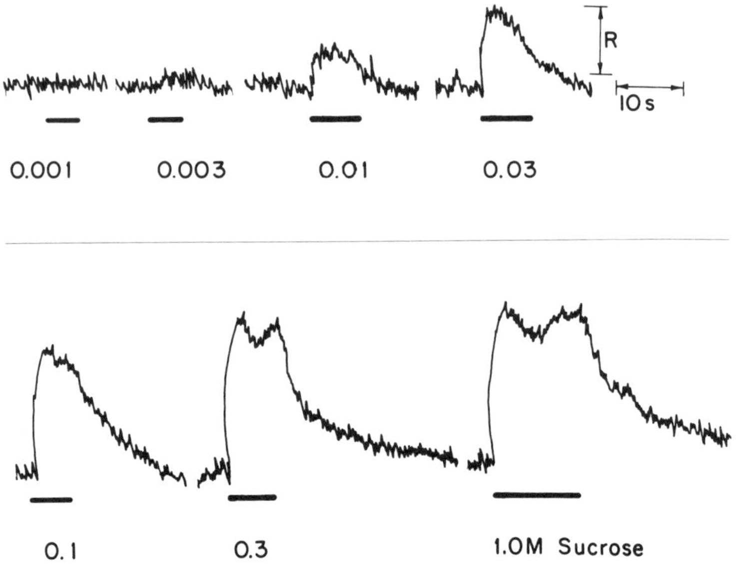

Figure 1. Integrated neural discharge from the gerbil's chorda tympani nerve in response to a series of increasing concentrations of sucrose applied to the tongue. The solid bar under the records indicate stimulus duration; R is the measure of response (from Jakinovich, 1976).

method of normalizing to a single concentration of sodium chloride, sucrose, or some other taste stimulant, since I have found more variability associated with the data when a single concentration is used as a standard, especially in the case of the sodium chloride response, which is sensitive to flow rate.

Summary of Evidence

Some of the evidence that led to the idea of a sucrose receptor site includes the following:

Structure-Activity Studies

1. Among disaccharides, monosaccharides, and polyols, sucrose is the most effective "natural sugar" to stimulate the gerbil's chorda tympani nerve.

2. The two most effective monosaccharide taste stimuli that have structural features in common with sucrose are methyl α-D-glucopyranoside and methyl β-D-fructofuranoside.

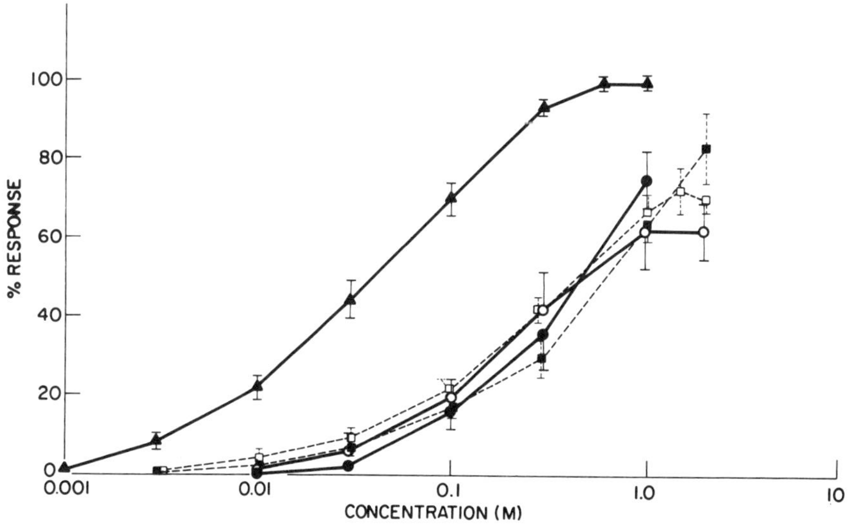

Figure 2. Comparison of mean integrated response of five disaccharides as gustatory stimulants. The maximum response of sucrose is considered to be 100% and the response of the other sugars are compared with the maximum sucrose response. Sucrose (▲ , n=37); maltose, (□, n=5); maltitol (■ , n=5); palatinose (● , n=4); and turanose (○, n=5). Bars indicate 95% confidence intervals (from Jakinovich, 1976).

3. The effectiveness of linear polyols as taste stimuli gradually increases with carbon chain length and plateaus at five carbons, a length approximately that of a sucrose molecule.

Kinetic Mechanism Studies

1. Based on the analyses of kinetic plots of the responses, sugars are thought to bind to the receptor site on a one-to-one basis. Moreover, because the electrophysiological response is saturable, it would appear that there are a finite number of receptor sites within the tase receptor cell membrane.

2. From pH experiments, this binding is believed to involve hydrogen bonding.

3. The taste responses to a mixture of methyl α-D-glucopyranoside and sucrose and a mixture of D-sorbitol and sucrose suggest that these sugars are all acting at a common receptor site.

4. In contrast, the taste responses to mixtures of sucrose and saccharin suggest that these two sweeteners are binding to separate receptor sites.

Table 1. *Stimulating Effectiveness of Some Disaccharides (Mean Values)*

Sugar	*Structure*[1]	*Thresh-old*[2]	CR_{50}	*Kd*	n^3	*Maximum response*	N^4
		M	M	M			
Sucrose	Glu α(1 ⟷ 2) Fru	0.001	0.042±0.005[5]	0.037	0.96	1.0	32
Turanose	Glu α(1 ⟶ 3) Fru	0.03	0.23 ±0.02	0.30	1.14	0.69±0.08	5
Palatinose	Glu α(1 ⟶ 6) Fru	0.03	—	0.49	1.0	—	4
Maltose	Glu α(1 ⟶ 4) Glu	0.01	0.24 ±0.05	0.29	1.00	0.75±0.06	5
Cellobiose	Glu β(1 ⟶ 4) Glu	0.01	—	0.33	1.2	—	5
Maltitol	Glu α(1 ⟶ 4) GluOH	0.03	—	0.34	0.86	—	5
Cellobiitol	Glu β(1 ⟶ 4) GluOH	0.03	—	0.50	1.27	—	5
Trehalose	Glu α(1 ⟷ 1) Glu	0.03	0.21 ±0.03	0.26	1.22	0.83±0.10	5
Lactulose	Gal β(1 ⟶ 4) Fru	0.01	0.18 ±0.02	0.23	0.98	0.88±0.08	5
β-Lactose	Gal β(1 ⟶ 4) Glu	0.01	—	0.31	0.89	—	5
Melibiose	Gal α(1 ⟶ 6) Glu	0.03	0.18 ±0.03	0.37	1.09	0.68±0.27	5
Lactitol	Gal β(1 ⟶ 4) GluOH	0.01	—	0.26	1.02	—	5
Melibiitol	Gal α(1 ⟶ 6) GluOH	0.03	—	0.23	1.00	—	5

1. Abbreviations: Glu, glucose; Gal, galactose; Fru, fructose; GluOH, glucitol.
2. Threshold is defined as the lowest concentration tested that elicited a measureable response in 50% of the animals (see text).
3. n, Number of molecules per receptor site.
4. N, Number of animals.
5. 95% confidence intervals are indicated.

5. Stimulation of the receptor cell by the sugar involves at least three steps: binding to the site, activation, and inactivation of the site.

Discussion of Results

Structure-Activity Studies

Disaccharide studies. The most effective "natural" disaccharide taste stimulus is sucrose (Figure 2 and Table I) (Jakinovich, 1976). The responses elicited by disaccharides in the Mongolian gerbil are consistent with the data obtained from the chorda tympani nerves of other mammals. The order of stimulatory effectiveness (sucrose > maltose = lactose) in the Mongolian gerbil, when compared at a single concentration (0.3 M), is generally consistent with the effectiveness of these disaccharides on other mammals tested, e.g., as taste stimuli, sucrose > maltose = lactose for both the rat and hamster (Hardiman, 1964), sucrose > lactose ≥ maltose in the human (Diamant et al., 1965), and sucrose > maltose > lactose in the dog (Andersen et al., 1962).

The variation in effectiveness of the disaccharides as taste stimuli is not easily explained by different chemical or physical properties. Whether a disaccharide is a reducing or nonreducing sugar does not seem to play a role in taste stimulation since neither type is necessarily more effective. Reducing sugars are those which reduce Fehling's or Tollen's reagent due to the presence of their free carbonyl groups. For example, maltose, cellobiose, lactose, turanose, palatinose, lactulose, and melibiose are all reducing sugars but poor stimuli compared with sucrose, a nonreducing sugar. On the other hand, the nonreducing disaccharides maltitol, cellobiitol, melibiitol, lactitol, and trehalose are no more effective than their reducing counterparts. Moreover, while increasing water solubility of sugars has long been associated with their stimulatory ability (Andersen et al., 1962), we have found that this is not always the case. For example, palatinose, cellobiose, and lactose, all poor stimuli, are barely soluble compared with sucrose. On the other hand, turanose is more soluble than sucrose but is a poorer stimulant. Therefore, solubility cannot be the sole determinant of effectiveness. The sugar responses are not due to osmotic pressure either, because sucrose is clearly a superior taste stimulus even though it has the same osmotic pressure as the rest of these sugars.

All of this would suggest, therefore, that the effectiveness of sucrose over the other disaccharides as taste stimuli is due to the presence of a specific sucrose receptor site. The other disaccharides may fit into this site but not as perfectly, as witnessed by their failure to stimulate as well. In addition, turanose and palatinose, the two fructosyl glucosides (the constitutional isomers of sucrose), are much less effective stimuli than sucrose. These two disaccharides, unlike sucrose, are reducing sugars that mutarotate and exist as a mixture of isomers. A paucity of one type of isomer that would be more complementary to the site could also account for the poor response.

Monosaccharide studies. In electrophysiological experiments most monosaccharides stimulate the gerbil's taste nerve in a manner similar to that of the disaccharides (Jakinovich and Goldstein, 1976). The order of effectiveness of reducing monosaccharides (fructose > mannose > glucose > galactose) at a single concentration (0.3 M) in producing taste responses in the chorda tympani nerve is the same in all mammals studied, including the human (Diamant et al., 1963), gerbils (Jakinovich and Oakely, 1975), rat (Noma et al., 1971), hamster (Noma et al., 1974), and dog (Andersen et al., 1963).

With regard to methyl glycopyranosides, the taste responses of some other species are consistent with those of the Mongolian gerbil. Specifically, in three other gerbil species (Jakinovich and Oakley, 1975) and the hamster (Noma et al., 1974), methyl α-D-glucopyranoside is a better taste stimulant than methyl β-D-glucopyranoside (Figure 3). This observation indicates that in these five species binding of a methyl D-glycopyranoside to its receptor site is enhanced if the sugar contains an axially oriented C-1 group. In addition, methyl α-D-glucopyranoside was found to be a more effective taste stimulant than either methyl α-D-mannopyranoside (C-2 epimer), methyl α-D-allopyranoside (C-3 epimer), or methyl α-D-galactopyranoside (C-4 epimer) (Jakinovich et al., 1981; Jakinovich, 1981a). This indicates that the equatorial hydroxyl groups at positions C-2, C-3, and C-4 are preferred to the axial for more effective binding.

The two methyl glycosides, methyl α-D-glucopyranoside and methyl β-D-fructopyranoside, which most closely resemble the two moieties of sucrose, were found to be the most stimulatory (Figure 3) (Jakinovich and Goldstein, 1976). This finding is consistent with our proposal of a sucrose receptor site, composed of two subsites, one that binds the glucopyranosyl and one that binds the fructofuranosyl portion of the sucrose molecule. At the same time, there is no reason to discount the idea of separate glucose and fructose sites responsible for the binding of these sugars. Research based on fly sugar responses has reported separate glucose and fructose receptor sites (Shimada et al., 1974). Furthermore, in various mammal single neuron experiments, fructose as well as some other sugars are reported to be better stimuli than sucrose (Pfaffmann, 1969), suggesting the possibility of other sugar receptor sites. Nevertheless, consistent with the sucrose site hypothesis, in the gerbil (Cheal et al., 1977) and other mammals (Noma et al., 1974; Bradley, 1970), many

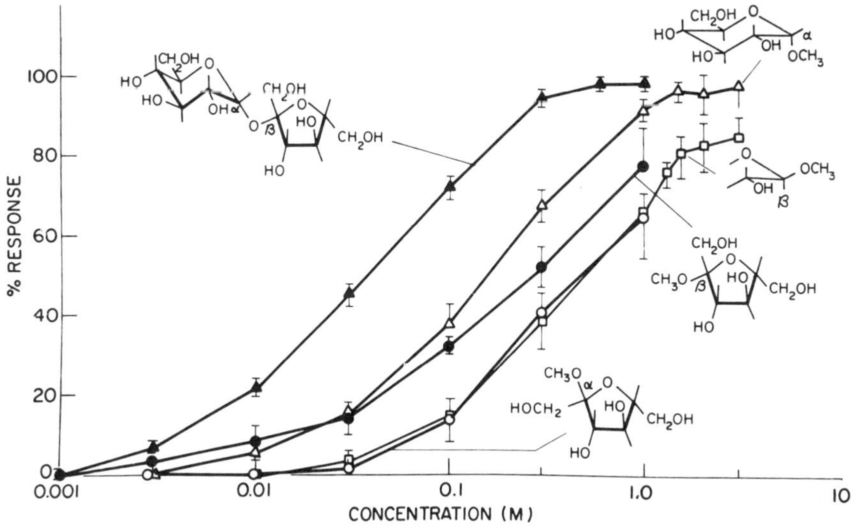

Figure 3. Mean integrated response of the chorda tympani nerve discharge in the gerbil to sucrose (▲ , $n=49$); methyl α-D-glucopyranoside (△ , $n=15$); methyl β-D-glucopyranoside (□, $n=5$); methyl β-D-fructofuranoside (●, $n=5$); and methyl α-D-fructofuranoside (○ , $n=5$). Responses are relative to the maximum sucrose response. Bars represent 95% confidence intervals (from Jakinovich and Goldstein, 1976).

single taste neurons that are responsive to sugars respond best to sucrose.

Polyol studies. The electrophysiological responses of the gerbil's taste nerve to polyols are similar to those observed for both disaccharides and monosaccharides (Jakinovich and Oakley, 1976). The polyols electrophysiologically stimulate the taste receptors of the dog (Andersson et al., 1950), monkey (Gordon et al., 1959), hamster (Hardiman, 1964), and rat (Noma et al., 1971). In the hamster, as in the gerbils, the order of stimulatory effectiveness of the few polyols tested was found to be mannitol > glycerol > ethylene glycol (Hardiman, 1964).

Like the monosaccharides, none of the polyols was as effective a stimulus as sucrose. We also found that the effectiveness of the polyols as taste stimuli gradually increases with chain length, plateaus at five carbons, and then remains the same for the six- and seven-carbon compounds (Figure 4). This leveling off may be interpreted to mean that the polyols are binding to a sucrose receptor site and their effectiveness is limited by their length.

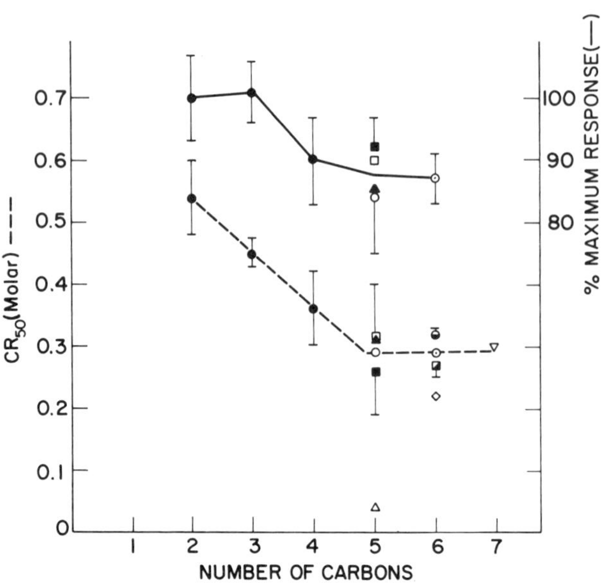

Figure 4. Relationship between the number of carbons in sugar alcohols and the concentration (left ordinate) which elicited a 50% response (CR_{50}) and the maximum response (right ordinate) relative to sucrose, maximum taken as 100%. Bars indicate 95% confidence interval. Ethylene glycol (2C, ●, n=5); glycerol (3C, ●, n=6); erythritol (4C, ●, n=6); D-ribitol (5C, ■, n=6); L-arabinitol (5C, ▲, n=6); D-arabinitol (5C, ○, n=6); D-xylitol (5C, □, n=5); D-sorbitol (6C, ○, n=5); D-galactitol* (6C, ◒, n=5); D-mannitol* (6C, ◪, n=5); myo-inositol* (6C, ◇, n=10); perseitol* (7C, ▽, n=6); sucrose (△, n=28). Asterisks indicate sugars whose insolubility prevented direct determination of maximum response. The CR_{50} for these compounds was estimated from K_ds (from Jakinovich and Oakley, 1976).

Kinetic Mechanism Studies

Theoretical model. Beidler (1954) was the first person to quantify taste responses and develop a theoretical model for the relationship between the stimulus concentration and the physiological response. According to Beidler, the binding of the stimulus, S, to the receptor site, R, can be measured as an increment of summated activity. This is represented by the following equation:

$$S + R \rightleftharpoons SR. \quad (1)$$

From this equation a dissociation constant (K_d) can be calculated

$$K_d = \frac{[S]\,[R]}{[SR]} \quad (2)$$

Assuming that (*a*) the response (*Resp*) measured was linearly related to the number of stimulus molecules bound, ($Resp \propto [SR]$), and (*b*) at a high concentration of *S*, when all receptors are occupied, the maximum response ($Resp_{max}$) is reached, ($Resp/Resp_{max} = 1$), then:

$$\frac{Resp}{Resp_{max}} = \frac{\text{sites filled}}{\text{total sites}} = \frac{[SR]}{[R] + [SR]} \qquad (3)$$

By substituting Eq. 2 into Eq. 3:

$$\frac{Resp}{Resp_{max}} = \cfrac{1}{1 + \cfrac{Kd}{[S]}} \qquad (4)$$

By rearrangement of Eq. 4 we get:

$$\frac{[S]}{Resp} = \frac{[S]}{Resp_{max}} + \frac{Kd}{Resp_{max}} \qquad (5)$$

Since $K_a = \frac{1}{K_d}$, Eq. 5 is essentially Beidler's taste equation (1954).

When $Resp/Resp_{max} = 1/2$, it follows that $CR_{50} = K_d = [S] = \frac{1}{K_a}$.

If more than one stimulus molecule combines with one receptor site, then Eq. 1 would become:

$$nS + R = RS_n \qquad (6)$$

and the response would be proportional to the amount of RS_n formed. Eq. 4 would then become:

$$\frac{Resp}{Resp_{max}} = \cfrac{1}{1 + \cfrac{Kd}{[S]^n}} \qquad (7)$$

By rearranging Eq. 7 and taking the log we get:

$$log \frac{Resp}{Resp_{max} - Resp} = n \log [S] - \log K_d \qquad (8)$$

This is identical to the Hill equation (1910). The Hill equation has been used before to quantify neural responses, e.g., Tateda and Hidaka (1966) using glycine in rats.

Our data from the gerbil's chorda tympani nerve responses to most disaccharides, monosaccharides, and polyols are consistent with Beidler's (1954) theory because they fit a straight line in the reciprocal plot (Figures 5A and

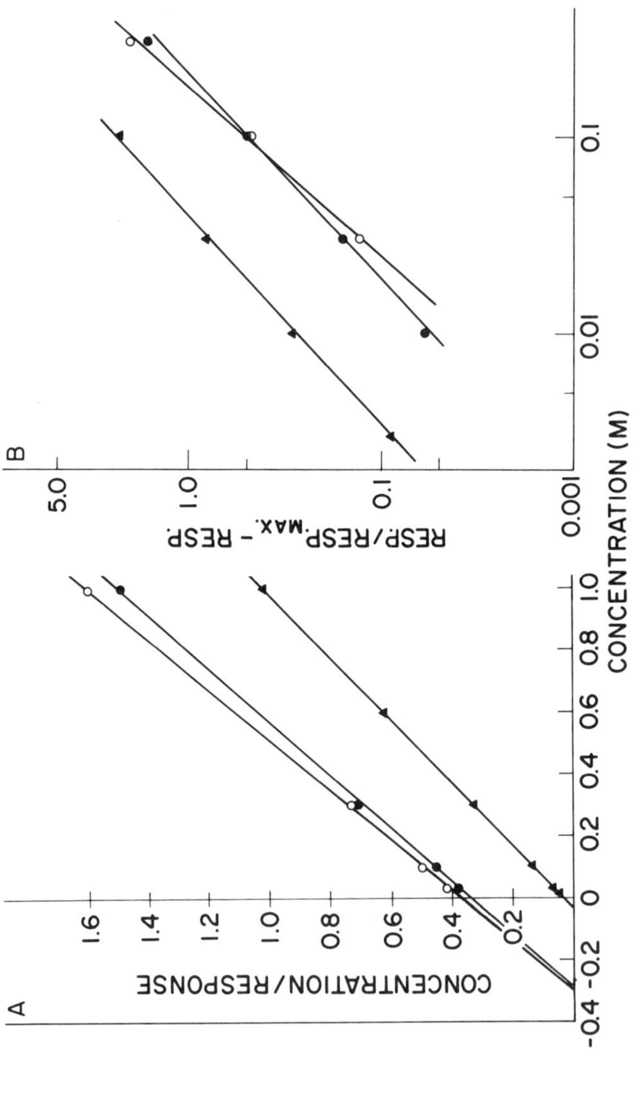

Figure 5. (A) Beidler plot of data from Figure 2. Resp, response (height of integrated nerve response); slope is 1Resp$_{max}$, Resp$_{max}$ being the maximum response and K_d/Resp$_{max}$ is the y-intercept. K_d is the dissociation constant and —K_d is the x-intercept. Sucrose (▲), turanose (O), and maltose (●) are presented. (B) Hill plot of data from A. The slope is N, N being the number of molecules interacting with each receptor site. Sucrose (▲ , N=0.96); turanose (O, N=1.14); maltose (● , N=1.00) (from Jakinovich, 1976).

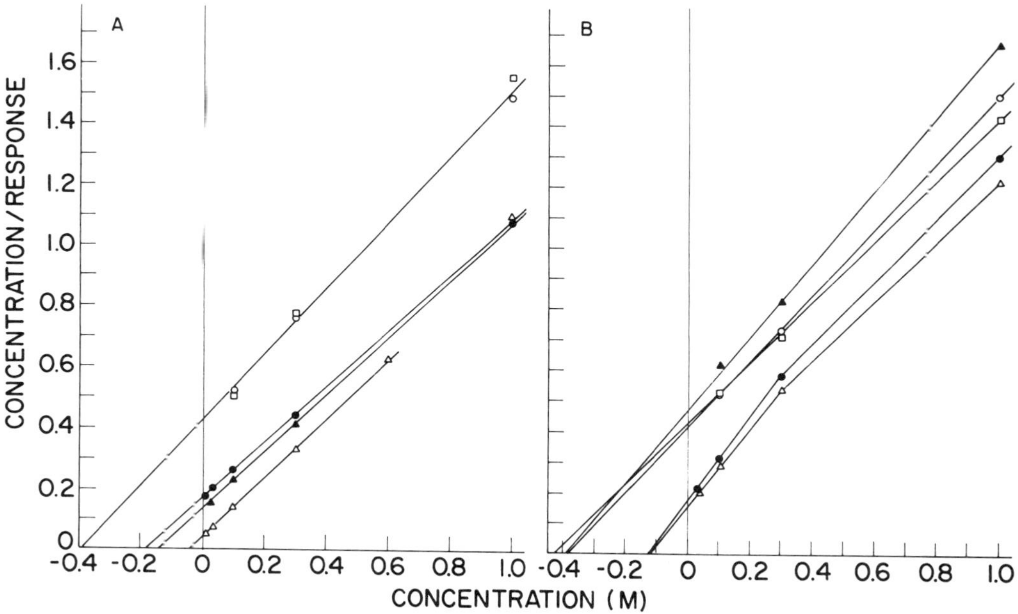

Figure 6. (A) Beidler plot of gerbil's integrated taste response to various glycosides. Sucrose (△); methyl α-D-xylopyranoside (●); methyl α-D-glucopyranoside (▲); methyl β-D-glucopyranoside (○); and methyl β-D-xylopyranoside (□). (B) Beidler plot of gerbil's integrated taste response to D-fructose (△) (equilibrated at 25°C) and the four methyl fructosides: methyl α-D-fructofuranoside (○); methyl β-D-fructofuranoside (●); methyl α-D-fructopyranoside (□); methyl β-D-fructopyranoside (▲) (from Jakinovich and Goldstein, 1976).

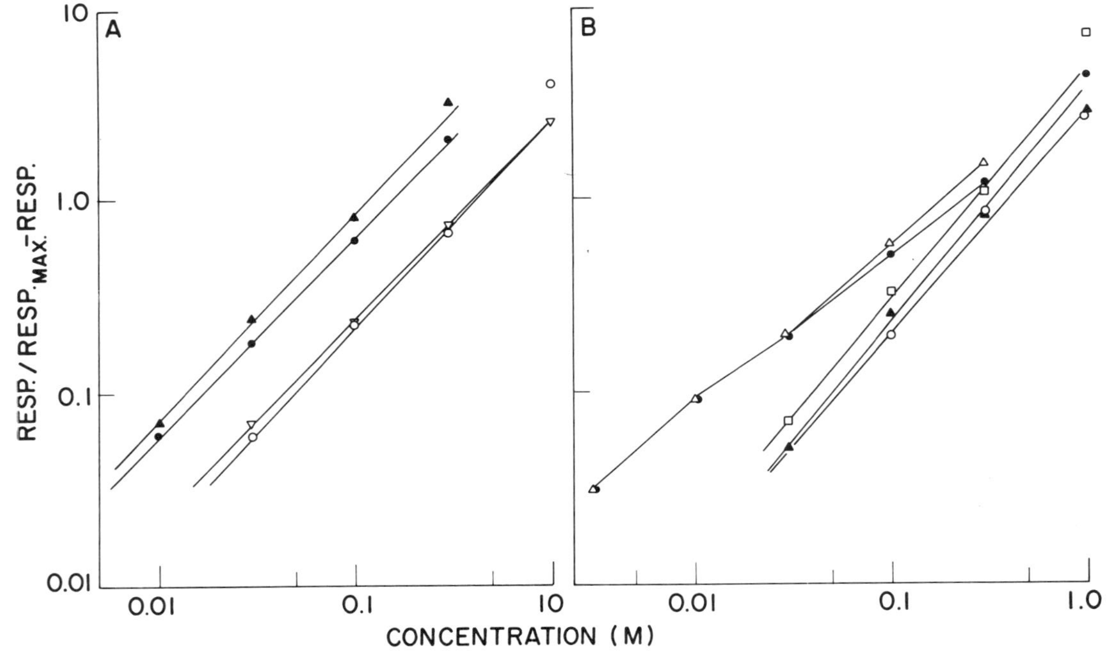

Figure 7. (A) Hill plot of data from Figure 6A. Methyl α-D-glucopyranoside (●, n=1.06); methyl α-D-xylopyranoside (▲, n=1.10); methyl β-D-glucopyranoside (▽, n=1.04); methyl β-D-xylopyranoside (○, n=1.07). For the meaning of Resp/$Resp_{max}$—Resp and N, see text. (B) Hill plot of data from Figure 6B. Methyl β-D-fructopyranoside (●); methyl α-D-fructofuranoside (○, n=1.14); methyl β-D-fructopyranoside (▲, n=1.18); methyl α-D-fructopyranoside (□, n=1.12); fructose (equilibrated, 25°C) (△) (from Jakinovich and Goldstein, 1976).

6A) and have a slope of 1 in a Hill plot (Figures 5B and 6B, Table I). Typically, sugar concentration-response curves are sigmoidal on semilog plots. For those sugars that are sufficiently soluble, concentration-response curves show saturation (Figures 2, 3, 4, and Table I).

In other animal species, including the human, responses to sucrose yield straight lines on a Hill or Beidler plot, indicating that for sucrose, at least, Beidler's monomolecular binding hypothesis holds. As one would expect, the concentration-response curves to sucrose are sigmoidal and saturate on a semilog plot. Although a few experiments failed to show saturation of the concentration-response functions done by others (Diamant et al., 1963), these discrepancies may be related to species' sensitivity and/or the failure of the researchers to test sufficiently high concentrations of sucrose.

We also found a few exceptions in our data. For example, the gerbil's responses to fructose and methyl β-D-fructofuranoside do not yield straight lines on a Beidler or a Hill plot (Figures 6 and 7). This finding could be interpreted to signify that the sugar receptor is composed of two subsites that may be occupied by a single disaccharide or two monosaccharide molecules. On the other hand, the variation in the aforementioned plots may indicate cooperativity. In the rat (Tateda and Hidaka, 1966) the formation of a multimolecular complex between stimulating molecules and the receptor site has been suggested. Another interpretation of plot non-linearity might be the presence of multiple fructose sites. However, regarding the data from the gerbil's taste responses to fructose and methyl β-D-fructofuranoside, we are still left without a satisfactory explanation.

Effect of pH. The effect of pH on the ionic state of organic molecules is well known and has proved to be a useful tool in exploring the binding mechanisms between drugs and their receptor sites (Ariens, 1964). One of the most widely accepted theories on sweetener binding mechanisms is Shallenberger and Acree's AH-B theory (1967). They believe that all sweeteners bind to their receptor sites by hydrogen bonding. However, in view of the fact that many of the sweeteners, in particular saccharin, will ionize, depending upon pH, we thought it possible that some of these compounds might be binding to the receptor site by ionic rather than hydrogen bonding. Therefore, we conducted pH experiments on gerbils with solutions of sucrose and saccharin, and found that, indeed, there seemed to be different binding mechanisms associated with these two sweeteners (Jakinovich, 1982b). For example (Figure 8), we observed that the optimal pH response range for saccharin is from pH 5–8. At pH 7 and 8, saccharin is fully ionized and, therefore, we believe it is held to its receptor site by ionic bonds. In contrast, the sucrose optimum pH response

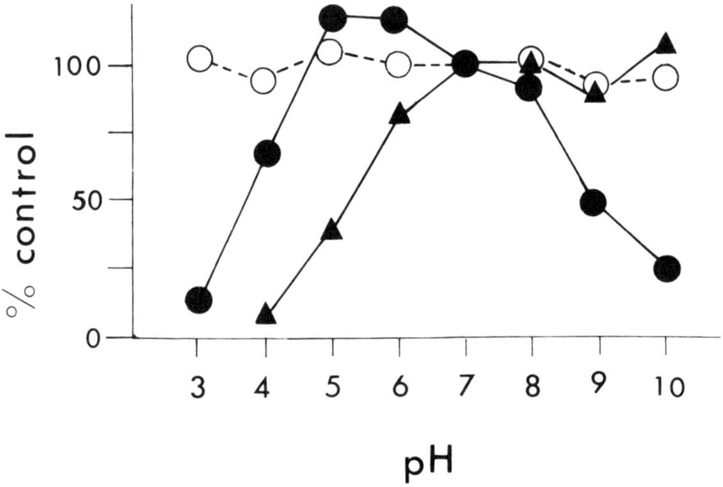

Figure 8. The effect of pH on the taste response to 0.1 M sodium chloride (○), 0.001 M H-saccharin (●), and 0.03 M sucrose (▲). The concentrations chosen produce nearly equivalent responses. The control response is the response of a substance dissolved in 0.1 M potassium phosphate buffer, pH 7 (from Jakinovich, 1982b).

range is found to be from pH 6–10. At pH 7 and 8, sucrose is completely un-ionized and, in this case, we believe that sucrose is hydrogen bonded to its receptor site, in accordance with the Shallenberger and Acree theory (1967). Our sucrose-pH results are borne out by work done on other mammals by Ogawa (1969) (working with whole nerves) and Hyman and Frank (1980) (working on single chorda tympani nerve fibers).

Mixtures. Taste responses to mixtures of sucrose and some other sugars suggest that these compounds are binding to a common receptor (Jakinovich and Goldstein, 1976; Jakinovich and Oakley, 1976). Ideally, this competition can be demonstrated when mixtures of sucrose and methyl α-D-glucopyranoside are applied to the gerbil's tongue (Figure 9), in which case taste responses to the mixtures closely fit theoretical curves calculated for competitive interaction of two substances for a single receptor site. Similar results were also obtained with another mixture of sucrose and D-sorbitol.

On the other hand, taste responses to mixtures of saccharin and sucrose do not easily fit the competitive interaction model as seen with the sugars (Jakinovich, 1982b). Instead, mixtures of these two compounds produce responses that suggest a different model involving independent but interacting receptor sites (Figure 10). It is notable that in the development of the theoretical curves, the two models differ very slightly, with the difference most easily discernible at the maximum response level. Even though the response differences between the single and two separate receptor site models are very small, those differences are clearly seen and, since all models are based on certain assumptions, any observed deviations therefrom must be considered carefully rather than ignored (Kenakin and Black, 1978).

Inhibitors. It has recently been revealed through the use of inhibitors that

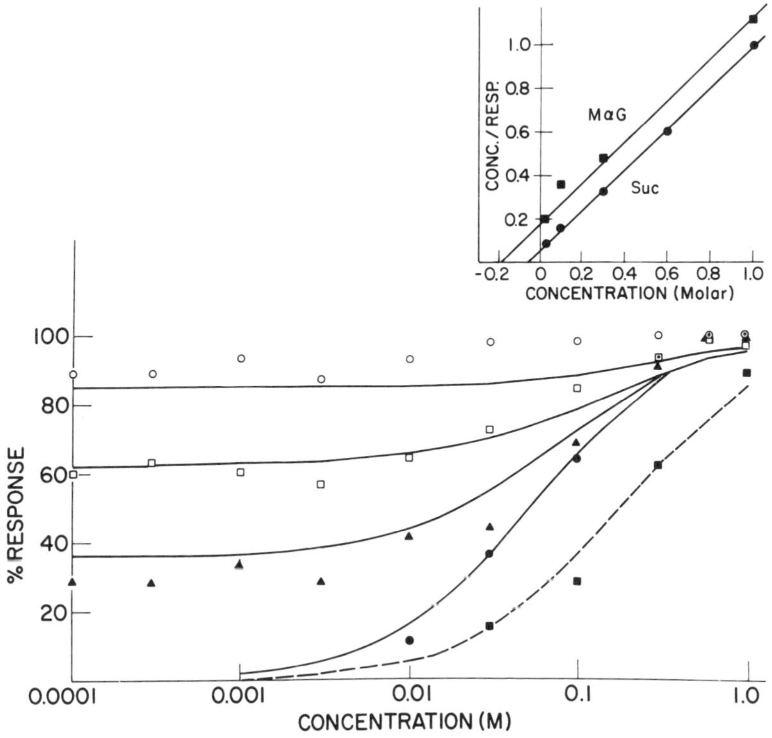

Figure 9. Concentration-response curve of sucrose in the presence of methyl α-D-glucopyranoside (MαG). The solid lines are theoretical curves obtained from an equation describing the competitive interaction of two substances with a single receptor site. Data points for sucrose alone (●); sucrose + 0.1 M MαG (▲); sucrose + 0.3M MαG) (□); sucrose + 1.0 M MαG (O); MαG alone (■). Dashed line is the theoretical curve drawn from the binding equation of MαG. K_d for sucrose (0.05 M) and K_d for MαG (0.18 M) were determined from the Beidler plot (see inset). CR_{50} for sucrose = 0.052 M and for MαG = 0.18 M (from Jakinovich and Goldstein, 1976).

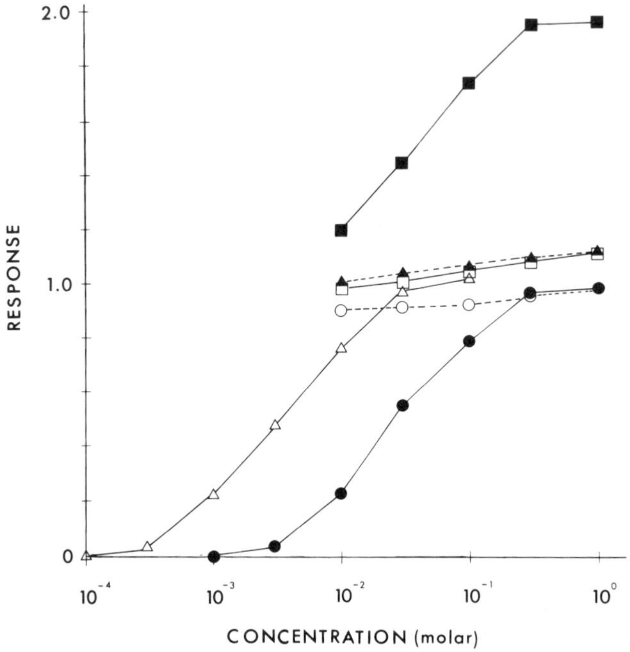

Figure 10. Concentration-response curve of sucrose in the presence of saccharin. The dashed lines are theoretical curves obtained from an equation describing (1) the competitive interaction of two substances with a single receptor site (○), (2) independent receptor sites (■), and (3) independent receptor sites with functional interaction (▲). Also shown are data points for sucrose alone (●), saccharin alone (△), sucrose + 3 X 10^{-2} M saccharin (□); N=5. To simplify figure standard error bars are not shown. The solvent is 0.1 M potassium phosphate buffer, pH 7 (from Jakinovich, 1982b).

the fly's sugar taste receptor cell involves at least three receptor sites (Shimada and Isono, 1978). Inhibitors have also been useful in studying enzyme-substrate binding sites (Dixon and Webb, 1964) as well as drug receptor sites (Ariens, 1964). Because of this, we have tested a number of substances as potential inhibitors of the gerbil's sucrose response. First we tested alloxan (unpublished) and gymnemic acid (Jakinovich and Oakley, 1975), two substances known to inhibit the sweet taste response in other animals (Hagstrom, 1957; Zawalich, 1973). However, these two compounds had no effect on the gerbil's sucrose response. Then we decided to test a number of sweeteners and/or derivatives that we know do not stimulate the gerbil's sweet taste receptor to see whether they would show any inhibition effects (Jakinovich,

1981b, 1983). These include: (*a*) L-aspartyl-L-phenylalanine methyl ester, (*b*) beryllium acetate, (*c*) furan acrylonitrile, (*d*) 4-(methoxymethyl), 1,4-cyclohexadiene-1-synoxine, (*e*) monellin, (*f*) 5-nitro-2-propoxyaniline, (*g*) perillartine, and (*h*) methyl 4,6-dichloro-4,6-dideoxy-α-D-galactopyranoside (DiCl-gal). Although these compounds do not stimulate, we felt it possible that they bind without activating the sucrose receptor. To test this hypothesis, we mixed each compound with sucrose and noted the taste response. To date, except for DiCl-gal, none of the above substances behave as inhibitors of the gerbil's sweet taste response. DiCl-gal, the quickly reversible inhibitor we have discovered, resembles part of the sweetener molecule, *chloro*-sucrose (4, 6-dichloro-4, 6-dideoxy-α-D-galactopyranosyl 1, 6-dichloro-1, 6-dideoxy-β-D-fructofuranoside).

In the gerbil, we have observed varied responses to DiCl-gal alone. Most of the time, it does not stimulate at all; at other times it barely stimulates, and, on a few occasions, it suppresses the baseline activity to a small degree. However, when mixed with sucrose or with saccharin, DiCl-gal suppresses the gerbil's taste response to these two sweeteners (Figure 11A). In contrast, when mixed with sodium chloride or hydrochloric acid, DiCl-gal has no effect on the taste responses produced by these compounds (Figure 12).

In our first report (Jakinovich, 1982a), the inhibition of these sweeteners by DiCl-gal was thought to be differential (competitive inhibition of the sucrose response and noncompetitive inhibition of the saccharin response). However, reanalysis of the data on reciprocal plots suggests the possibility that both sweeteners might be inhibited competitively by DiCl-gal (Figure 11B) (Jakinovich, 1983). This is a puzzle because it is difficult to understand how DiCl-gal can be acting simultaneously as a competitive inhibitor with both sucrose and saccharin. The three compounds fail to exhibit any known common molecular features, and in addition, the two sweeteners are believed to bind to separate receptor sites. One possible solution to this dilemma is that DiCl-gal is interacting with sucrose and saccharin and, by reducing the stimulus concentration, is lowering the taste response. Another explanation of the kinetic plots is that DiCl-gal may be acting as a type of allosteric inhibitor that is only diminishing the sweetener binding (Lehninger, 1975). Its allosteric site could be the sucrose site itself. The latter possibility, which takes into account the molecular similarities between DiCl-gal and sucrose, suggests that when DiCl-gal binds to the sucrose site, not only is DiCl-gal a competitive inhibitor at that site, but, at the same time, it allosterically affects the saccharin binding site.

The finding that DiCl-gal may be a competitive inhibitor of the sugar response required modification of Beidler's theoretical model with regard to

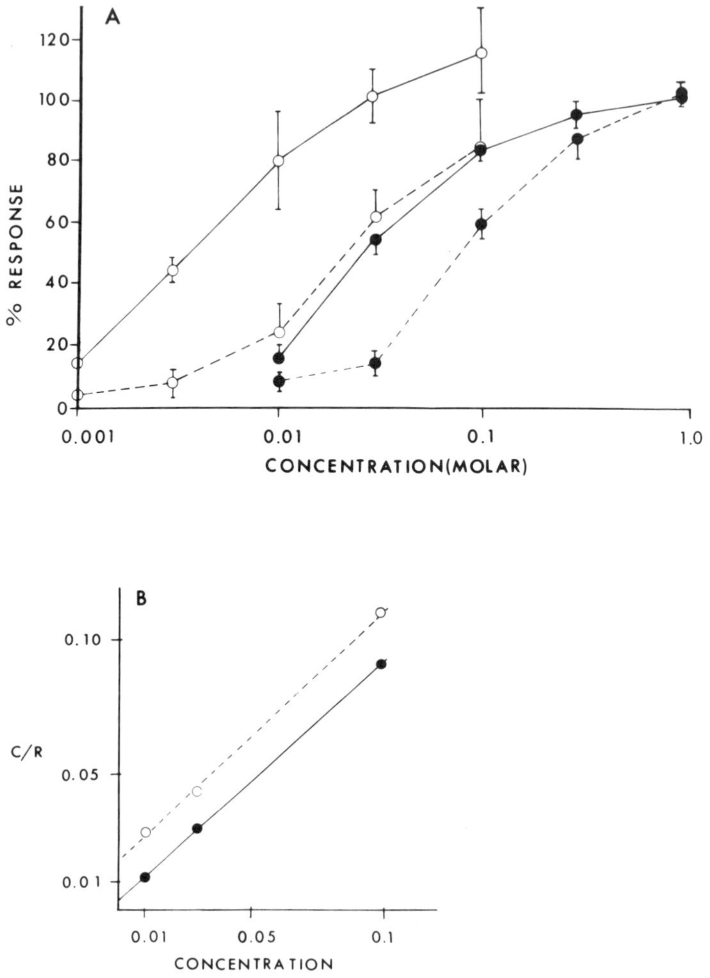

Figure 11. (A) A comparison of the mean integrated nerve responses to sweetener solutions with the responses to mixtures of the sweetener and DiCl-gal. The sucrose solutions were prepared with deionized water: ●——●, sucrose (N=5); ●- - -●, mixture of sucrose and 0.1 M DiCl-gal (N=5). The H-saccharin solutions were prepared with 0.1 M potassium phosphate buffer. pH 7: ○——○, H-saccharin (N=4); ○- - -○, mixture of H-saccharin and 0.1 M DiCl-gal (N=4). The bars indicate ±2 SE. (B) A Beidler plot of integrated nerve responses to H-saccharin (●) and mixtures of H-saccharin and 0.1 M DiCl-Gal (○); C represents concentration and R is response (from Jakinovich, 1983).

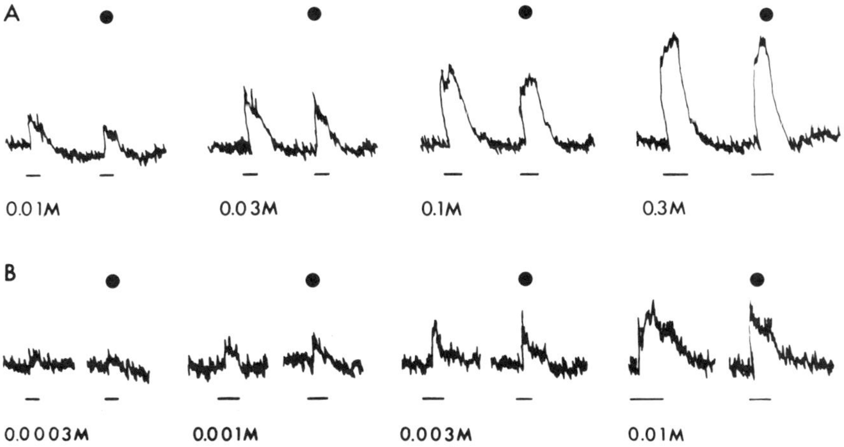

Figure 12. (A) A comparison of the integrated nerve responses to sodium chloride with the responses to mixtures of sodium chloride and 0.1 M DiCl-gal (●). The bars indicate the addition of the stimuli to the flow system. (B) A comparison of the integrated nerve responses to hydrochloric acid with the responses to mixtures of hydrochloric acid and 0.1 M DiCl-gal (●) (from Jakinovich, 1983).

activation and inactivation of the receptor. In his original theory, Beidler assumed that the receptor became activated when it bound a stimulus molecule and, further, that the taste response was proportional to the number of receptor-stimulus complexes formed.

$$S + R \rightleftharpoons SR. \qquad (1)$$

However, in the case of sugars at least, it would appear that this basic assumption must be changed because DiCl-gal is binding to the receptor, forming receptor complexes, but is not activating the receptor cell. On the other hand, when sucrose binds to that receptor, activation follows. At this point, it is evident that at least two separate states must be involved, an inactivated (SR) and an activated (SR^*) state.

$$S + R \rightleftharpoons SR \rightleftharpoons SR^*. \qquad (2)$$

To take this one step further, to account for the adaptation that occurs in the taste receptor cell during continuous exposure to sucrose (Ozeki, 1971) and the diminished receptivity of the taste receptor cells shortly after removal of the stimulus, still another state of the receptor (SR') must be considered to exist. Therefore, we propose the following modifications:

$$\begin{array}{ccc} S + R & \rightleftharpoons & SR. \\ \updownarrow & & \updownarrow \\ SR' & \leftarrow & SR^* \end{array} \qquad (3)$$

This scheme for binding, activation, and adaptation of the sucrose receptor is similar to the one proposed by Katz and Thesleff (1957) for acetylcholine receptors.

In summary, the stimulus molecule (*S*) binds to its receptor (*R*), forming a stimulus-receptor complex (*SR*). If the stimulus is uniquely suited to that receptor, activation should occur and taste response result. But, if the complexing is less than structurally perfect, binding may occur but not activation. In the case of the activated receptor, should the stimulus be continuously applied, as often happens with taste receptor cells, a third stimulus-receptor complex may be formed, (*SR′*), at which state the response diminishes—to recover slowly after removal of the stimulus.

Concluding Remarks

Putative sucrose receptor site. We believe that this sucrose site (Figure 13) is located in the protein of the taste receptor cell membrane and consists of two subsites, a "superficial" α-glucopyranosyl and a "deep" β-fructofuranosyl site. All the data accumulated so far on the many sugars we have tested, with the exception of one (*chloro*-sucrose), can be explained by our putative sucrose receptor site model. The exception will be discussed later.

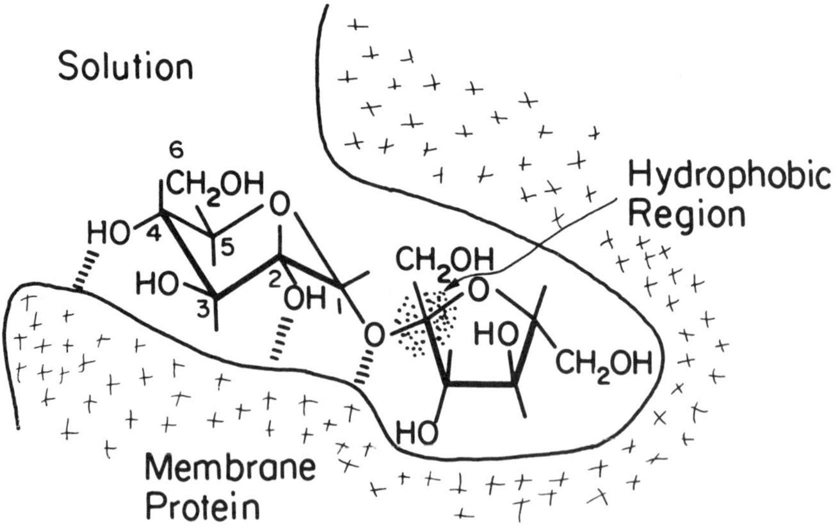

Figure 13. Proposed model for the "sucrose site" in the membrane of the gerbil gustatory cell. See text for explanation (from Jakinovich and Oakley, 1976).

The deep β-fructofuranosyl subsite is believed to be in the shape of a five-membered ring, complementary to the fructose moiety of sucrose, which could easily bind molecules of similar size and shape, for example, methyl β-D-fructofuranoside and the fructose moiety of sucrose itself. However, the requirement for a five-membered ring alone does not satisfy since methyl α-D-fructofuranoside is a less effective stimulant than methyl β-D-fructofuranoside (Jakinovich and Goldstein, 1976). In addition, since fructose in solution can assume five- or six-membered rings, molecules of both types may bind in varying degrees to the deep subsite. This type of specificity may account for the failure of some other fructose derivatives to stimulate very well, for example, turanose (3-O-α-D-glucopyranosyl-D-fructose), palatinose (6-O-α-D-glucopyranosyl-D-fructose), methyl α-D-fructopyranoside, and methyl β-D-fructopyranoside. The first two compounds mutarotate; the other two do not, but are bulky six-membered rings. Both these conditions contribute to the inability of all four compounds to bind as efficiently as sucrose.

With regard to the superficial β-glucopyranosyl subsite, we believe that this site binds sugars whose glucose moiety contains a pyranosyl ring. Our response data suggest that the various structural elements required for maximum response may also account for the differences in specificity between the two aforementioned subsites. For example, with regard to the OH groups on the glucose moiety, equatorial orientation on C-2, C-3, and C-4 are essential, whereas the C-5 CH_2OH is not (Jakinovich and Goldstein, 1976). Moreover, response to the axial OH on C-1 is favored over the equatorial OH. In addition, our observations that methyl α-D-glucopyranoside (with its methyl group at C-1) is a better stimulus than α-D-glucopyranose (without that methyl group) leads us to believe that there may be a hydrophobic region in the vicinity of the C-1 position, within which the methyl group may bind to enhance the taste response. On the other hand, the methyl group may merely prevent the mutarotation of the glucose molecule.

The combined specificity of the deep and the superficial subsites may account for the relatively weak responses of some disaccharides tested as gustatory stimulants. For example, large bulky sustituent groups at the C-1 position of the α-D-pyranosides (e.g., maltose, turanose, trehalose, etc.) may prevent effective interaction between the sugar and the deep site. Therefore, these sugars may have to interact mostly with the superficial site, resulting in weaker responses. We have also noted that disaccharides containing α-D-glucopyranoside are better stimuli than those with the β-D-glucopyranoside (Jakinovich, 1976), which agrees with our model. A seeming exception is the case of disaccharides containing the β-D-galactopyranoside eliciting a better response than those with the α-D-galactopyranoside. However, we feel this can be explained by visualizing a superimposition of the β-galactopyranoside molecule over the glucopyranosyl receptor site such that the C-4 OH group of galactose is in the same orientation as the C-1 OH group of glucose. When this occurs, there are one axial and three equatorial OH groups in the optimum stimulatory postions; this cannot be accomplished with the α-D-galactopyranoside.

With regard to the previously mentioned exception, *chloro*-sucrose, according to our model, when we tested this disaccharide of unusual chemical structure, we might have expected that it would not stimulate, because generally whenever the OH groups on a sugar molecule are removed and replaced by other substituent groups, the result is usually diminution of sweet taste (Birch, 1976). However, not only did *chloro*-sucrose stimulate the gerbil's chorda tympani nerve, but it was about 100 times more potent than sucrose (Jakinovich, 1981b). In addition, it probably tastes like sucrose, since gerbils trained to avoid sucrose in conditioned taste aversion experiments, generalized this avoidance to *chloro*-sucrose as well (Jakinovich, 1981b). Therefore, it would appear that more information is needed to explain the differences in response to *chloro*-sucrose vs. sucrose. One possible explanation is the solubility factor. *Chloro*-sucrose, in contrast to sucrose, is soluble in both water and lipid solvents, which may reaffirm our belief that the sucrose receptor also contains a hydrophobic region where the chlorine atoms may interact. On the other hand, the difference in lipid solubility may indicate the *chloro*-sucrose is able to partition into the lipid portion of the cell membrane and is, therefore, bound in higher concentrations, resulting in the greater than expected response to that compound.

Comparison with human behavior. A quantitative comparison of the gerbil's electrophysiological sugar taste responses with human behavioral experiments reveals that both species may have sugar receptor sites with similar specificity properties. This comparison is permissible because in humans a correlation has been established (Zotterman, 1971) between chorda tympani nerve responses and subjective taste intensity. Furthermore, the comparison is reasonable since taste aversion experiments suggest that the gerbil's behavioral taste responses to sugars parallel (in general) those of humans (Jakinovich, 1982c). With regard to our quantitative comparison, the following similarities were observed in both species:

1. Sucrose is the best taste stimulant among naturally occurring disaccharides (Fabian and Blum, 1943; Cameron, 1947; Jakinovich, 1976).

2. The synthetic disaccharide, *chloro*-sucrose, is many times more effective

a taste stimulant than sucrose (Hough and Phadnis, 1976; Jakinovich, 1981b).

3. Disaccharides that contain α-D-glucopyranoside, such as maltose and maltitol, are better taste stimuli than disaccharides that contain β-D-glucopyranoside, e.g., cellobiose and cellobiitol (Jakinovich, 1976; Lee, 1977).

4. Lactose, which contains the β-D-galactopyranoside, is a better stimulus than melibiose, the α-D-galactopyranoside (Jakinovich, 1976; Lee, 1977).

5. The taste intensity of methyl glycosides is affected by the orientation of individual hydroxyl or methoxy groups. For example, methyl α-D-glucopyranoside is a more effective taste stimulus than either methyl β-D-glucopyranoside or methyl α-D-mannopyranoside (Birch, 1976; Jakinovich and Goldstein, 1976).

6. Replacing the sugars' hydroxyl groups with hydrogen atoms at various positions (deoxy sugars) results in less intense taste stimuli (Birch, 1976; Jakinovich and Goldstein, 1976).

7. In humans, among reducing monosaccharides, fructose has a lower taste threshold (0.02 M) than glucose (0.04 M) (Fabian and Blum, 1943). In the gerbil the electrophysiological threshold of fructose (0.003 M) is also less than the glucose threshold (0.03 M) (Jakinovich and Goldstein, 1976).

8. In both species, the sucrose taste responses are suppressed by lowering the pH of a sucrose solution (Jakinovich, 1982b; Stone et al., 1969).

There are also a few dissimilarities:

1. In the human, β-D-fructopyranose was believed to be the sweetest fructose isomer (Shallenberger and Acree, 1967). In the gerbil, however, using comparable but not exactly the same experimental conditions and compounds, methyl β-D-fructofuranoside was found to produce a stronger electrophysiological response than methyl β-D-fructopyranoside (Jakinovich and Goldstein, 1976).

2. The linear polyols taste like sucrose to both man (Moskowitz, 1971) and the gerbil (Jakinovich, 1982c), but human taste behavior neither plateaus nor exhibits correlation between increasing chain length and taste responses, a phenomenon that is seen in the gerbil's electrophysiological responses (Jakinovich and Oakley, 1976).

3. Many artificial sweeteners (e.g., aspartame) that presumably stimulate man's taste receptors do not stimulate those of the gerbil (Jakinovich, 1981b).

4. Some compounds, such as cyclamate, which are sweet to humans, stimulate the gerbil's taste nerve but do not produce taste avoidance generalization in animals taught to avoid sucrose (Jakinovich, 1981b).

5. The sweet-taste inhibitor gymnemic acid does not abolish the electro-

physiological taste responses of sucrose in the gerbil (Jakinovich and Oakley, 1975) as it does in the human (Zotterman, 1971).

One possible explanation to account for these dissimilarities is species differences in sweet taste receptor specificity. Single site theories of sweet taste reception such as the "AH-B taste theory" may only apply to humans or closely related primates (Glaser et al., 1978). Slight differences between the human sweetener receptor and that of the gerbil may account for nonstimulation in this animal by the sweetener aspartame, which may be binding to but not activating its taste receptor. However, this possibility seems to be ruled out because aspartame and other nonstimulating sweeteners do not diminish the gerbil's sucrose response in mixture experiments (Jakinovich, 1981b). Another possibility is that the sweetener site is slightly modified in the gerbil so that some, but not all, sweeteners can interact there. A further alternative is that taste reception to sweeteners involves a number of, rather than one, specific sweet receptor sites, each with individual requirements, two examples being a sucrose receptor site (Jakinovich and Oakley, 1976) and a saccharin site (Jakinovich, 1982b).

At present, we are not completely certain of the mechanism of sweet taste reception. The multiple receptor hypothesis is attractive because there is evidence suggesting the presence of more than one sweetener site in mammals. For example, in electrophysiological experiments using dogs, gerbils, and monkeys, single taste neurons respond differentially to sugars (Andersson et al., 1950; Pfaffmann, 1969; Cheal et al., 1977). In adaptation experiments with the human, a given sweetener does not completely cross-adapt to all sweeteners (McBurney, 1972). Mixture experiments in the gerbil, using sucrose and saccharin together, provide evidence that more than one receptor is involved in the stimulation of the chorda tympani nerve (Jakinovich, 1982b). Finally, in the rat, alloxan will inhibit the sucrose electrophysiological response but not the saccharin response (Zawalich, 1973). To really understand the mechanism of the sweet taste response, and especially the multiple receptor hypothesis, further investigation is needed using specific inhibitors.

Acknowledgments

This work was supported in part by grants from the National Institutes of Health and PSC-CUNY. Thanks to V. Vlahopoulos for her contribution in both the electrophysiological experiments and synthesis, to D. Sugarman for the editorial comments, and to A. Ward for the typing.

References

Andersen, H. T., M. Funakoshi, and Y. Zotterman. 1962. Electrophysiological investigation of the gustatory effect of various biological sugars. *Acta Physiol. Scand.* 56:362–375.

———. 1963. Electrophysiological responses to sugars and their depression by salt. In *Olfaction and Taste* (Proceedings of the First International Symposium). Y. Zotterman, ed. Oxford: Pergamon Press Ltd., pp. 177–192.

Andersson, B., S. Landgren, L. Olsson, and Y. Zotterman. 1950. The sweet taste fibres of the dog. *Acta Physiol. Scand.* 21:105–119.

Ariens, E. J., ed. 1964. *Molecular Pharmacology*, Vol. I. New York: Academic Press, Inc., p. 363.

Beidler, L. M. 1953. Properties of chemoreceptors of tongue of rat. *J. Neurophysiol.* (*Bethesda*). 16:595–607.

———. 1954. A theory of taste stimulation. *J. Gen. Physiol.* 38:133–139.

———. 1969. Innervation of rat fungiform papilla. In *Olfaction and Taste III* (Proceedings of the Third International Symposium). C. Pfaffmann, ed. New York: The Rockefeller University Press, pp. 352–369.

Birch, G. G. 1976. Structural relationships of sugars to taste. *Crit. Rev. Food Sci. Nutr.* 8:57–95.

Bradley, C. M. A study of rat chorda tympani fiber discharge patterns in response to lingual stimulation with a variety of chemicals. Ph.D. thesis, The Florida State University, 1970.

Cameron, A. T. 1947. The taste sense and the relative sweetness of sugars and other sweet substances. New York: Sugar Research Foundation, Inc. Scientific Report Series no. 9, 74 pp.

Cheal, M., W. P. Dickey, L. B. Jones, and B. Oakley. 1977. Taste fiber responses during reinnervation of fungiform papillae. *J. Comp. Neurol.* 172:627–646.

Diamant, H., M. Funakoshi, L. Strom, and Y. Zotterman. 1963. Electrophysiological studies on human taste nerves. In *Olfaction and Taste* (Proceedings of the First International Symposium). Y. Zotterman, ed. Oxford: Pergamon Press Ltd., pp. 193–203.

Daimant, H., B. Oakley, L. Strom, C. Wells, and Y. Zotterman. 1965. A Comparison of neural and psychophysical responses to taste stimuli in man. *Acta Physiol. Scand.* 64:67–74.

Dixon, M., and E. C. Webb. 1964. *Enzymes*. 2nd ed. New York: Academic Press, Inc.

Fabian, F. W., and H. B. Blum. 1943. Relative taste potency of some basic food constituents and their competitive and compensatory action. *Food Res.* 8:179–193.

Frank, M. 1973. An analysis of hamster afferent taste nerve response functions. *J. Gen. Physiol.* 61:588–618.

Glaser, D., G. Hellekant, J. N. Brouwer, and H. van den Wel. 1978. The taste response in primates to the proteins thaumatin and monellin and their phylogenetic implications. *Folia Primatol.* 29:56–63.

Gordon, G., R. Kitchell, L. Strom, and Y. Zotterman. 1959. The response pattern of taste fibres in the chorda tympani of the monkey. *Acta Physiol. Scand.* 46:119–132.

Hagstrom, E. C. Nature of taste stimulation by sugar. Ph.D. thesis, Brown University, 1957.

Hardiman, C. W. Rat and hamster chemoreceptor responses to a large number of compounds and the formulation of a generalized chemosensory equation. Ph.D. thesis, Florida State University, 1964. University Microfilms, Ann Arbor.

Hill, A. V. 1910. The possible effects of the aggregation of the molecules of haemoglobin on its dissociation curves. *J. Physiol.* (*Lond.*). 40:iv–vii.

Hough, L., and S. P. Phadnis. 1976. Enhancement in the sweetness of sucrose. *Nature* (*Lond.*). 263:800.

Hyman, A. M., and M. E. Frank. 1980. Effects of binary taste stimuli on the neural activity of the hamster chorda tympani. *J. Gen. Physiol.* 76:125–142.

Jakinovich, W., Jr. 1976. Stimulation of the gerbil's gustatory receptors by disaccharides. *Brain Res.* 110:481–490.

———. 1981a. Gustatory responses of the Mongolian gerbil to methyl α-D-allopyranoside. The Association for Chemoreception Sciences, 3rd Annual Meeting. Abstract.

———. 1981b. Stimulation of the gerbil's gustatory receptors by artificial sweeteners. *Brain Res.* 210:19–81.

———. 1982a. Inhibition of the gerbil's electrophysiological sweetner taste response by methyl 4,6-dichloro, 4,6-dideoxy-α-D-galactopyranoside. The Association for Chemoreception Sciences, 4th Annual Meeting. Abstract.

———. 1982b. Stimulation of the Gerbil's gustatory receptors by saccharin. *J. Neurosci.* 2:49–56.

———. 1982c. Taste aversion to sugars by the gerbil. *Physiol. Behav.* 28:1065–1071.

———. 1983. Methyl 4,6-dichloro-4,6-dideoxy-α-D-galactopyranoside: an inhibitor of sweet taste responses in gerbils. *Science* (*Wash. DC*) 219:408–410.

Jakinovich, W., Jr., and I. J. Goldstein. 1976. Stimulation of the gerbil's gustatory receptors by monosaccharides. *Brain Res.* 110:491–504.

Jakinovich, W., Jr., and B. Oakley. 1975. Comparative gustatory response in four species of gerbilline rodents. *J. Comp. Physiol.* 99:89–101.

———. 1976. Stimulation of the gerbil's gustatory receptors by polyols. *Brain Res.* 110:505–513.

Jakinovich, W. Jr., D. Sugarman, and V. Vlahopoulos. 1981. Gustatory responses of the cockroach, housefly and gerbils to methyl glycosides. *J. Comp. Physiol.* 141:297–301.

Katz, B., and S. Thesleff. 1957. A study of the 'desensitization' produced by acetylcholine at the motor end-plate. *J. Physiol.* (*Lond.*). 138:63–80.

Kenakin, T. P., and J. W. Black. 1978. The pharmacological classification of practolol and chloropractolol. *Mol. Pharmacol.* 14:607–623.

Kimura, K., and L. M. Beidler. 1961. Microelectrode study of taste receptors of rat and hamster. *J. Cell. Comp. Physiol.* 58:131–139.

Lee, C. 1977. Structural functions of taste in the sugar series: taste properties of sugar alcohols and related compounds. *Food Chem.* 2:95–105.

Lehninger, A. L. 1975. *Biochemistry*. New York: Worth, p. 238.

McBurney, D. H. 1972. Gustatory cross adaptation between sweet-tasting compounds. *Percept. Psychophys.* 11:225–227.

Moskowitz, H. R. 1971. The sweetness and pleasantness of sugars. *Am. J. Psychol.* 84:387–405.

Noma, A., J. Goto, and M. Sato. 1971. The relative taste effectiveness of various sugars and sugar alcohols for the rat. *Kumamoto Med. J.* 24:1–9.

Noma, A., M. Sato, and Y. Tsuzuki. 1974. Taste effectiveness of anomers of sugars and glycosides as revealed from hamster taste responses. *Comp. Biochem. Physiol. A: Comp. Physiol.* 48: 249–262.

Ogawa, H. 1969. Effects of pH on taste responses in the chorda tympani nerve of rats. *Jpn. J. Physiol.* 19:670–681.

Ozeki, M. 1971. Conductive change associated with receptor potentials of gustatory cells in rat. *J. Gen. Physiol.* 58:688–699.

Pfaffmann, C. 1969. Taste preference and reinforcement. In *Reinforcement and Behavior*. J. T. Tapp, ed. New York: Academic Press Inc., pp. 215–241.

Shallenberger, R. S., and T. E. Acree. 1967. Molecular theory of sweet taste. *Nature (Lond.)* 216:480–482.

Shimada, I., and K. Isono. 1978. The specific receptor site for aliphatic carboxylate anion in the labellar sugar receptor of the fleshfly. *J. Insect Physiol.* 24:807–811.

Shimada, I., A. Shiraishi, H. Kijima, and H. Morita. 1974. Separation of two receptor sites in a single labellar sugar receptor of the fleshfly by treatment with *p*-chloromercuribenzoate. *J. Insect Physiol.* 20:605–621.

Stone, H., S. Oliver, and J. Kloehn. 1969. Temperature and pH effects on the relative sweetness of suprathreshold mixtures of dextrose fructose. *Percep. Psychophys.* 5:257–260.

Tateda, H., and I. Hidaka. 1966. Taste respoonses to sweet substances in rat. *Mem. Fac. Sci. Kyushu Univ. Series E Biol.* 4:137–149.

Zawalich, W. S. 1973. Depression of gustatory sweet response by alloxan. *Comp. Biochem. Physiol. A: Comp. Physiol.* 44:903–909.

Zotterman, Y. 1971. The recording of the electrical response from human taste nerves. In *Handbook of Sensory Physiology*, vol. 4. *Chemical Senses*, pt. 2. L. M. Beidler, ed. Berlin: Springer-Verlag, pp. 102–115.

4. Trophic Competence in Mammalian Gustation

The taste neuron is frequently cited as the classic example of a nerve cell that maintains the normal morphological, physiological, and biochemical features of its end organ (see reviews by Smith and Kreutzberg, 1976; Rosenthal, 1977). I consider here the neurotrophic maintenance of mammalian taste buds, addressing in turn the role of axonal transport and the kinds of competent tissue and the types of neurons that will support taste buds. I close by considering briefly some advantages of neurotrophic maintenance.

Although the detailed mechanism of the trophic influence of innervation upon taste buds is still unknown, recent research has demonstrated that acute localized treatment of a gustatory nerve (IXth), either with colchicine or by cooling, impairs axonal transport and simultaneously reduces taste responses (Oakley et al., 1981). Additionally, when a taste nerve is chronically treated with a Silastic nerve cuff containing colchicine, taste responses fail and taste buds degenerate (Sloan et al., 1983). (See Figure 1.) This recent work has resolved a long-standing problem by showing that the action of colchicine derives from impairment of axonal transport and not from direct effects on cells at the taste bud, such as the blockage of cell division and turnover. The most critical evidence was that the small concentrations of colchicine that reached the tongue (< 2 pM) were at all times equal on the two sides, whereas the loss of taste responses and taste buds was strictly unilateral (Sloan et al., 1983).

Given the necessity of axonal transport, it is appropriate to examine the view that taste buds are maintained by a diffusible trophic chemical transported down the taste axons to the tongue where the released chemical triggers differentiation of epithelial cells into taste receptor cells. Apart from the concern that taste axons could function by removing an inhibitory factor from the gustatory epithelium, there is a more fundamental issue. In any neurotrophic system, axonal transport may produce its effects indirectly. For example, the direct effect of impaired axonal transport may be to disrupt the axon

---

Bruce Oakley, Department of Zoology
University of Michigan, Ann Arbor, Michigan 48109

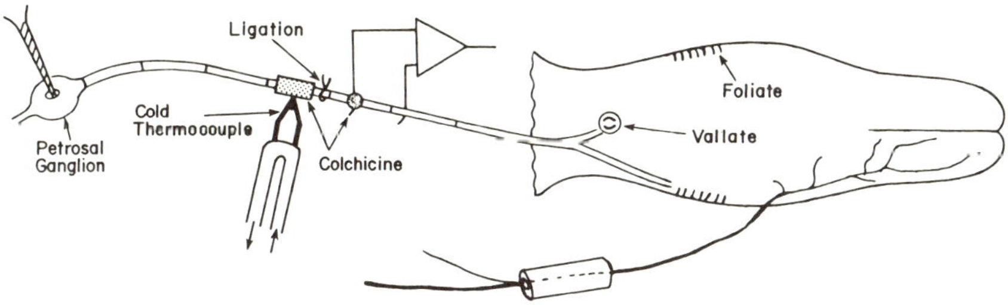

Figure 1. Schematic drawing of the gerbil's tongue and treatments of the IXth and lingual nerve used to evaluate the role of axonal transport. Protein, labeled by injecting tritiated leucine into the petrosal ganglion, was transported along IXth nerve axons to the tongue until the transport was blocked by ligation, colchicine, or local cooling of the IXth nerve as shown. Recording electrodes monitored the effect of these treatments upon taste impulse discharges. The pair of parallel arrows indicates cold methanol flowing at adjustable rates through a U-shaped section of hypodermic tubing used to cool a projecting thermocouple. The thermocouple cooled the nerve locally and registered its temperature. A Silastic nerve cuff (1.5 x 0.4 mm I.D.) was chronically implanted around the lingual–chorda tympani nerve. The chronic effects of colchicine or lumicolchicine mixed with the Silastic (1% by weight) were evaluated by recording taste responses and counting taste buds.

terminals, which secondarily disturbs membrane interactions and the trophic support of the target tissue. That the trophic interaction normally involves the release of a diffusible trophic agent is conjectural. (In the retina a requirement for actual cellular contact has been suggested for enzyme induction [Moscona and Linser, 1983].) At the present time one can simply state that the integrity of mammalian taste buds depends in some manner upon axonal transport.

The primary objective of this review is to consider the kinds of axons and epithelial tissues competent to engage in the trophic interactions leading to taste buds.

Competent Epithelial Tissue for Taste Bud Formation

Normal Distribution of Taste Buds

Mammalian taste buds develop under the influence of gustatory fibers at limited sites in stratified squamous epithelium and pseudostratified ciliated columnar epithelium of the oral cavity. These sites include the fungiform, foliate, and vallate papillae of the tongue, the nasoincisor ducts, the tongue-cheek margin, the base of the tongue near the ducts of the sublingual glands, the soft palate, the pharynx, and the epiglottis (e.g., Miller, 1977). To determine whether other tissues are competent to develop into taste buds, four approaches have been taken: evaluation of taste bud regeneration after papilla destruction, transplantation of foreign tissue onto the tongue, reinnervation of regions of the tongue between taste papillae, and explantation of tongue tissue to a foreign site.

Destruction of Taste Papillae

Fungiform, foliate, or vallate papillae have been cauterized in dogs, rabbits, and rats. The absence of most taste buds after tissue healing would imply that the remaining tongue epithelium was not competent to support taste bud formation. In these experiments the adequacy of tissue regeneration has ranged from a smooth tongue virtually lacking papillae and taste buds (Arey, 1942) to substantial regeneration of both normal papillae and taste buds (Whiteside, 1926). Three recent investigations have examined the consequences of excising the vallate papilla. The rat has a single vallate papilla on the midline of the posterior part of the tongue. A horseshoe-shaped trench or crevasse surrounds the vallate papilla, except on the anterior margin. Beneath this trench lie the salivary glands of von Ebner, the ducts of which empty into the bottom and side of the trench. Hundreds of taste buds line both sides of the trench. More frequently in young animals than in adults, one finds a few taste buds on the dorsal surface of the vallate papilla (Zalewski, 1969). Taste buds are also occasionally found deep in the rootlike extensions of the trench as part of the ducts of the von Ebner's glands (Gómez-Ramos et al., 1979, their Figure 1). Fifteen days after excision of the vallate papilla, Sanchez et al. (1978) observed an abnormally small and asymmetrical regenerated papilla. With longer recovery periods the papillae and crevasses became less prominent (Sanchez et al., 1978; Zalewski, 1970). After complete removal of the vallate papilla, as defined by removal of the "bottom of the sulcus," Sanchez et al. (1978) found a few regenerated taste buds only in squamous epithelium which lay in those crevasses having a von Ebner's gland duct at their base (Sanchez et al., Figure 10). The regenerated taste buds were normal in shape, size, and histochemical reaction to alkaline phosphatase (Sanchez et al., 1978; Zalewski, 1970). The taste cells had synaptic contacts and microvilli and were of the usual three types (dark, light, and intermediate). When the excision did not completely remove the base of the trench, papilla regeneration was improved and more taste buds were seen (Sanchez et al., 1978)

The vallate papilla is a particularly favorable choice for obtaining ample reinnervation because the bilateral overlap of the IXth nerve allows either of the IXth nerves to maintain 80–90% of the vallate taste buds (Whiteside, 1927; Oakley, 1970; State, 1977a). Thus, it seems highly likely that the combined regeneration of both IXth nerves over a short distance would have reformed many taste buds, had the tissue that remained after excision been competent. However, a maximum of nine regenerated taste buds has been reported after excision of the rat vallate papilla (Zalewski, 1970). That is, over 98% of the taste buds failed to re-form after vallate removal. The wound healed in a matter of days and there was no morphological indication that the healing thwarted reinnervation of this area. In fetal sheep after excision of vallate papillae a healed surface was found which bore several taste buds but no papilla (Mistretta et al., 1979).

From this class of experiments one can conclude that when the gustatory epithelium is extensively destroyed few, if any, taste buds re-form. The virtual elimination of taste buds by excision of the vallate papillae suggests that excision has removed the vast majority of stem cells for taste buds. The occasional regenerated taste bud may derive from a few remaining stem cells such as those found ventral to the trench in the deep roots of the crevasses that are continuous with the ducts of the von Ebner's glands. This is not to exclude the possibility that the competence to respond to taste axons may have been reactivated in a few indifferent cells by the trauma of excision and healing (Singer and Mutterperl, 1963). But clearly, with more that 98% of the taste buds permanently lost, the general tongue epithelium cannot reasonably be considered to be competent to form taste buds; those few taste buds which are found must derive either from residual stem cells or token dedifferentiation and respecification.

The conclusion emerging from these experiments that tissue competence and not axon abundance is the normal factor limiting the number and distribution of taste buds is supported by evidence of a different kind. The rat chorda tympani nerve normally innervates an average of 109 taste buds. However, when it is directed to innervate the posterior portion of the tongue via the IXth nerve, it re-forms in the foliate papillae more than one and one half times as many taste buds as it normally innervates (Oakley, 1970). In all instances where foreign chemosensory nerves have been used to innervate the tongue (to be described) the new taste buds are always restricted to their normal gustatory papillae. The experiments described below in the next three sections also support the uniqueness of the gustatory epithelium.

Transplantation of Foreign Epithelium to the Tongue

The competence of skin has been tested by grafting developing hairy skin from the face to the tongue in fetal sheep. The skin graft was successfully innervated and maturation continued, as most conspicuosly indicated by continued hair follicle development. However, no taste buds were formed in hairy skin, whereas in control experiments when tongue epithelium was lifted and replaced in situ, taste buds developed in the fungiform papillae (Mistretta and Bradley, 1977).

Reinnervation of the Tongue along Pathways Directing Regenerating Taste Axons to Nongustatory Epithelium

The capacity of tongue epithelial tissue lying between the fungiform papillae to form taste buds has been tested by suturing the chorda tympani nerve to the lingual nerve so that some regenerated taste axons were directed to areas between the fungiform papillae. Notwithstanding, taste buds appeared exclusively in fungiform papillae (Oakley, 1970), suggesting again that the mere presence of taste axons is insufficient; tissue incompetence excludes taste bud formation from much of the tongue surface.

Transplantation of Epithelium to the Eye

Taste buds failed to develop in ear skin when this epithelium and the vagal nodose ganglion (which normally innervates pharyngeal and epiglottal taste buds) were cotransplanted to the anterior chamber of the eye (Zalewski, 1972).

An explant of the vallate papilla to the anterior chamber of the eye not uncommonly becomes encysted and has an excessively keratinized and atrophied papilla (Zalewski, 1976; Gómez-Ramos et al., 1979). Four of 241 structures identified as taste buds by ATPase reactivity appeared in keratinized marginal lobules that resembled large filiform papillae (Zalewski, 1972). In view of the excessive keratinization, tissue atrophy, and unusual morphology associated with these vallate explants, the improved explanting technique introduced by Zalewski (1976), which significantly increases the yield of well-formed unencysted papillae, should be used to show convincingly that filiform papillae can reliably support taste buds with normal cellular morphology.

Taste buds have been reported on the dorsal surface of the vallate papilla after testosterone treatment (Zalewski, 1969) or explantation of the vallate papilla into the anterior chamber of the eye (Zalewski, 1972). Since taste buds are also found in this location in normal neonates and sometimes in adults, this locus must be included as gustatory epithelium.

In vallate papillae explanted to the eye, taste buds were found not only on the dorsal surface and in trenches but also in circularized squamous epithelium lying beneath or sometimes attached to trench walls (Zalewski, 1976). Abrasion during dissection to remove muscle and salivary glands from the ventral surface of the explant before insertion into the eye may have played a role in promoting the development of the form and position of these circular epidermal structures, which were probably derived from stratified squamous epithelium of the deep trench/ductal region. Similar deeply situated taste buds are found in trench/ductal areas in normal vallate papillae (Gómez-Ramos et al., 1979, Figure 1). Thus, there is no convincing evidence such taste buds in explants occur in nongustatory epithelium.

A straightforward conclusion emerges from these studies on the problem of tissue competence in mammalian gustation. In experimental manipulations, taste buds only appear in gustatory papillae and adjacent squamous epithelium where they are normally found. Experiments that find taste buds after purporting to have destroyed the gustatory epithelium are subject to the criticism that bits of competent tissue remained. Recent investigations have found that destruction of the vallate papilla permanently eliminates nearly all taste buds. No compelling evidence has been published that taste buds can be found in nongustatory epithelium at novel sites that have never previously supported taste buds.

Nerve Specificity in Gustation

In this section I shall consider two questions. Is innervation necessary for the development of taste buds? What kinds of nerves are able to re-form taste buds in adults? There are three classes of experiments designed to test for an absolute requirement for nerves or the capacity of foreign nerves to re-form taste buds: (a) reinnervation of (gustatory) epithelium with a (foreign) nerve, (b) transplantation of gustatory epithelium to a foreign site, and (c) transplantation of both gustatory epithelium and a ganglion to a foreign site. Thus, one can transplant the nerve, the epithelium, or both to a new site.

The Necessity for Nerves in Taste Bud Formation

In considering the neurotrophic dependence of taste buds one must be cognizant of the possibility that the processes controlling development and regeneration may differ. Since the development of taste buds occurs at or shortly after the papillae become innervated (Farbman, 1965), it is tempting to presume that the axons initiate taste bud development. However, among developmental biologists there has been a long history of dissenting opinion. Ross Harrison (1904), for example, believed the invasion of taste papillae by nerve fibers and the formation of taste buds was merely coincidental. He believed the role of the nerve was to maintain taste buds that had already developed by their inherent capacity to differentiate. In contrast, on the basis of indirect observations using cauterized or explanted vallate papillae, Torrey (1940) concluded that axons probably are required for taste bud development, but he was unable to perform the critical experiment of early denervation. Hosley and Oakley (1982) have recently carried out these experiments successfully in rats. Hosley found that not only does the initial formation of taste buds depend upon the presence of axons, but there is also a critical period during which axons must interact with the epithelium or most taste buds will never form, even if the tissue is innervated at a later time. Perhaps taste axons are required during a critical period to commit precursor cells permanently to taste bud formation.

When the gustatory epithelium of fetal sheep tongue is homografted to the external cheek, taste buds fail to develop to maturity (Mistretta and Bradley, 1977; Mistretta et al., 1979). The trigeminal cutaneous axons were unable to replace the normal gustatory innervation in development. Hence, in development, specific gustatory innervation seems to be necessary for taste bud formation.

Reinnervation of Gustatory Epithelium with Foreign Nerves

Several investigations have shown that neither motor axons nor general cutaneous sensory axons have gustatory trophic capacities (e.g., Guth, 1958; Oakley, 1974; State, 1977b). For example, although the auriculotemporal (sensory), hypoglossal (motor), and mylohyoid (mixed) nerves will innervate the vallate and foliate papillae, they will neither re-form, maintain, nor functionally innervate existing taste buds (Oakley, 1974). Cross-regenerated gustatory nerves, however, will readily re-form taste buds in appropriate locations and will make functional connections (Oakley, 1967 and 1970). The chemical responses of the taste buds are tissue-specific, not nerve-specific (Oakley, 1967), and the several branches from a single axon to different taste buds innervate closely similar functional types of receptors (Oakley, 1975). Other gustatory nerves (e.g., vagus) not normally found in the tongue are able to re-form tongue taste buds (Guth, 1958). In addition, fibers of the carotid branch of the glossopharyngeal nerve and of the main descending trunk of the vagus nerve are also able to re-form vallate taste buds (Dinger et al., 1981; Zalewski, 1981). In these experiments, the responsible axons are likely to be chemosensory fibers associated with the carotid and aortic bodies, respectively. It is reasonable to conclude that only chemosensory fibers are able to reinnervate the tongue and re-form taste buds in the gustatory papillae.

Transplantation of Gustatory Papillae to the Field of a Foreign Nerve

When the rat vallate papilla is transplanted to the anterior chamber of the eye, the vallate taste buds die. After a few weeks, small taste buds re-form (Zalewski, 1976; Gómez-Ramos et al., 1979). It is not necessarily the chemical milieu or revascularization (Smith and Wolpert, 1975) that is responsible for taste bud regeneration, since these explants are invaded by myelinated and unmyelinated axons of unknown origin (Farbman, 1971; Zalewski, 1976; Gómez-Ramos et al., 1979). It would be interesting to know whether the fibers contacted the taste cells in morphologically typical synapses, for it would suggest a neurotrophic role and certify the reciprocal nature of the trophic interactions that cause both target cell and axon ending to differentiate. In summary, taste buds reappear in gustatory epithelium that has been explanted to the anterior chamber of the eye. The responsible factors have not been analyzed.

Transplantation of Both Gustatory Epithelium and a Test Ganglion to a Foreign Field

The model that typifies this paradigm is the joint explantation of the vallate papilla and the vagal nodose ganglion into the anterior chamber of the rat's eye. Under the most favorable conditions, 5–10% of the normal number of vallate taste buds regenerate. One can only guess whether the low yield of taste buds reflects the poor health of the explant, the modest density of innervation, or the changes in the cells that survive in the explanted tongue fragment and ganglion. Taste buds also form in the vallate papilla when it is co-explanted with lumbar spinal ganglia (Zalewski, 1973). Unfortunately, such explantation experiments are confounded by the prompt invasion of fibers from the intrinsic innervation of the eye. As Farbman (1974) has pointed out, whether a foreign ganglion would have the capacity to re-form taste buds cannot be determined from this model system since the contribution of the ganglion toward taste bud re-formation cannot be separated logically from the trophic capacity of the intrinsic innervation to re-form taste buds. For example, the particular contribution of the lumbar ganglion might be similar to that of testosterone, which can augment the number of taste buds in the tongue, yet cannot independedntly induce taste buds in a denervated papilla (Allara, 1952; Zalewski, 1969). The origin of the intrinsic axons of the eye that innervate the vallate explants is unclear. The removal of the superior cervical ganglion does not eliminate the myelinated and unmyelinated fibers from the explant (Farbman, 1971). However, since the vallate is implanted in the eye through a corneal incision, it would not be surprising if the corneal chemosensory fibers trophically innervate the vallate papilla. From changes in cholinesterase staining, it is likely that foreign ganglia innervate the co-explanted vallate papilla, although this source of axons has not been directly shown. It might be argued that the presence of cholinesterase-staining fibers in the explants reinnervated by the nodose ganglion of the vagus nerve and their absence in explants reinnervated by intrinsic innervation indicates that there are two types of fibers that influence taste bud regeneration (Zalewski, 1976). However, it could be that a non-cholinesterase-staining subpopulation of vagal fibers contributes to the formation of taste buds. Until a system is developed that is free of contaminating sources of innervation, it will not be possible to evaluate unambiguously the neurotrophic capacity of added ganglia upon explanted gustatory epithelium. For this reason there are no convincing demonstrations that mammalian taste buds can be independently induced by nonchemosensory fibers.

The anterior chamber of the eye is the most complex of those model systems used to study the neurotrophic capacity of various nerves and epithelium, because both the tissue and ganglia are altered by removal from their normal sites and by explantation to a third site, where the tongue epithelium is permeated by a new, local vascular bed chemical environment, and local innervation. Taste buds in vallate papillae explanted to the eye may form because the chemical environment is stimulatory, because the eye has nerve fibers with neurotrophic capacity which invade the explant, or because the cells in the explant become more responsive to trophic influences. The introduction of a foreign ganglion adds a fourth potentially stimulatory factor. Although common chemosensory fibers directed principally to the cornea may be responsible for taste bud formation in this model system, it would be no less interesting if an increased capacity of the explanted epithelium to respond were a contributing factor. In salamanders it has been shown that transplantation of the limb blastema reduces the number of axons required to trigger limb regeneration (Singer and Mutterperl, 1963). In spite of uncertainty about the factors responsible for taste bud regeneration in this complex eye model system, it has some important strengths which have been cleverly

Table I. *Regeneration of Mammalian Taste Buds in Situ*

Motor ( )	*Sensory* (−)	Chemosensory (+)
Hypoglossal (XII)	Lingual (V)	Chorda tympani (VII)
Mylohyoid (V)	Mylohyoid (V)	IXth—lingual branch
	Auriculotemporal (V)	Xth—superior laryngeal branch
		IXth—carotid branch
		Xth—main descending

Only chemosensory nerves have the trophic capacity to elicit regeneration of mammalian taste buds in situ.

exploited. For example, using immunologically deficient nude mice, Zalewski and co-workers (1977) have shown that the formation of taste buds can be stimulated in donor tissue from another species (rat/mouse combinations). Table I, based largely upon experimental reinnervation of the tongue, summarizes the present evidence on nerve specificity in taste bud regeneration by demonstrating that chemosensory neurons are required. It would be valuable to test the generality of these conclusions by more systematic examination of the trophic dependence of taste buds in locations other than the vallate papilla.

Summary and Conclusions

The structure and function of mammalian taste buds are trophically maintained by processes requiring axonal transport in gustatory nerves. That taste buds will re-form only in gustatory epithelium is shown by experiments demonstrating the lack of trophic responsiveness of hairy skin, of nongustatory tongue epithelium, and of tongue epithelium after excision of gustatory papillae. In embryonic development recent experiments indicate that gustatory innervation is required for the initial formation of taste buds. In adults only chemosensory axons have been convincingly shown to be competent to re-form taste buds. Innervation by local, possibly chemosensory, axons of the eye may be responsible for the presence of taste buds in rat vallate papillae explanted to the anterior chamber of the eye. These local axons of the eye contaminate the explanted tongue fragments and make it impossible to evaluate what independent contribution a co-explanted foreign ganglion might make toward taste bud re-formation. On the basis of the available evidence the most prudent conclusion is that taste buds are only known to be re-formed by chemosensory fibers and only in gustatory epithelium.

Nerve specificity in the neurotrophic formation of taste buds ensures that nearby nontaste axons such as warm, cold, nociceptive, mechanosensory, and motor axons are prevented from making spurious connections to taste receptor cells. Tissue specificity in the formation of taste buds provides for a constant number and distribution of taste buds. The locations of taste buds near salivary ducts and other passages must have functional significance, and those taste bud locations and numbers are reliably maintained when only gustatory stem cells are competent to re-form taste buds. The existence of a neurotrophic dependency for taste buds may assist in healing after injury and limit tissue vulnerability. Consequently, a neurotrophic requirement for gustatory axons and gustatory epithelium ensures the development, preservation, and meaningful reconstitution of gustatory sensory information.

References

Allara, E. 1952. Sull'influenza esercitata dagli ormoni sessauli sulla struttura dell formazioni gustative di *Mus rattus albinus*. *Riv. Biol.* (Perugia). 44:209–229.

Arey, L. B. 1942. The regeneration of lingual papillae and taste buds after cautery. *Q. Bull. Northwest. Univ. Med. Sch.* 16:100–104.

Dinger, B., L. J. Stensaas, and S. Fidone. 1981. Taste buds reinnervated by carotid sinus nerve fibers. *Neurosci. Lett.* 27:285–289.

Farbman, A. I. 1965. Electron microscope study of the developing taste bud in rat fungiform papilla. *Dev. Biol.* 11:110–135.

———. 1971. Differentiation of foetal rat tongue homografts in the anterior chamber of the eye. *Arch. Oral Biol.* 16:51–57.

———. 1974. Taste bud regeneration in organ culture. In *Trophic Functions of the Neuron*. D. B. Drachman, ed. New York: New York Academy of Sciences, pp. 350–354.

Gómez-Ramos, P., E. Leon-Felíu, and E. L. Rodríguez-Echandía. 1979. Taste buds in vallate papillae grafted to the anterior chamber of the eye. *Anat. Embryol.* 156:217–224.

Guth, L. 1958. Taste buds on the cat's circumvallate papilla after reinnervation by glossopharyngeal, vagus and hypoglossal nerves. *Anat. Record.* 130:25–37.

Harrison, R. G. 1904. Experimentelle Untersuchungen über die Entwicklung der Sinnesorgane der Seitenlinie bei den Amphibien. *Arch. f. mikr. Anat.* 63:35–149.

Hosley, M. A., and B. Oakley. 1982. Reduced innervation during an early sensitive period prevents development of taste buds in the rat vallate papilla. *Soc. Neurosci. Abstr.* 8:757.

Miller, I. J. 1977. Gustatory receptors of the palate. In *Food Intake and Chemical Senses*. Y. Katsuki, *et al.* eds. Baltimore: University Park Press, 173–185.

Mistretta, C. M., and R. M. Bradley. 1977. Interactions between nerves and epithelia during taste bud and papilla development in fetal sheep. *Soc. Neurosci. Abstr.* 3:81.

Mistretta, C. M., R. M. Bradley, and H. M. Stedman. 1979. Development of tongue epithelium grafts in fetal sheep. *Soc. Neurosci. Abstr.* 5:130.

Moscona, A. A., and P. Linser. 1983. Developmental and experimental changes in retinal glia cells: cell interactions and control of phenotypic expression and stability. *Curr. Top. Dev. Biol.* 18:155–188.
Oakley, B. 1967. Altered temperature and taste responses from cross-regenerated sensory nerves in the rat's tongue. *J. Physiol. (Lond.)*. 188:353–371.
———. 1970. Reformation of taste buds by crossed sensory nerves in the rat's tongue. *Acta Physiol. Scand.* 79:88–94.
———. 1974. On the specification of taste neurons in the rat tongue. *Brain Res.* 75:85–96.
———. 1975. Receptive fields of cat taste fibers. *Chem. Sens. Flav.* 1:431–442.
Oakley, B., J. S. Chu, and L. B. Jones. 1981. Axonal transport maintains taste responses. *Brain Res.* 221:289–298.
Rosenthal, J. 1977. Trophic interactions of neurons. *Handbk. Physiol.* 1(2):775–802.
Sanchez, B. F., M. Rodrigo Angulo, J. Cano Garcia, and E. L. Rodríguez-Echandía. 1978. Regeneration of the vallate papilla in the rat with special reference to the origin of the taste bud-bearing epithelium. *Cell Tissue Res.* 193:399–411.
Singer, M., and E. Mutterperl. 1963. Nerve fiber requirements for regeneration in forelimb transplants of the newt *Triturus*. *Dev. Biol.* 7:180–191.
Sloan, H. E., S. E. Hughes, and B. Oakley. 1983. Chronic impairment of axonal transport eliminates taste responses and taste buds. *J. Neurosci.* 3:117–123.
Smith, A. R., and L. Wolpert. 1975. Nerves and angiogenesis in amphibian limb regeneration. *Nature (Lond.)*. 257:224–225.
Smith, B. H., and G. W. Kreutzberg. 1976. Neuron-target cell interactions. *Neurosci. Res. Program Bull.* 14:211–453.
State, F. A. 1977a. Histological changes following unilateral reinnervation of the circumvallate papilla of rat. *Acta Anat.* 98:343–352.
———. 1977b. Circumvallate papilla of dog following suture of the hypoglossal and glossopharyngeal nerves. *Acta Anat.* 98:413–419.
Torrey, T. W. 1940. The influence of nerve fibers upon taste buds during embryonic development. *Proc. Natl. Acad. Sci. USA.* 26:627–634.
Whiteside, B. 1926. The regeneration of the gustatory apparatus in the rat. *J. Comp. Neurol.* 40:33–45.
———. 1927. Nerve overlap in the gustatory apparatus of the rat. *J. Comp. Neurol.* 44:363–377.
Zalewski, A. A. 1969. Neurotrophic-hormonal interaction in the regulation of taste buds in the rat vallate papilla. *J. Neurobiol.* 1:123–132.
———. 1970. Regeneration of taste buds in the lingual epithelium after excision of the vallate papilla. *Exp. Neurol.* 26:621–629.
———. 1972. Regeneration of taste buds after transplantation of tongue and ganglia grafts to the anterior chamber of the eye. *Exp. Neurol.* 35:519–528.
———. 1973. Regeneration of taste buds in tongue grafts after reinnervation by neurons in transplanted lumbar sensory ganglia. *Exp. Neurol.* 40:161–169.
———. 1976. The neural induction of taste buds in the salivary ducts of the lingual gland of von Ebner. *Exp. Neurol.* 52:565–580.
———. 1981. Regeneration of taste buds after reinnervation of a denervated tongue papilla by a normally nongustatory nerve. *J. Comp. Neurol.* 200:309–314.
Zalewski, A. A., G. F. Creswell, H. G. Goshgarian, and T. H. Oh. 1977. The nude mouse: an *in vivo* model for demonstrating cross-species trophic nerve function. *Exp. Neurol.* 54:397–402.

Neurophysiological Coding of Taste Information

5. On the Neural Code for Sweet and Salty Tastes

Introduction

An across-neuron pattern theory of taste quality coding was proposed initially, because single peripheral neurons were found to respond to stimuli of very different taste qualities (Pfaffmann, 1955; Erickson, 1963), but, subsequently, because single cells *throughout* the gustatory nervous system were found to do so (e.g., Doetsch and Erickson, 1970; Perrotto and Scott, 1976; Sott and Erickson, 1971). These neural recordings also suggested that peripheral gustatory neurons could not be divided into a few types; instead, each neuron appeared to have a unique, individual response profile across stimuli applied to the tongue (Erickson and Schiffman, 1975). Much of the neural data on which this theory was based was obtained from rats; the quality attributed to stimuli was based largely on human judgments (e.g., Smith and McBurney, 1969; McBurney and Shick, 1971; McBurney, 1972; McBurney and Bartoshuk, 1973), but also on animal preferences and discriminations (Erickson, 1963; Morrison, 1967). It did not seem possible that, considering the neural data available, taste qualities described by people as sweet, salty, sour, and bitter could be mediated by separate peripheral neural channels.

People could, however, quite reliably describe their taste sensations as comprising four components. Figure 1 depicts data from a psychophysical study (McBurney and Shick, 1971) in which people described taste sensations elicited by application of a large number of chemical compounds to the anterior of the tongue. In the figure, the heights of the four bars represent the mean magnitude estimates of sweetness, saltiness, sourness, and bitterness for 20 people for the 9 compounds indicated. Concentrations were chosen so that overall taste intensities were equal. An asterisk above a bar means 10 or more of the 20 people attributed that quality to the compound. Note that some chemicals, particularly the sweet sucrose, sodium saccharin, and fructose, as well as the bitter quinine hydrochloride, are predominantly described by one

Marion E. Frank, Department of Oral Biology
University of Connecticut Health Center, Farmington, Connecticut 06032

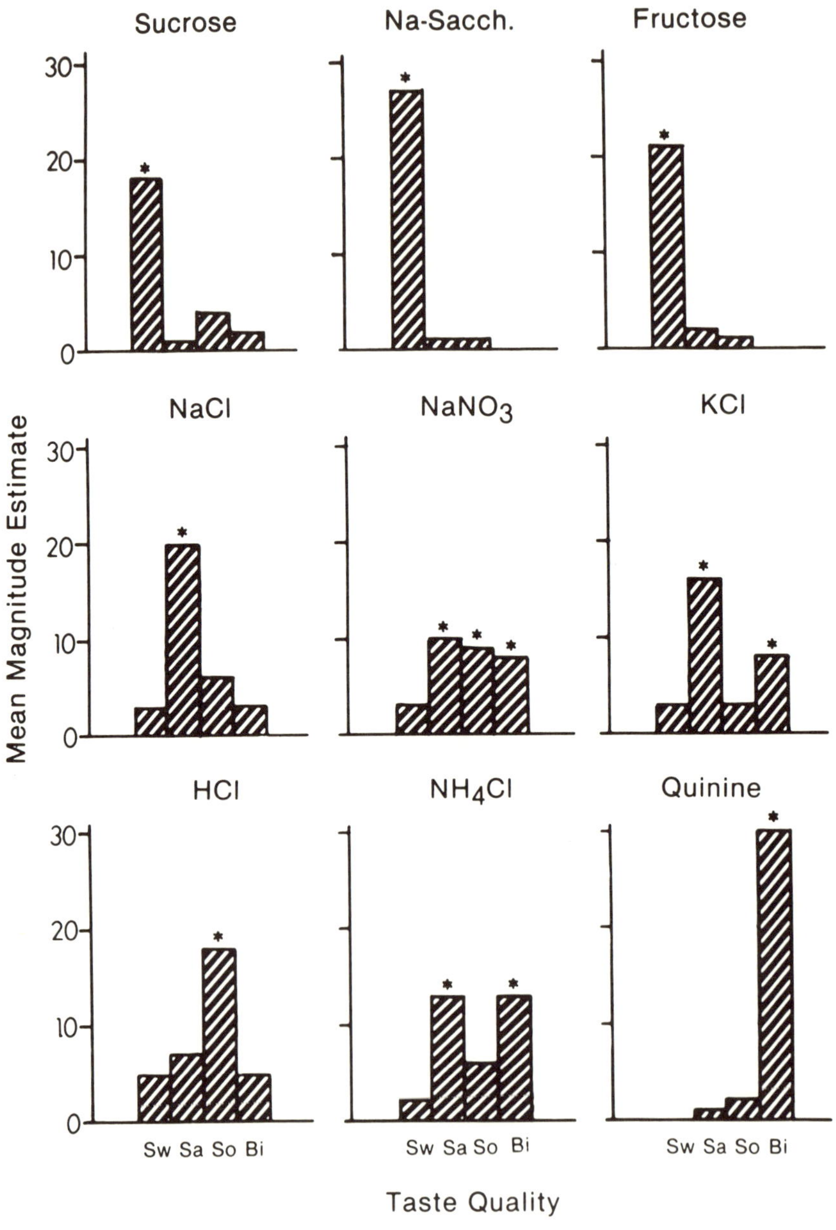
Sucrose
Na-Sacch.
Fructose
NaCl
$NaNO_3$
KCl
HCl
NH_4Cl
Quinine
Mean Magnitude Estimate
0
10
20
30
Sw Sa So Bi
Taste Quality

quality adjective; the salts, however, are often described by two or three adjectives: salty, sour, and bitter (Bartoshuk, 1975; Murphy et al., 1981). The amount of saltiness attributed to the equally intense salts is ordered NaCl > KCl > NH_4Cl > $NaNO_3$, and all the salts are more salty than HCl.

Sucrose, NaCl, HCl, and quinine hydrochloride have frequently been used in studies of neural sensitivities as prototypes of the four qualities, although it is possible that the salty prototype, NaCl, and the sour prototype, HCl, elicit a mixture of qualities. Psychophysical evidence suggests that HCl has significant salty and bitter components besides the dominant salty (Bartoshuk, 1975). The purest salty metal halide salt, NaCl (Murphy et al., 1981), is described as either salty or sour-bitter depending where on the tongue it is applied: salty when the fungiform papillae are activated but sour-bitter when the circumvallate papillae are activated (Sandick and Cardello, 1981). Apparently, NaCl is capable of stimulating more than one kind of taste receptor. It also should activate more than one neural quality-specific channel. This would help explain the "nonspecificity" observed in neural recordings, which would result from a nonspecificity of prototypal probe stimuli; it would not, however, explain why peripheral gustatory neurons could not be classified into types.

The Salty Taste

Figure 2 shows photographs of oscilloscope traces of recordings obtained from the hamster chorda tympani nerve. The chorda tympani contains nerve fibers that serve the taste buds on the front of the mammalian tongue. The hamster chorda tympani was studied because it shows responsiveness to sweet taste stimuli greater than, as well as a responsiveness to salts and acids similar to, that of the rat chorda tympani (Beidler et al., 1955; Fishman, 1957). Study of the rat chorda tympani, with its strong sensitivities to sour and salty stimuli, may have suggested an across-fiber pattern theory because the salty and sour stimulus prototypes were nonspecific; but study of the hamster chorda tympani made a test of the theory possible because there are pure sweet stimuli.

Figure 1. Mean magnitude estimates of humans for sweetness (Sw), saltiness (Sa), sourness (So), and bitterness (Bi) of 0.18 M sucrose, 0.003 M Na saccharin, 0.18 M fructose, 0.1 M NaCl, 0.46 M $NaNO_3$, 0.11 M KCl, 0.0032 M HCl, 0.051 M NH_4Cl, and 0.0001 M quinine HCl. Asterisks indicate that half or more of the 20 subjects used an adjective to describe a taste. Data are from McBurney and Shick, 1971.

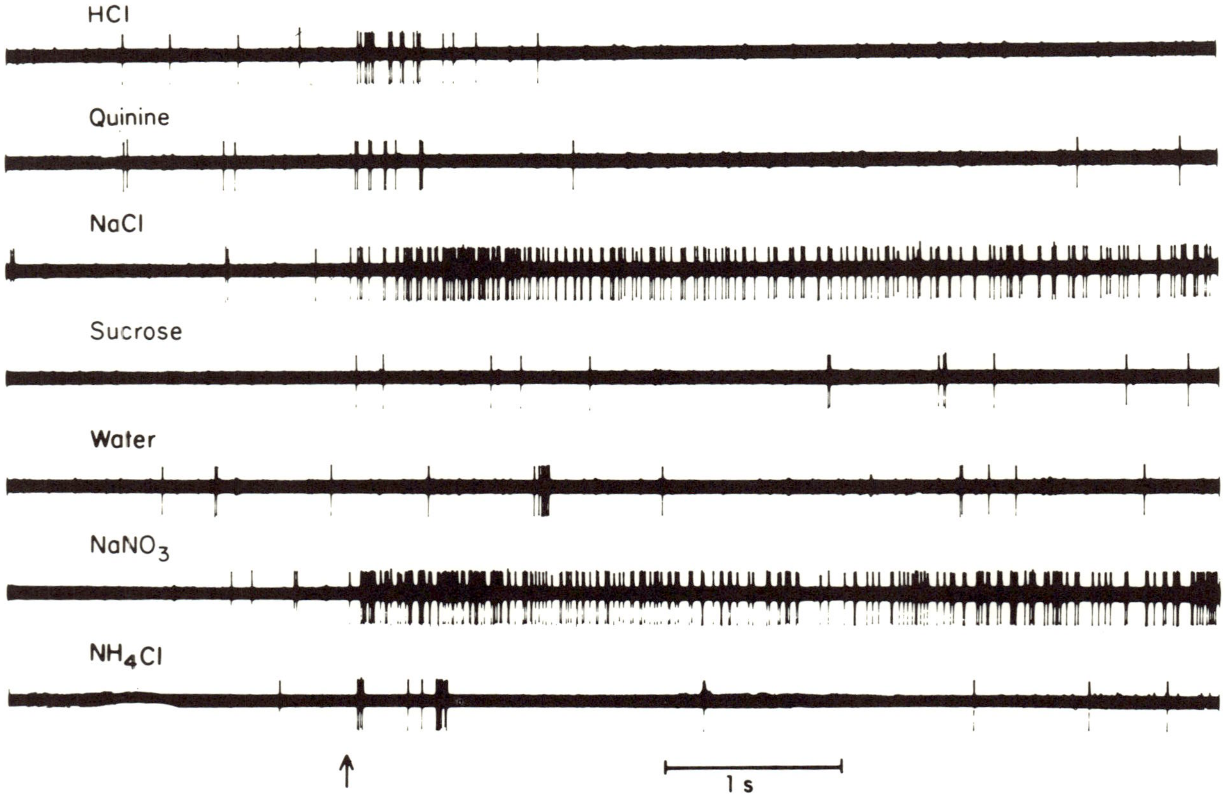

Figure 2. Photographs of oscilloscope traces of recordings from a strand of the chorda tympani nerve of the hamster showing action potentials of a single fiber sensitive to application of salts to the tongue. Stimulus solutions were 0.003 M HCl, 0.001 M quinine HCl, 0.03 M NaCl, 0.1 M sucrose, distilled water, 0.03 M $NaNO_3$, and 0.03 M NH_4Cl. Solutions flowed continuously over the tongue for the 5 s of response shown, beginning at the arrow.

The recordings show the effect of taste quality prototypes, HCl, quinine hydrochloride, NaCl, and sucrose, as well as $NaNO_3$, NH_4Cl, and the solvent water on a single preparation. Recordings were obtained by the traditional technique of teasing apart the desheathed nerve after it had been cut distal to the cell bodies of its fibers. The concentrations of the stimuli applied to the anesthetized animal's tongue were in the middle of the effective range of concentrations measured by the multiunit integrated response of the hundreds of fibers in the whole nerve (Frank, 1973). HCl, NaCl, $NaNO_3$, and NH_4Cl were equally intense in that they elicited whole nerve responses of nearly equal size, but the response to sucrose (or sodium saccharin) was about 20% smaller and the response to quinine (or KCl), about 30% smaller. These latter chemicals were not quite as effective as the former four; the saturated response rate to quinine, for example, was noticeably lower.

Note the strong, sustained response of this fiber to the two sodium salts; there are also shorter transient responses apparent in the first second to HCl, quinine HCl, and NH_4Cl, but they are not sustained through the 5 s of steady stimulation shown. Sucrose does not stimulate this fiber any more than the solvent, water, does. The total elicited activity in the 5 s response period shown has been used as a measure of sensitivity of a fiber to a stimulus. There is some "spontaneous activity" occuring in the 5 s before stimulus application; this can be subtracted from the response. The total number of action potentials in 5 s is a simple measure that overlooks interesting complexities seen in the train of impulses during stimulation, but it has been very helpful in describing sensitivities of different fibers and allows them to be classified into types. The type of fiber illustrated has been called an NaCl-best or salt-best fiber because it is most responsive to NaCl of the four quality prototypes and it has been seen in the chorda tympani of the hamster (Frank, 1973), rat (Frank et al., 1983), goat (Boudreau et al., 1982), macaque (Sato et al., 1975), and squirrel monkey (Pfaffmann et al., 1976), but not in the chorda tympani of carnivores such as the cat (Boudreau and Alev, 1973) or dog (Boudreau and White, 1978).

Figure 3 shows response profiles for 8 of 21 salt-best fibers sampled from the chorda tympani of the hamster. Profiles of the four most sensitive fibers, showing the largest responses (number of impulses in 5 s) to NaCl, and profiles of four fibers that were less sensitive, showing responses that were the largest below the mean response to NaCl, are depicted. Points for each unit's responses are connected by lines to trace its response profile. The fibers sampled varied nearly 10-fold in absolute sensitivity, but responses of each single fiber to NaCl and $NaNO_3$ are nearly equal; there is little or no response

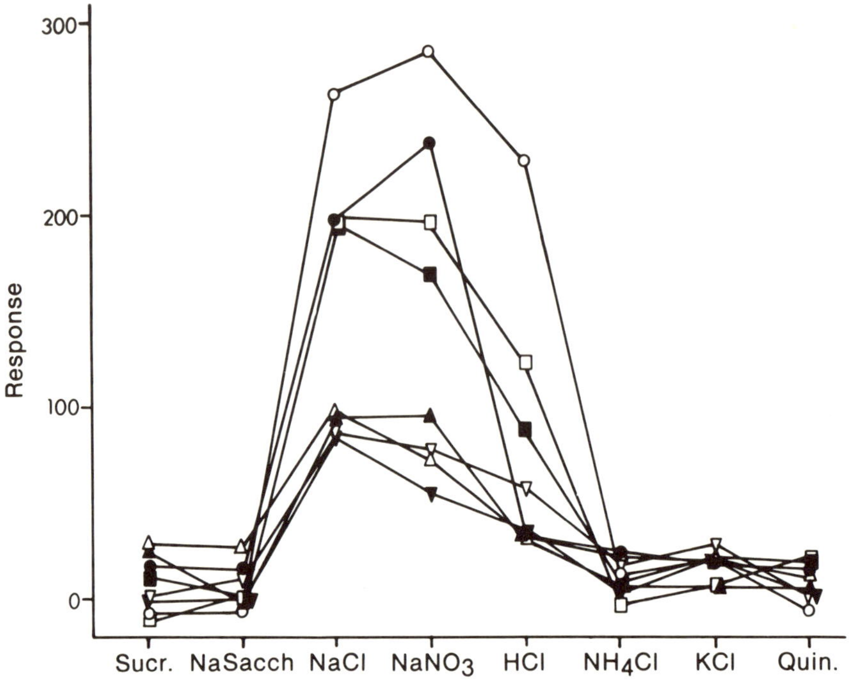

Figure 3. Response profiles for eight chorda tympani fibers of the hamster that responded more to NaCl than to sucrose, HCl, or quinine HCl. The total number of action potentials elicited in 5 s of solution flow minus the total elicited with the flow of distilled water over the tongue is plotted for each stimulus. Concentrations of sucrose, NaCl, $NaNO_3$, HCl, NH_4Cl, and quinine HCl are as noted in the legend of Figure 2; 0.001 M Na saccharin and 0.1 KCl were used. Points for responses of an individual fiber are connected to trace the response profile, not to imply any function.

to sucrose, sodium saccharin, NH_4Cl, KCl, or quinine HCl in any fiber, although each compound was applied to the tongue at a concentration that produced as significant effects on the whole nerve as did NaCl and $NaNO_3$. Some of the fibers respond strongly and others respond weakly to HCl. In fact, responses to HCl in these salt-best fibers can be any percentage of the NaCl response, varying from $< 10\%$ percent to $> 90\%$ of it. This variation of the relative size of the response to HCl in salt-best fibers of hamsters may result from a "sensitization" such as shown for peripheral taste units of the cat and rat. Geniculate ganglion, group II, amino acid-sensitive units increase their already substantial responsiveness to 1.0 M NaCl by about 50% after the tongue is stimulated by a series of phosphate-citrate buffers (Kruger and Boudreau, 1972). Salt-best fibers of the rat chorda tympani increase their slight responsiveness to 0.01 M HCl by 300%, on the average, during the course of an experiment; the tongue was bathed with a concentration series of NaCl, HCl, sucrose, and quinine-HCl solutions in the 30–40 min between HCl stimulations (Frank et al., 1983). Thus, the variation in acid-sensitivity of salt-sensitive fibers shown in Figure 3 could result from differing amounts of such a sensitization for different fibers which depends upon the point at which, during an experiment, a fiber was isolated. In this case, the response to HCl might be large in units isolated soon after the tongue had been stimulated by sensitizing chemicals. If such sensitization is the explanation for the variable size of responses to HCl in salt-sensitive fibers of the hamster chorda tympani, responses to the nonsodium salts KCl and NH_4Cl do not appear to be as easily sensitized; they are small in most units. Some of the variability in the relative effects of different compounds on salt-best units might also be developmental; it has been suggested that the membranes of taste cells change their sensitivities to salts and acids as they age (Hill et al., 1982; Mistretta and Bradley, 1983). A given chorda tympani salt-best fiber from which recordings are obtained may be innervating younger or older taste receptor cells that have an average life-span of about 10 d in rats (Beidler and Smallman, 1965). In the hamster salt-best units, stimuli of midrange intensity have the following order of effectiveness: NaCl = $NaNO_3$ > HCL > KCl = NH_4Cl. This order does not compare well with the amount of saltiness attributed to the stimuli by people who find NaCl and KCl the most salty. It is possible, however, that the salt-best fibers of the hamster carry a specific quality message to the brain that is not associated with stimuli in exactly the same way saltiness is for humans. It would be helpful, if possible, to determine the amount of "saltiness" in various stimuli for hamsters.

Figure 4 is the result of an atempt to do just that. Taste aversions were conditioned in groups of 12 hamsters to each of the stimuli indicated at concentrations similar to those used in the neural studies (Nowlis and Frank, 1977; Nowlis et al., 1980). After conditioning, aversions were tested by determining the amount of 0.1 M sucrose, 0.1 M NaCl, 0.01 M HCl, and 0.001 M quinine HCl that hamsters would drink relative to the amount a control group of hamsters would drink. The extent of the demonstrated aversion to the four prototypal test stimuli was taken as a measure of the similarity of a conditioning stimulus and the test stimuli. The height of each bar in Figure 4 shows the mean percent suppression of drinking attributable to the learned aversion; on top of each bar, standard errors of the mean are indicated. From

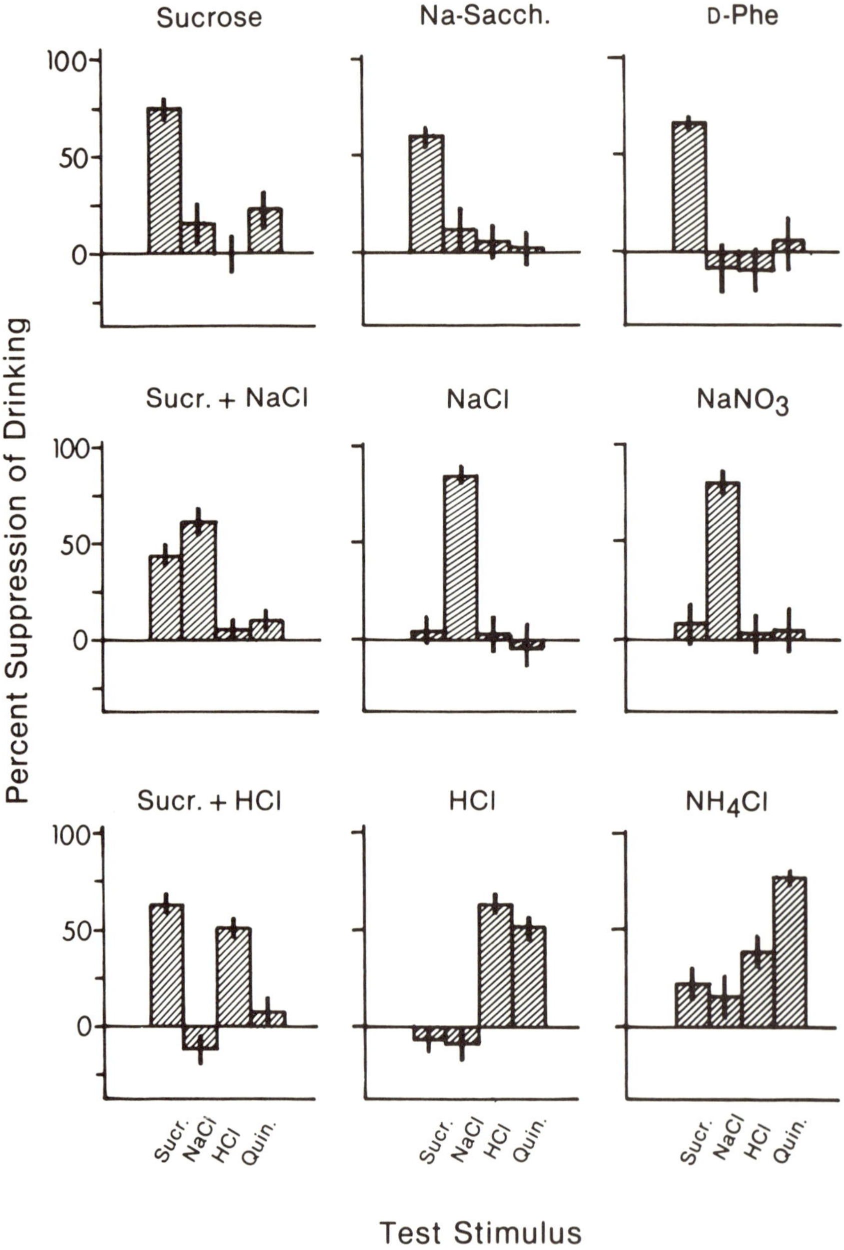

Sucrose
Na-Sacch.
D-Phe
Sucr. + NaCl
NaCl
$NaNO_3$
Sucr. + HCl
HCl
NH_4Cl
100
50
0
Percent Suppression of Drinking
Sucr.
NaCl
HCl
Quin.
Test Stimulus

these experiments, it can be concluded that sodium saccharin tastes like the sweet sucrose, as does D-phenylalanine, which is also sweet to humans. For these three stimuli, there is a great similarity between the way hamsters and humans classify the chemicals by taste. NaCl, $NaNO_3$, and NH_4Cl, however, are not classified similarly by the two species. NaCl and $NaNO_3$, which taste quite different to humans (cf. Figure 1), apparently taste identical to hamsters; and NH_4Cl, which has a strong salty taste as well as a strong bitter taste to humans, does not elicit a significant NaCl-like taste response in hamsters; an aversion to KCl yielded a similar cross-generalization profile, including no aversion to NaCl. In the hamster, both NH_4Cl and KCl, in fact, appear more similar to HCl than to NaCl; aversions to all three compounds generalize to quinine and HCl. These results with conditioned taste aversions equate NaCl and $NaNO_3$ in taste and indicate that the taste of NH_4Cl or KCl has little in common with the taste of sodium salts in hamsters.

The possibility that NH_4Cl, which elicits a mixture of taste qualities in humans, actually has a strong NaCl-like taste to hamsters that is not detected by the method used for testing aversions is not likely: hamsters are capable of showing aversions to both components of a binary mixture after an aversion is established to a mixture. The results for mixtures of sucrose and NaCl and sucrose and HCl are shown in Figure 4: aversions are shown to both components as they were for the other four binary mixtures of the prototypal stimuli (Nowlis and Frank, 1981).

Although the lack of effect of NH_4Cl and KCl on salt-best units is correlated with behavioral classification of taste stimuli by hamsters, the effect of HCl on some salt-best units does not parallel behavioral assessments: aversions to NaCl and HCl do not cross-generalize. It is noteworthy that many salt-best units of the hamster do not respond or respond only slightly to HCl; only 6 of the 21 fibers sampled, some of which are illustrated in Figure 3, responded at rates of 10 impulses/s or higher to HCl, whereas, 19 of the 21 responded that

---

Figure 4. Patterns of generalization of learned taste aversions in hamsters across four taste stimuli representing human taste qualities: 0.1 M sucrose (sweet), 0.1 M NaCl (salty), 0.01 M HCl (sour), and 0.001 M quinine HCl (bitter). Aversions were established to 0.1 M sucrose, 0.001 M Na saccharin, 0.04 M D-phenylalanine, 0.1 M sucrose + 0.1 M NaCl, 0.1 M NaCl, 0.1 M $NaNO_3$, 0.1 M sucrose + 0.01 M HCl, 0.01 M HCl, and 0.3 M NH_4Cl. Mean percent suppression of drinking plus and minus one standard error of the mean are indicated for the four test stimuli in a four-bar pattern for aversions established to the compounds indicated above each pattern. (Data are from Nowlis et al., 1980, and Nowlis and Frank, 1977.)

well to NaCl, and 18 of 21 did to $NaNO_3$. Perhaps the sensitization of the acid response that occurs in the experimental recording preparation creates a sensitivity to acids in these salt-best fibers that does not occur in animals during behavioral experiments.

However, the lack of an observed cross-generalization of taste aversions to NaCl and HCl in hamsters is not only curious because of the strong activation of a minority of salt-best units by HCl but also because NaCl reliably activates another group of electrolyte-sensitive taste afferents in hamsters: the HCl-best or acid-best. Figure 5 shows response profiles of six hamster chorda tympani nerve fibers that were more responsive to HCl than to other quality prototypes (data from Hyman, 1978). These were the only acid-best fibers isolated for which responses to 0.03 M NaCl were obtained. Although the fibers vary greatly in absolute sensitivity, the three electrolytes differentially affect all six acid-best fibers in the same way: responses to 0.003 M HCl $>$ 0.05 NH_4Cl M $>$ 0.03 M NaCl. The HCl and NH_4Cl elicited about equal responses from the whole chorda tympani nerve; the NaCl elicited about 25% more activity. A small response in these acid-best fibers and a larger response in salt-best fibers were also elicited by 0.01 M NaCl. Data points for 0.03 M were plotted since it was the concentration used to characterize salt-best fibers whose profiles are shown in Figure 3. Note that neither sucrose nor D-phenylalanine had much effect on these fibers.

Although NH_4Cl is not described as very sour by humans, hamsters show an aversion to HCl after a conditioned aversion to NH_4Cl is established, which is consistent with the strong activation of acid-sensitive fibers by the two stimuli in the hamster chorda tympani. However, the significant activation of these acid-sensitive taste fibers by NaCl clearly has no parallel in behavior (Figure 4): an aversion to NaCl does not establish any aversion to HCl or vice versa.

Apparently, when an aversion is established to a stimulus with a mixture of qualities of unequal intensity, an aversion may not be shown to the weaker quality (Nowlis et al., 1980). This can be shown by manipulating the learning conditions. If an animal has a stimulus of one taste quality present in his drinking water for several days, that quality will not serve as a cue for conditioning. Using such a pre-exposure technique, Nowlis has shown, for example, that there is a quininelike taste component to sodium saccharin not seen in the results in Figure 4 for which there was no pre-exposure. Therefore, NaCl may elicit an acidlike taste quality in hamsters, as it can in humans (Sandick and Cardello, 1981), which would parallel its effect on acid-best units of the hamster chorda tympani, but, because NaCl is the purest salty stimulus

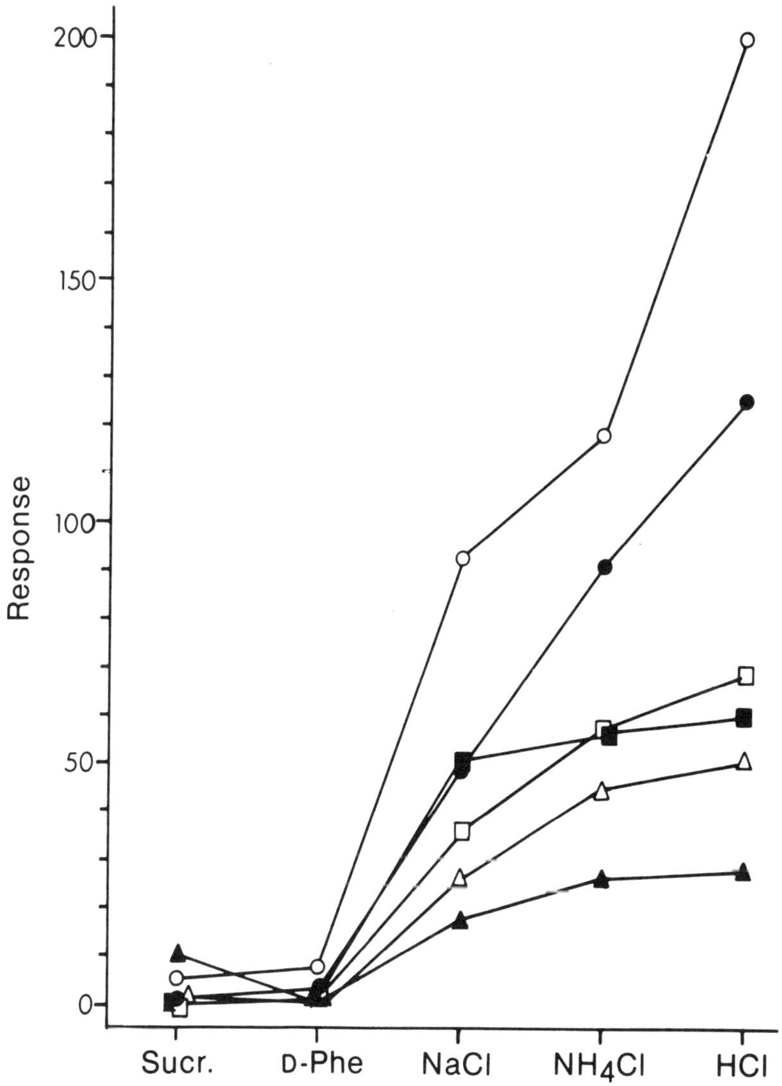

Figure 5. Response profiles for six chorda tympani fibers of the hamster that responded more to HCl than to sucrose or NaCl. Response was measured by counting the total number of action potentials for the initial 5 s of a response to stimulus flow over the tongue. Points for responses of an individual fiber to 0.1 M sucrose, 0.07 M D-phenylalanine, 0.03 M NaCl, 0.05 M NH_4Cl, and 0.003 M HCl are connected by lines to trace the profile. Data are from Hyman, 1978.

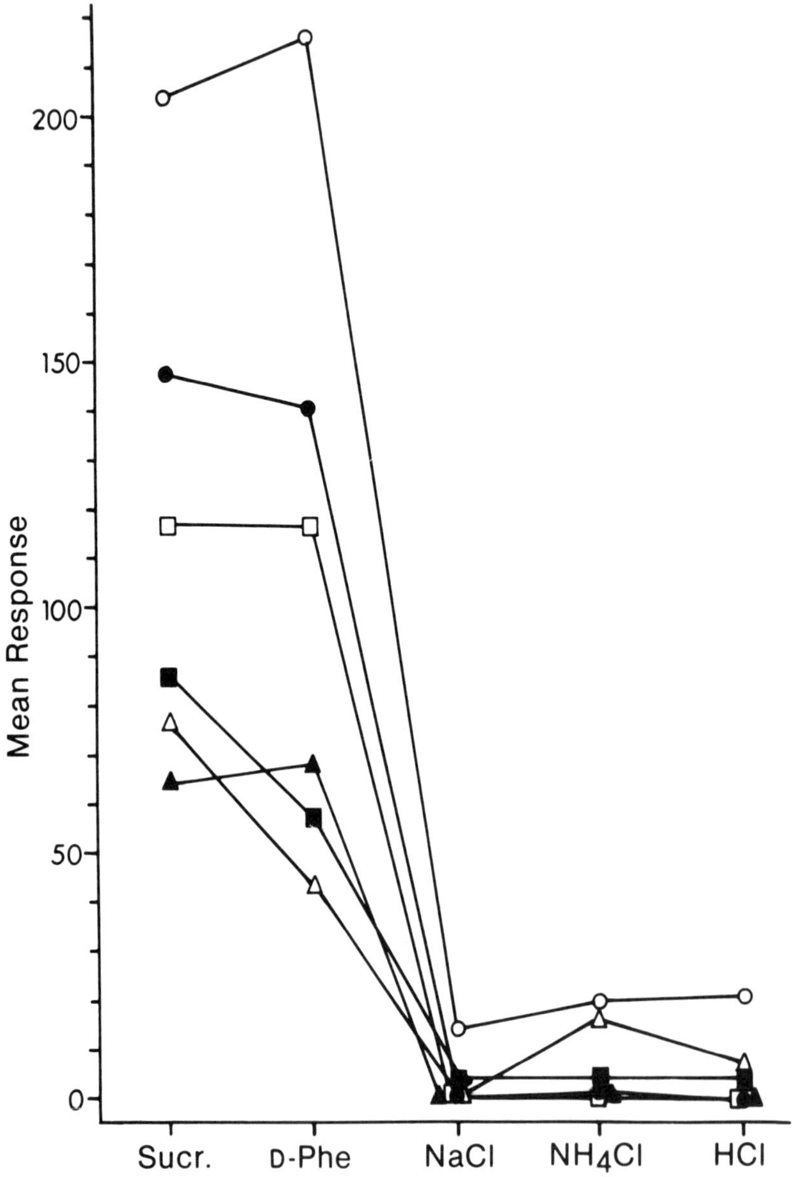

200
150
100
50
0
Mean Response
Sucr.
D-Phe
NaCl
NH4Cl
HCl

known, it would not be possible to pre-expose animals to a salty-quality stimulus and test NaCl for aversions to other qualities.

The two types of peripheral taste nerve fibers described above have recently been characterized for the rat chorda tympani (Frank et al., 1983). However, given the variable acid response in salt-sensitive fibers and the sensitivity of acid-sensitive fibers to salts, it is not surprising that investigators first observing responses of taste fibers in the rat chorda tympani would not see either that there were types of fibers or that each might be specialized for one taste quality. The initial observations were even more confusing than can be seen from the response profiles in Figures 3 and 5 because of another factor adding variability to the relative sensitivity across stimuli of acid-sensitive fibers. If a fairly strong acid stimulus (for example, 0.01 M HCl—in the earlier studies this or even 0.03 M HCl was used [Pfaffmann, 1955; Erickson, 1963]) is applied to the tongue, responses to other stimuli (e.g., NaCl and quinine HCl) that follow are suppressed even though the tongue is rinsed with water for several minutes after acid application (Frank et al., 1983). Similarly, acid-sensitive units of the geniculate ganglion show reduced responses to NaCl after a series of phosphate-citrate buffers are applied to the tongue (Kruger and Boudreau, 1972). Perhaps the effect is not seen in the fibers whose profiles are shown in Figure 5 because a weaker acid stimulus was used (0.003 M HCl). If a suppression had been produced, profiles would show some units having atypically small responses to one or the other salt depending upon which followed acid in the test series.

The Sweet Taste

Figure 6 shows response profiles for six hamster chorda tympani fibers that are more responsive to sucrose than to the other stimulus prototypes (data from Hyman, 1978). Mean responses for two presentations of all stimuli are plotted. The six fibers characterized in Figure 6 across stimulus compounds are the only ones isolated that were also completely characterized for the effects of two other concentrations of each stimulus and binary mixtures. This type of fiber is not found in great numbers in the rat chorda tympani; it was the

---

Figure 6. Response (5 s) profiles for six chorda tympani fibers of the hamster that responded more to sucrose than to NaCl or HCl. Concentrations of sucrose, D-phenylalanine, NH_4Cl, and HCl are as noted in the legend of Figure 5; 0.01 M NaCl was used. Points for an individual fiber are connected by lines to trace the profile.

discovery of its specificity in the hamster (Frank, 1973, 1977) that revived the possibility that there could be quality-specific peripheral channels in taste. Sucrose and D-phenylalanine, at concentrations with equal effect on the whole nerve, affect the units about equally; but NH_4Cl, NaCl, and HCl, at concentrations chosen on the same basis, hardly affect these fibers at all. These sucrose-best fibers vary considerably in absolute sensitivity to the effective stimuli. This large, nearly 10-fold variation in sensitivity is a general characteristic of peripheral taste neurons, whether recorded from nerve fibers severed from their cell bodies, as shown here, or recorded from intact neurons near their cell bodies, as shown by Boudreau and his his colleagues (e.g., Boudreau and White, 1978; Boudreau et al., 1982) and is therefore not determined by the experimental method of nerve dissection. However, a small decrease in overall elicited impulse rate is seen during the course of an experiment. The average elicited rate to 0.03 M NcCl in 19 salt-best fibers isolated early in a 6-h recording session was about 1.5 times greater than the rate for 23 salt-best fibers isolated later, but a great range of elicited rates was obtained from fibers throughout the experiment. Perhaps the greater part of the variation in absolute responses is related to the dynamic nature of taste receptor cells which turn over about biweekly, requiring new synaptic connections between innervating fibers and receptor cells to be formed repeatedly, or the variation in the number of taste buds innervated by the fiber under examination (Pfaffmann, 1970; Miller, 1971). These two phenomena might, in fact, be related; a fiber with a low absolute sensitivity may make functional contact with one or two taste buds at the time of recording but may have contacted more at a previous time when the cells in its path were more active. Apparently such a large variation in absolute reactivity does not occur in peripheral neurons of the auditory system (Kiang et al., 1965; Kim and Molnar, 1979), for example, where such a turnover in receptor cells is not present.

The sucrose-best chorda tympani fibers are quite "specific," in contrast to the two types of electrolyte-sensitive fibers seen in the hamster (Figures 3 and 5) or rat (Frank et al., 1983); they respond nearly exclusively to two stimuli that are sweet to humans and to which learned aversions cross-generalize behaviorally in hamsters (Figure 4). It was the observation of these "specific" sweet-sensitive fibers that sparked an interest in going back to look at the electrolyte-sensitive taste fibers and to the discovery of some of the sources of their variability in response profile.

Up to this point, responses of taste nerve fibers to a single concentration of a number of different chemicals have been considered. That concentration was chosen because it had an overall intensity, based on the response of the whole nerve, that was equal or midrange for all chemicals applied to the tongue. Figure 7 is a plot of response (number of impulses in 5 s) as a function of log stimulus concentration for the six sucrose-best fibers whose response profiles are shown in Figure 6. Although fibers vary considerably in absolute response rate to strong concentrations and response would probably saturate at different response levels, the functions appear to be approaching a common threshold, with the most responsive fibers' response rates dropping off more rapidly than those of the less responsive fibers, as concentration is decreased. It is as if each sucrose-best fiber obtained the same kind of input from receptors but some fibers obtained more and others less of it. This is consistent with the most-responsive fibers "contacting" more receptor cells or more receptor cells in their prime.

In Figure 8, mean responses of the 6 sucrose-best fibers to 3 concentrations of sucrose, D-phenylalanine, and HCl, NaCl, and NH_4Cl, are plotted as open symbols; mean responses of 11 electrolyte-sensitive fibers (NaCl-best and HCl-best) to sucrose and D-phenylalanine are plotted as closed symbols. There was no significant response in sucrose-best fibers to three concentrations of HCl, NaCl, or NH_4Cl that spanned a 10-fold increase in molarity, the midpoint of which produced a response equal to the response to 0.1 M sucrose in the whole nerve. Nor was there a significant response to 3 concentrations of sucrose or D-phenylalanine in the electrolyte-sensitive fibers for a comparably effective 10-fold increase in molarity (Hyman and Frank, 1980a). The singular quality of sweetness of sucrose and D-phenylalanine to humans and the cross-generalization of learned aversions to them in hamsters (Figure 4) suggest these sucrose/D-phenylalanine sensitive fibers carry information to the brain about "sweetness"; the presence of these two "sweet" chemicals on the tongue does not affect other taste fibers.

Another issue that may be addressed with recordings from peripheral taste fibers is how many independent sites of interaction exist on receptor cell membranes that are sensitive to "sweeteners." One way to address this issue is in the study of mixtures. If two different compounds (e.g., the sugar, sucrose, and the sweet amino acid D-phenylalanine) were applied to the tongue, mixed at concentrations that produced equal responses when presented alone, the result should be equivalent to doubling the concentration of either compound if they both interacted with the same receptor sites. Ideally the response to the mixture could be predicted from $R = f(C)$, with R = response magnitude and C = stimulus molarity, for either compound. As an illustration, using Figure 8, the response to a mixture of 0.07 M D-phenylalanine and 0.1 M sucrose was predicted to be 118.5 impulses in 5 s; the concentration of sucrose that would

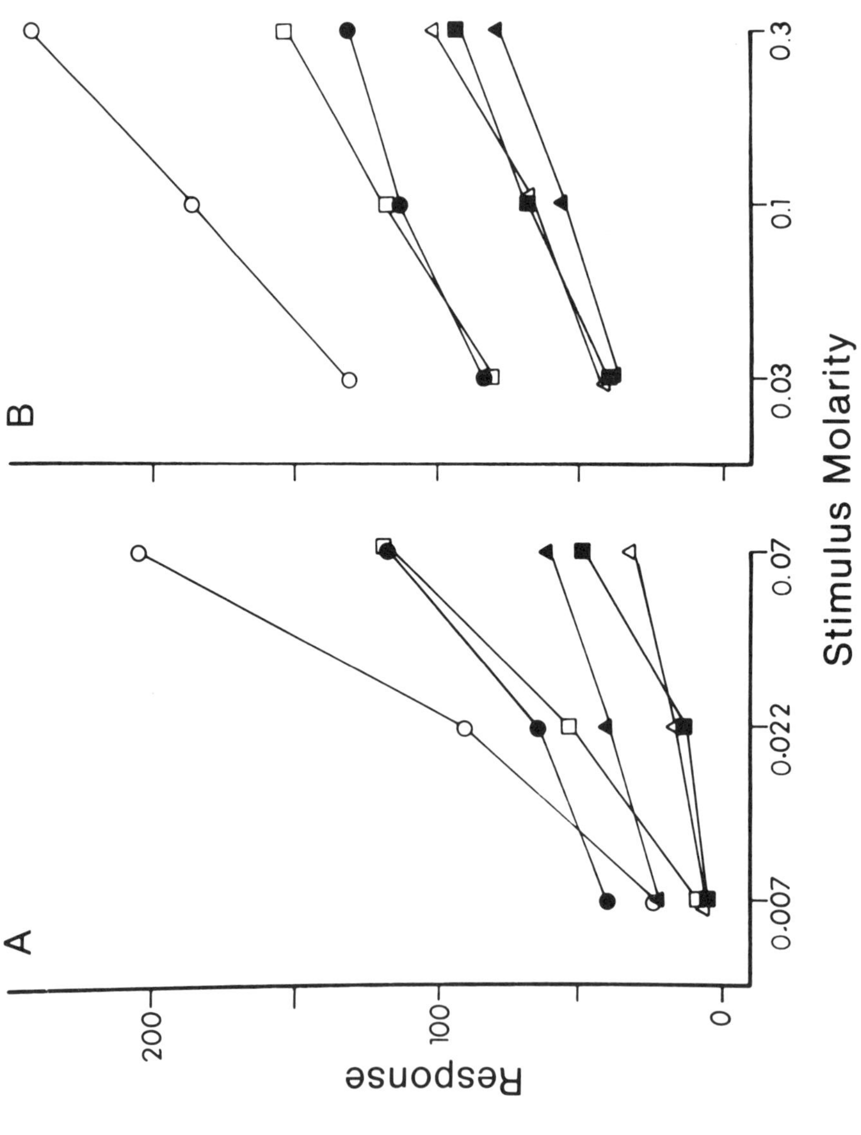

Figure 7. Response (5 s) as a function of concentration of (A) D-phenylalanine and (B) sucrose of the six fibers whose response profiles are shown in Figure 6. Points for responses to the two compounds for a fiber are indicated by the same symbol used in Figure 6.

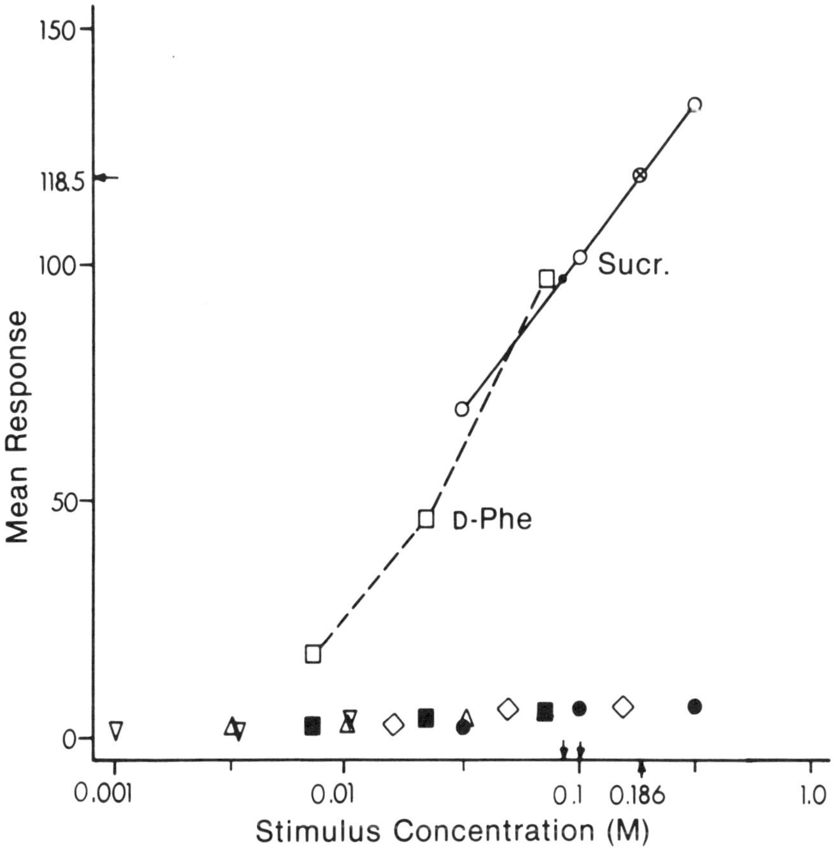

Figure 8. Mean responses (5 s) of nerve fibers of the chorda tympani of the hamster to three concentrations of sucrose and D-phenylalanine as well as to HCl, NaCl, and NH_4Cl in fibers sensitive to sucrose and D-phenylalanine. Mean responses of the six fibers whose profiles are shown in Figure 6 are indicated with open symbols: sucrose (circles), D-phenylalanine (squares), HCl (inverted triangles), NaCl (triangles), and NH_4Cl (diamonds). Closed symbols indicate the mean responses of 11 electrolyte-sensitive fibers to sucrose (circles) and D-phenylalanine (squares). Other points and arrows are explained in the text. Data are from Hyman, 1978.

produce this response was 0.186 M, slightly weaker than a doubled 0.1 M because the response to 0.07 M D-phenylalanine is slightly smaller than the response to 0.1 M sucrose. The small black dot on the sucrose function in Figure 8 is at the level of the 0.07 M D-phenylalanine response, and the downward pointing arrows along the abscissa point to the two sucrose concentrations that should be added to elicit a response equal to that from the mixture (indicated by an encircled × on the sucrose function). That is, a mixture of 0.1 M sucrose and 0.07 M D-phenylalanine should elicit a response of about 120 impulses in 5 s if the two compounds interact with the same receptive site. The D-phenylalanine response function could not be used for a prediction; it was not defined at response levels as high as the one elicited by 0.1 M sucrose because of a solubility limit.

Figure 9 shows the results of a study on the effects of binary mixtures on the six sucrose-best fibers of the hamster chorda tympani nerve whose response profiles are shown in Figure 6. The three dark bars at the left represent mean responses to 0.1 M sucrose, 0.07 M D-phenylalanine, and the mixture of the two; the asterisk above the bar representing the response to the mixture indicates that it is larger than either response to a component (Hyman and Frank, 1980b), but it does not approach a sum of responses to components (~ 200 impulses in 5 s), a result that would be consistent with independent receptor sites for the two sweeteners. The white bar to the right of the dark bar for the sucrose/D-phenylalanine mixture indicates the prediction, obtained as described above, for a mixture of two chemicals that interact with the same receptive site. There is little difference between the mixture response and this predicted response.

The second set of three bars in Figure 9 represents responses of these fibers to 0.01 M NaCl and to mixtures of NaCl and 0.1 M sucrose or 0.07 M D-phenylalanine. The addition of NaCl does not add to responses to the two sweeteners. The third set of three bars represents responses of the same fibers to 0.05 M NH_4Cl and to mixtures of NH_4Cl and sucrose or D-phenylalanine. The asterisk above the bar indicates that the response to the mixture of NH_4Cl and sucrose is smaller than the response to sucrose alone. The fourth set of three bars on the right of Figure 9 represents the sucrose-best fibers' responses to 0.003 M HCl and HCl mixed with sucrose and D-phenylalanine. Responses to both sucrose and D-phenylalanine were suppressed by addition of HCl, as is indicated by the asterisks above the two bars (Hyman and Frank, 1980b).

These peripheral neural effects of mixtures correspond nicely to psychophysical observations on humans. Common psychophysical phenomena are "compression" for mixtures of chemicals with the same taste quality (Barto-

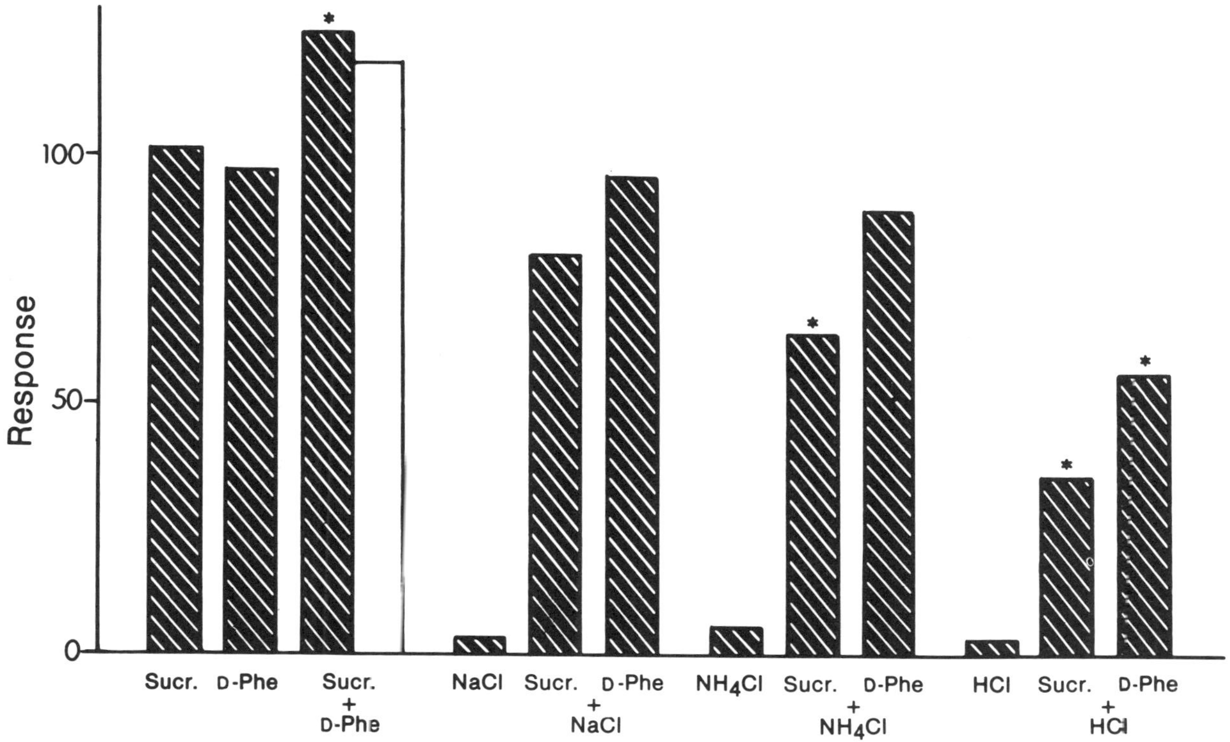

Figure 9. Responses of hamster sucrose-sensitive fibers to binary mixtures and their components. Mean (5 s) responses of the six fibers whose profiles are shown in Figure 6 are indicated by dark bars. Concentrations of solutions, in mixtures or components, are those indicated in the legend to Figure 6. The light bar to the right of the bar for the response to the sucrose/D-phenylalanine mixture is the response predicted for a mixture if stimuli are activating the same receptor sites. Asterisks indicate that responses to mixtures are larger (sucrose + D-phenylalanine) or smaller (sucrose + NH_4Cl, sucrose + HCl, D-phenylalanine + HCl) than a response to the more effective component. Data are from Hyman, 1978.

shuk and Cleveland, 1977) and "suppression" for mixtures of different qualities (Bartoshuk, 1975, 1977). "Compression" or a nonlinear summation is probably seen in mixtures of compounds of the same quality because the two chemicals are interacting with a common, saturable receptive system, and responses follow a single response function that shows less than a doubling of response with a doubling of concentration. For mixtures of different quality, the components of which have a positive effect on totally different sets of peripheral nerve fibers, the "suppression" can have a peripheral basis, as seen in the reduction of the responses to sucrose and D-phenylalanine. NaCl, NH_4Cl, and HCl reduce the effectiveness of the sweeteners at the sites of stimulus interaction with receptors or sites of interaction of receptors with innervating nerve fibers. Incidentally, the similar ordering of suppression by the three electrolytes (HCl $>$ NH_4Cl $>$ NaCl) on the effect of sucrose and D-phenylalanine is exactly the order of their effectiveness on acid-best chorda tympani fibers (Figure 5). Their lesser effect on the response to D-phenylalanine than on that to sucrose could be consistent with a single sweet receptor site, if the electrolytes' effect is an interference with receptor binding and if this interference is, for some reason, greater for the dissaccharide's binding than for the amino acid's.

Conclusion

There are sharp similarities between the stimuli that elicit the sensations of sweet and salty in people, or similar reactions that may be assumed to occur in animals, and the stimuli that activate two types of peripheral taste neurons innervating taste buds on the anterior tongue of animals. In fact, two very different sweetener molecules, a sugar and an amino acid, appear to be interacting with a single receptor site. Therefore, it no longer is necessary to invoke an across-fiber pattern theory of taste quality coding. In addition, since the essential quality distinctions appear to be made at the taste bud in the tongue, neural processing in the central nervous system probably has other functions than simply assisting taste quality discrimination as defined by distinctions between chemicals that are sweet or salty.

Acknowledgment

This work was supported by a grant from the National Science Foundation (BNS 81-12180).

References

Bartoshuk, L. M. 1975. Taste mixtures: is mixture suppression related to compression? *Physiol. Behav.* 14:643–649.

———. 1977. Psychophysical studies of taste mixtures. In *Olfaction and Taste VI* (Proceedings of the Sixth International Symposium). J. Le Magnen and P. MacLeod, eds. London: IRL Press, pp. 377–384.

Bartoshuk, L. M., and C. T. Cleveland. 1977. Mixtures of substances with similar tastes: a test of a psychophysical model of taste mixture interactions. *Sensory Processes.* 1:177–186.

Beidler, L. M., I. Y. Fishman, and C. W. Hardiman. 1955. Species differences in taste responses. *Am. J. Physiol.* 181:235–239.

Beidler, L. M., and R. L. Smallman. 1965. Renewal of cells within taste buds. *J. Cell Biol.* 27:263–272.

Boudreau, J. C., and N. Alev. 1973. Classification of chemoreceptive tongue units of the cat geniculate ganglion. *Brain Res.* 54:157–175.

Boudreau, J. C., J. J. Oravec, and N. K. Hoang. 1982. Taste systems of goat geniculate ganglion. *J. Neurophysiol. (Bethesda).* 48:1126–1242.

Boudreau, J. C., and T. D. White. 1978. Flavor chemistry of carnivore taste systems. *ACS (Am. Chem. Soc.) Symp. Ser.* 67:102–128.

Doetsch, G. S., and R. P. Erickson. 1970. Synaptic processing of taste quality information in the nucleus tractus solitarius of the rat. *J. Neurophysiol. (Bethesda).* 33:490–507.

Erickson, R. P. 1963. Sensory neural patterns and gustation. In *Olfaction and Taste* (Proceedings of the First International Symposium). Y. Zotterman, ed. Oxford: Pergamon Press, Ltd., pp. 205–213.

Erickson, R. P., and S. S. Schiffman. 1975. The chemical senses: a systematic approach. In *Handbook of Psychobiology.* M. S. Gazzaniga, and C. Blakemore, eds. New York: Academic Press, Inc., pp. 393–426.

Fishman, I. Y. 1957. Single fiber gustatory impulses in rat and hamster. *J. Cell. Comp. Physiol.* 49:319–334.

Frank, M. 1973. An analysis of hamster afferent taste nerve response functions. *J. Gen. Physiol.* 61:588–618.

———. The distinctiveness of responses to sweet in the chorda tympani nerve. In *Taste and Development: The Genesis of Sweet Preference.* J. M. Weiffenbach, ed. Washington, DC: U.S. Government Printing Office, pp. 25–41.

Frank, M. E., R. J. Contreras, and T. P. Hettinger. 1983. Nerve fibers sensitive to ionic taste stimuli in the chorda tympani of the rat. *J. Neurophysiol. (Bethesda).* 50:941–960.

Hill, D. L., C. M. Mistretta, and R. M. Bradley. 1982. Developmental changes in taste response characteristics of rat single chorda tympani fibers. *J. Neurosci.* 2:782–790.

Hyman, A. M. The effects of taste stimuli in mixtures on electrophysiological activity of the hamster chorda tympani nerve and its single fibers. Ph.D. dissertation, The Rockefeller University, New York, 1978.

Hyman, A. M., and M. E. Frank. 1980a. Effects of binary taste stimuli on the neural activity of the hamster chorda tympani. *J. Gen. Physiol.* 76:125–142.

———. 1980b. Sensitivities of single nerve fibers in the hamster chorda tympani to mixtures of taste stimuli. *J. Gen. Physiol.* 76:143–173.

Kiang, N. Y.-S., T. Watanabe, E. C. Thomas, and L. F. Clark. 1965. *Discharge Patterns of Single Fibers in the Cat's Auditory Nerve.* Cambridge, MA: MIT Press.

Kim, D. O., and C. E. Molnar. 1979. Population study of cochlear nerve fibers: comparison of spatial distributions of average-rate and phase-locking measures of responses to single tones. *J. Neurophysiol. (Bethesda)*. 42:16–30.
Kruger, S., and J. C. Boudreau. 1972. Responses of cat geniculate ganglion tongue units to some salts and physiological buffer solutions. *Brain Res.* 47:127–145.
McBurney, D. H. 1972. Gustatory cross-adaptation between sweet tasting compounds. *Percept. Psychophys.* 11:225–227.
McBurney, D. H., and L. M. Bartoshuk. 1973. Interactions between stimuli with different taste qualities. *Physiol. Behav.* 10:1101–1106.
McBurney, D. H., and T. R. Shick. 1971. Taste and water taste of twenty-six compounds for man. *Percept. Psychophys.* 10:249–252.
Miller, I. J., Jr. 1971. Peripheral interactions among single papilla inputs to gustatory nerve fibers. *J. Gen. Physiol.* 57:1–25.
Mistretta, C. M., and R. M. Bradley. 1983. Neural basis of developing salt taste sensation: response changes in fetal, postnatal, and adult sheep. *J. Comp. Neurol.* 215:199–210.
Morrison, G. R. 1967. Behavioral response patterns to salt stimuli in the rat. *Can. J. Psychol.* 21:141–152.
Murphy, C., A. V. Cardello, and J. G. Brand. 1981. Tastes of fifteen halide salts following water and NaCl: anion and cation effects. *Physiol. Behav.* 26:1083–1095.
Nowlis, G. H., and M. Frank. 1977. Qualities in hamster taste: behavioral and neural evidence. In *Olfaction and Taste VI* (Proceedings of the Sixth International Symposium). J. Le Magnen and P. MacLeod, eds. London: IRL Press, pp. 241–248.
———. 1981. Quality coding in gustatory systems of rats and hamsters. In *Perception of Behavioral Chemicals*. D. M. Norris, ed. Amsterdam: Elsevier/North Holland Biomedical Press, pp. 59–80.
Nowlis, G. H., M. E. Frank, and C. Pfaffmann. 1980. Specificity of acquired aversions to taste qualities in hamsters and rats. *J. Comp. Physiol. Psychol.* 94:932–942.
Perrotto, R. S., and T. R. Scott. 1976. Gustatory neural coding in the pons. *Brain Res.* 110:283–300.
Pfaffmann, C. 1955. Gustatory nerve impulses in rat, cat and rabbit. *J. Neurophysiol. (Bethesda)*. 18:429–440.
———. 1970. Physiological and behavioural processes of the sense of taste. In *CIBA Foundation Symposium on Taste and Smell in Vertebrates*. G. E. Wolstenholme and J. Knight, eds. London: J. & A. Churchill, Ltd., pp. 31–50.
Pfaffmann, C., M. Frank, L. M. Bartoshuk, and T. C. Snell. 1976. Coding gustatory information in the squirrel monkey chorda tympani. *Prog. Psychobiol. Physiol. Psychol.* 6:1–27.
Sandick, B., and A. V. Cardello. 1981. Taste profiles from single circumvallate papillae: comparison with fungiform profiles. *Chem. Senses.* 6:197–214.
Sato, M., H. Ogawa, and S. Yamashita. 1975. Response properties of macaque monkey chorda tympani fibers. *J. Gen. Physiol.* 66:781–810.
Scott, T. R., and R. P. Erickson. 1971. Synaptic processing of taste quality information in the thalamus of the rat. *J. Neurophysiol. (Bethesda)*. 34:868–884.
Smith, D. V., and D. H. McBurney. 1969. Gustatory cross-adaptation: does a single mechanism code the salty taste? *J. Exp. Psychol.* 80:101–105.

6.
Definitions: A Matter of Taste

Perusal of any modern chapter in taste will show that the basic and guiding idea is that there are four tastes. With the exception of the sections on taste neuroanatomy, almost all materials are presented as discussions of, or in terms of, four tastes.

I have been concerned about the meaning and validity of this core concept for some time. If there can be any doubts raised about its validity, and there can, there could scarcely be a more important topic for discussion at this time. Thus, this chapter will be devoted to the issue of the four tastes. My purpose could easily be misinterpreted; I do not intend to try to prove the four-taste position wrong. But as with other sciences, which have benefited so immensely by examination of their basic concepts, it can only be instructive to examine our primary rule. This process will either strengthen our understanding of the four-taste position, or lead us in more profitable directions.

The idea of taste primaries is basically a psychophysical concept, but of necessity it has had a strong impact on other areas of taste, especially neural coding. Most other research areas are less intimately bound up with this concept, so this chapter will be limited to psychophysical and neural issues.

Psychophysics

The topic of taste primaries in the psychophysical realm suffers in two ways. The first is that the concept of "a primary taste" is not clearly defined. It would seem essential to know what it means since it influences so strongly what we do in our research and how we interpret the data.

Dictionary definitions are seldom useful in cases of this nature. However, one definition of "primary" (Random House) gives some perspective: "not dependent on or derived from something else." This would place taste primaries in the category of the separate sensory "modalities" of Helmholtz, as used by Öhrwall (1891). This is probably the definition most used today, although the use is not always explicit.

The second difficulty is the question of whether there are taste primaries at

---

Robert P. Erickson, Department of Psychology
Duke University, Durham, North Carolina 27706

all. In other words, instead of constituting separate sensory modalities, as vision vs. audition vs. touch, perhaps they fit better into Helmholtz's other category, that of "qualities"; here various tastes would relate as members of a continuum as advocated by Henning (1916), analogous to the differences among colors.

Workers in this area have always assumed the separate modality definition as reflected in the fact that the question has always been "how many tastes are there?"; this is well documented in Boring's histories (1942) and other recent general treatments of taste.

Relevant History

We cannot resolve the issue of whether there are four tastes by referring to past authorities in the area because our psychophysical techniques are certainly improved over theirs. However, a brief look at their work turns out to be very instructive in understanding our present acceptance of four tastes. Aristotle seems to have set the mold; by posing the question in terms of *how many* tastes there are, he seems to have largely prevented the question of *whether there are* primary tastes at all.

Henning, the "father" of the four-taste position, was totally opposed to that idea; a reading of what Henning said (1916) is very surprising. He felt that taste was organized in a continuous or serieslike fashion, analogous to colors, emphatically and explicitly ruling out the limitation of tastes to four. He strongly felt that the edges and surfaces of his tetrahedron were composed of "innumerable transitional" taste sensations between the basic stimuli at the corners; these many tastes were unique or singular in the sense that they were not reducible to, or analyzable into the primary four (see Erickson and Covey, 1980). Although he mainly discussed single chemical stimuli, he also felt that stimulus mixtures produced singular sensations. For example, he states that concerning sweet-salty tastes, "One does not sense a sugary separated from a salty taste just as one does not see an orange color as dark red at one moment and as bright yellow the next. If one is presented a chemical combination of table salt and cane sugar, one experiences a uniform sensation of fusion."

Henning was not entirely alone in this many-taste position. He was anticipated by Wundt (e.g., 1907), a well-known experimental psychologist. Wundt felt that the most "sharply differentiated sensations" in taste were few in number (six, including metallic and alkaline which he later withdrew from the list, leaving our primary four). But then he commented on these few tastes as follows: "In saying this, we do not mean to assert that these are the only taste-sensations possible at all. It is clear that, e.g., by combining sweet and bitter, we can produce a taste which is neither sweet nor bitter, although it has something of both qualities." His position was precisely that of Henning; that there are not just the four tastes, but also intermediaries between them.

Von Skramlik (1926) took Henning to task for his rejection of four tastes. Von Skramlik felt that all tastes, including those of mixtures, consisted of and were reducible to the primary four. These could be placed along the edges and surfaces of Henning's tetrahedron, but only as complex tastes analyzable into various amounts of the still-distinguishable tastes represented at the corners. But a reading of the original publication turns out to be rather surprising as a defense of the four-taste position.

First, von Skramlik seems to have agreed that there is a great range and diversity of tastes and that the number of taste sensations is "naturally infinitely large." He also commented several times that each of these many sensations, including tastes of mixtures, appears to be *unitary* (singular) and unique, not *obviously* composed of various degrees of several of the primary four. He repeatedly asserted that this initial impression of unity in most tastes can be dissected into the primary components only by thoroughly trained observers (also noted earlier by Öhrwall); obviously, this training could bias the results depending on the nature of the training, which is also a possible problem in modern studies. Even with trained observers, he agrees that real fusions occur between four of the six possible mixtures of the primaries (i.e., all except bitter-salty and bitter-sour), such as Wundt's earlier example of the fusion of sweet and bitter, into the "pleasant unitary sensation" of chocolate. He also mentions that a rather convincing cancellation of tastes (analogous to the grey resulting from mixing color complements) can occur in some cases. Of course, most of what he said was strongly in support of only four tastes. However, these comments intermingled with the rest make his defense of the quadripartite position equivocal.

But it was the earlier work of Öhrwall (1891) that set the stage for the modern four-taste position (McBurney, 1974, 1978; Bartoshuk, 1978; McBurney and Gent, 1979). His logic had two stages. First, he argued that there were no variations among each of the primary tastes. For example, he felt that various sugars, such as sucrose, lactose, and dextrose, could not be distinguished from each other; this is discussed further below. These conclusions were based on very little data from a total of about four subjects, including Öhrwall, and were not supported in later work. Second, he concluded that since there were no variations within each taste (in terms of Henning's later tetrahedron, there were no tastes except exactly at the corners), then there certainly could not be continua between them (the edges and surfaces of the tetrahedron). On these grounds, the four tastes met Helmholtz's definition of separate modalities, being as separate from each other as vision is from audition. I believe that is the position taken today.

It should be clear that our historical four-taste roots are shaky; when Boring (1942) wrote that Henning's taste tetrahedron "settled the problem to everyone's satisfaction" that there were four tastes, the issues were not as well resolved as he implied. A close reading of Boring shows that he was discussing the issue of *how many* tastes there are rather than asking the more fundamental questions of how taste is organized, such as "are there primaries?" We are left with at least the following possibilities: tastes defining the limiting cases, as the corners of the tetrahedron (von Skramlik and Henning), the fact that some tastes are more salient than others, for example, those that Henning put at the corners of his tetrahedron (Wundt, and probably all the others), taste continua or series (Henning, Wundt, and Kiesow), and only four tastes as separate modalities (Öhrwall and von Skramlik).

In general, taste scientists have tended to select or interpret data in support of the four-taste position. The data, however, do not unequivocally support that position. A very simple and typical example is that of the distribution of the various tastes across the surface of the tongue. Does this fact support the position that there are four separate tastes? No. By analogy, the fact that four tones produce maximum activity at different parts of the basilar membrane does not prove that each tone is a separate modality.

But what if the sensitivity to sour, bitter, sweet, and salty were distributed equally or similarly across the tongue? This should not deter anyone interested in the four-taste position. For example, both cold and touch can be experienced across the total body surface; but this does not prevent them from being clearly separate modalities. In other words, such data are *not relevant* to the issue of four tastes. Perhaps these data are used to support the four-taste position because of a historical or psychological pressure to prove the existence of four tastes.

This example, of the distribution of tastes over the tongue, is a fairly trivial one, but in a like manner the various other supports for four primary tastes may be shown to be equivocal (Schiffman and Erickson, 1980). I also suspect that any data that might detract from the four-taste position, by showing similarities among tastes, are not emphasized. It seems that there may have been, and may continue to be, an active, if unconscious, search for and selective attention to those data that might support the idea of four tastes. Neither the originators of this idea (Öhrwall, von Skramlik) nor the data commonly used in defense of it are completely convincing.

Recent Data Relevant to the Existence of Four Tastes

Some data are more clearly relevant to the issue of four tastes; I would like to briefly point out several recent findings of this nature. In general, they favor but do not prove the continuous model of Henning rather than the four-taste model of Öhrwall.

Are There Variations within Each of the Primary Tastes?

It should be recalled here that Öhrwall's distinction among the four primary tastes as being separate modalities rested primarily on the idea that there were no variations within each taste. Is there evidence that this is true?

Sweet. Öhrwall felt that he could not distinguish among sucrose, lactose, dextrose, etc. Current research does not support this finding. For example, Schiffman et al. (1979) found that different sweeteners have *qualitatively different* sweet tastes. Kuznicki and Ashbaugh (1979, 1982), in addition to agreeing with Schiffman's results, found that in mixtures with NaCl, sucrose attains a taste qualitatively different from that produced when sucrose is presented alone. This suggests that tastes may vary from pure cases of the primary four (variations in the vicinity of "sweet" in this case), and thus that the four primaries may not be exhaustive descriptors. Kuznicki and Ashbaugh put this point in the following manner: "It is suggested that taste mixtures can be considered as single tastes even though component qualities can be reliably identified using other techniques." This is a precise analogue of the data in color-naming discussed below.

In neural studies congruent with these findings, Beidler (Chapter 2 of this volume) suggests the existence of more than one kind of sugar receptor from recordings of single chorda tympani fibers in the rat, Shimada and his co-workers (1974, 1978) find evidence for two classes of sugar receptors in the fly, and Hellekant (1975) finds evidence for several classes of sweet receptor mechanisms in neural recordings in various mammals.

As another example of variations in the area of "sweet," mixtures of saccharine and glucose have a hedonic property, as shown in consumption measures (Valenstein et al., 1967; Smith and Foster, 1980; Smith et al., 1980, 1982), greater than that of either of its components, and therefore somewhat difficult to explain in terms of two separate tastes (Smith et al., 1982). One rather direct explanation is that the two components synthesize a new taste, different from and more rewarding than either of the sweet components.

Bitter. Öhrwall felt that strychnine, quinine, and morphine were indistinguishable from each other. But it seems possible that there are variations within that area too (e.g., Schiffman and Erickson, 1971; see Bartoshuk, 1978, for discussion). For example, caffeine, phenylthiocarbamoyl (PTC), quinine, and urea may all function differently, but we have only one term available for each—"bitter."

Sour. Öhrwall concluded that HCl, H_2SO_4, and acetic, tartaric, and oxalic acids were all indistinguishable, controlling, of course, for odor. These acids are quite discriminable from each other (e.g., Schiffman and Erickson, 1971).

Salty. Öhrwall had considerably more trouble with the salty taste, in that most salty stimuli had one or more "side-tastes." However, he came to the conclusion that within these complex tastes, the salty quality was the same in each. He did not specify how he came to this conclusion.

In general, it is clear that most salts are very complex and easy to distinguish from each other. Concerning the salty component, however, McCutcheon and Brown (in press) have shown that sodium-deficient rats will drink more of a NaCl-sucrose mixture than of sucrose alone, tending to suggest that they could identify the sodium. But the fact that they would drink less of the mixture than of NaCl alone suggests that the "salty" taste was distorted by mixture with another taste. In other words, there may be a series of different tastes in the "salty" area.

Kuznicki and Ashbaugh, in the studies mentioned above, found that NaCl has different tastes in mixtures with other stimuli.

Concerning all these tastes, Schiffman's studies on the multidimensional arrangements of stimuli in taste cast strong suspicion on the adequacy of four tastes in accounting for the full range of possibilities; these have been discussed extensively elsewhere (Schiffman and Erickson, 1980).

It should be clear that the assumption that there are no variations within each primary taste is at best questionable.

Can the Components of a Taste Be Separately Sensed?

As discussed above, it seems quite possible that there may be variations within each primary taste. An issue even more essential to the four primary taste position is whether we can sense the individual components of a taste (even if each is somewhat variable). The four-taste position requires that we can.

From the beginning, it was obvious that the four primary components of a complex taste are not unequivocally discernible. Even von Skramlik and Öhrwall pointed out that the components of a taste are not at all obvious, and that thorough training is necessary to perceive them. O'Mahoney and Buteau (1982) have directly addressed this issue and agree that very extensive training is necessary to evaluate tastes in terms of the four primaries. Even with the most thorough training, analysis of the components of a complex taste is very equivocal. This suggests that the components do not have clear and separate perceptual existences. (Their procedures on the identification of the actual components used are somewhat different from those on the identification of perceptions discussed below, but the similarities are interesting enough to warrant mention.)

In most modern psychophysical experiments, practice is essential, but the question of whether a taste is unitary or multiple (Erickson, 1977, 1982a; Erickson and Covey, 1980) may be different from most psychophysical procedures in its extreme simplicity. With simple stimuli it requires no practice to determine whether the experience is unitary or not in vision or audition. When presenting subjects with one to four tones, there is absolutely no difficulty in knowing when there is only one tone or more than one (although it takes practice to be able to accurately state whether there are two or three or four tones, and to identify each). I have gone so far as to jeopardize my reputation as a responsible and meaningful scientist by asking subjects whether each of a variety of hues was one or more than one; the answers were always unequivocally "one," even for colors intermediate to red, yellow, green, or blue. No practice was required. The subjects were puzzled that we should ask such a simple question because all hues are so clearly seen as "one." On only two occasions did subjects state that a hue was "two," because they knew it would take a mixture of two stimuli to produce that color; perhaps these were our more "practiced" subjects (Erickson, 1977; Erickson and Covey, 1980).

In taste we found that the situation is quite different from that in vision and audition. With unpracticed subjects, it is never clear that there is more than one sensory component present even when primaries are mixed, or when "complex" stimuli (with "side-tastes") are presented (discussed further below). This is not to say that, with practice, they cannot *rate* these stimuli in terms of sweetness, sourness, etc.

Thus, it seems possible that with von Skramlik and Öhrwall, as well as in modern studies, the reports on four tastes depend on the nature of the instructions and the training required.

In another vein, psychophysical data suggest that separate tastes, such as the primary four, may not remain separate, but may combine in ways in which the original components are lost, and thus tastes other than the primary four must be produced. For example, there are recent as well as old data (see Kiesow, 1896; von Skramlik, 1926, on "psychic inhibition" and "compensation") on taste mixtures and adaptation effects which show that the components of a taste are lost in a "fusion" process. This process follows in surprising detail the synthetic processes of color vision. (The analogy to color vision was introduced by Kiesow [1896, 1898], but his emphasis was on contrast, a topic then in vogue but now not considered particularly relevant in taste.) This process would require a continuum of tastes rather than just four. This is currently an active area (e.g., Beebe-Center et al., 1959; Kamen et al., 1961; Pangborn, 1961, 1962; Indow, 1969; Moskowitz, 1972; Bartoshuk, 1975; Lawless, 1977, 1979, 1982; Kroeze, 1982; Gillan, 1982; see also Bartoshuk, 1978, for review), and the effects are termed "mixture suppression."

For suppression, in general, the components of a mixture of dissimilar stimuli appear weaker (suppressed) than when presented alone. That is, a mixture of NaCl and HCl will not be rated as highly on saltiness and sourness than when each is presented singly. In addition, prior adaptation with one component releases the suppression of the other in the mixture; in the previous example, the sourness rating of the mixture will not be suppressed if the tongue is first adapted to NaCl. If the tongue is first adapted to the mixture, there is a partial suppression of subsequent responses to the components presented alone.

That there is any interaction at all runs counter to the idea of four primaries being separate modalities in Öhrwall's sense.

All these effects are exactly analogous to events in color vision, substituting green and blue, for example, for the two tastes. First, it is important to note that subjects can reliably rate any color, including mixtures, in terms of a few primaries (Boynton et al., 1964; Erickson, 1977), just as it is possible in taste. This alone does not prove that only those taste primaries exist since in color it is clear that we are dealing with a continuum, rather than with a few sensations that are present in various degrees in any given color. When blue and green are mixed, a subject can rate the ensuing blue-green on both blue and green. His rating of each in the mixture decreases, however, just as in taste. This is *not* because he still sees each color separately, but more dimly, as is required in "suppression" in the four-taste model; instead, he sees a new, unique color. This is in accord with the findings of Henning and Wundt, and accounts for the fact that von Skramlik's subjects always initially perceived the taste mixtures as unitary. Also, just as in color, prior adaptation to salt will restore the intensity of the sour in the salt-sour mixure, as well as reduce or eliminate the sensation of salt. (Color afterimages would be analogous to the various water tastes, and, also as in color, occasional enhancement of the taste of the second of a pair of stimuli might occur; for example, sweet followed by sour [see Bartoshuk, 1978].)

These data on taste suppression flow exactly and simply from a continuous, synthetic model (in which "fusions" occur; see Erickson, 1982b). In fact, the data on suppression seem to be good demonstrations of synthesis. However, this does not prove the system to be synthetic; the answer may lie between this and the four-taste position, or elsewhere. Development of the concept of "suppression," however, does illustrate the problem of where our assumptions may take us. In this example, *if* four separate tastes are assumed, the addition of a concept of "suppression" (not needed in color vision) is required. This brings with it various attendant issues such as whether the component stimuli block or inhibit alien receptors, or whether the "sweet fibers" produce inhibitory postsynaptic potentials (IPSPs) in "salt" fibers at some peripheral or central locus, etc. Thus, although it is convenient, the acceptance of the four-primary model may have adverse effects; it may lead us into areas of study, the premises of which are problematic.

The color vision model calls into question another set of data that provide some of the strongest support for the identifiablility of individual tastes in the four-taste model. Nowlis and Frank (1977) have shown that if an animal is shocked for drinking a mixture of two primary stimuli after the animal has learned that one of them is "safe," the avoidance will appear only to the other stimulus. This suggests that the two tastes are separate; that the animal knew that one of them was safe, and thus that the other was perceived as the unsafe component. However, the data may be interpreted in a different way; assume that the system functioned as does color vision, and that an animal was nonrewarded for a mixture of two stimuli (e.g., blue-green) after learning that one was rewarded (e.g., blue). Guttman (1965) has shown that in this situation (which differs from the taste paradigm mainly in the use of nonreward rather than punishment), the depression of responding extends beyond the nonrewarded stimulus (blue-green); that is, if the nonrewarded stimulus were blue-green following reward on blue, the responses to green would be depressed below those for the blue-green. This a very close approximation to the taste experiments of Nowlis and Frank, and suggests that even here the synthetic, continuous color vision model may be as appropriate in taste as the four-taste model.

As another example in which mixtures produce tastes alien to the primary four, DiLorenzo et al. (1982) found that for rats, the taste of ethyl alcohol was like a mixture of quinine hydrochloride (QHCl) and sucrose, whereas it did not generalize to either QHCl or sucrose alone, or to any of the other primaries. This suggests that QHCl and sucrose add together to taste like something entirely different from bitter or sweet, a possible case of the synthesis of a new taste which is not described by any of the four primaries.

Data from my laboratory supporting the idea that individual taste components can be identified are (Erickson and Covey, 1980; Erickson, 1982a) clear in one respect and problematical in another. First, they are quite clear in showing that the so-called complex tastes are *not* perceived as two, three, or four stimuli, as required in four-taste analytic theory. Henning and von Skramlik were close to being correct in stating that these tastes are perceived as unitary; on the other hand, they were not completely correct because our data showed that nothing was perceived as completely unitary, as we would expect from synthetic continuous theory. The so-called "primary" stimuli were *rather* more unitary than the rest, but never totally unitary. Yet many of the mixtures were as unitary as many of the "single" stimuli. Mixtures of NaCl with several salts and acids were as unitary or more so than the primary stimuli alone, as was a mixture of QHCl and HCl. Mixtures of sugars with other stimuli gave the best examples of stimuli with possibly more than one component taste, but even these were equivocal. The general import of these data is very forceful in suggesting that we do not clearly perceive just four tastes, but it falls short of complete and unequivocal support of synthetic theory.

Summary: Four-taste Theory in Psychophysics

Concerning these contrasting points of view in taste, the four-primary analytic position, following Öhrwall and von Skramlik, is probably simplest to understand, but is cumbersome with certain portions of the data. On the other hand, most areas of taste data are easily fit by the continuous synthetic position, in agreement with Henning. Still, a degree of analysis may occur between realms of taste stimuli that are really very different. One example might be between the sugars on the one hand, and most other stimuli on the other. This particular possibility, suggested by our own neural and psychophysical studies, is contrary to the fusions of sweet and bitter mentioned by Wundt and von Skramlik, and the synthesis or "suppression" between sucrose and other stimuli reported by others. But the degrees of analysis and synthesis in taste will not be discovered unless further research on this topic is undertaken without preconceptions that lead to the selection or analysis of data in a manner favoring one particular position.

So the meaning of four primary tastes is not always made clear, and thus this model is hard to evaluate in its particular usages. I believe that this position should be analyzed more specifically before being used as a general theoretical or working framework.

Neural Mechanisms and Four Tastes

Neural Codes

Does the question of whether there are four separate primary tastes have any bearing on neural coding mechanisms? Any discrete and separate sensory process (especially in Öhrwall's sense of four tastes as four separate modalities) would require a separate neural process for each. This is the intent of the "labeled-line" (LL) theory (Frank and Pfaffmann, 1969; Frank, 1973, 1974; Pfaffmann, 1974, Pfaffmann et al., 1976; Nowlis et al., 1980), in which a given neuron unequivocally signals the presence of one of the primary tastes. These neurons constitute four groups, each member of which signals the same information as the others in that group.

On the other hand, any continuously variable sensory process would be incompatible with a few types of labeled lines. The across-fiber pattern (AFP) theory (Pfaffmann, 1955, 1959; Erickson, 1963; Marshall, 1968; Dethier, 1971, 1973, 1977, 1978, 1982; Scott and Erickson, 1971; Scott, 1974; reviewed in Erickson [1982b]) is designed to mediate such realms of sensory events.

Some examples of both types of coding may be seen in olfaction. Pheromones are certainly discrete stimuli, and labeled lines would serve them well, probably in the vomeronasal system. Shepherd (Chapter 13 of this volume) has identified an olfactory area possibly specialized to mediate suckling behavior in rats; the afferents to this area and the area itself are good candidates as labeled lines, since the behavior is specific to one odorant situation. On the other hand, AFP coding has been advocated for general olfactory coding (O'Connell and Mozell, 1969; O'Connell, 1975, 1981; MacKay-Sim et al., 1982), wherein the concept of "primary" seems not to be required or used.

In taste, neural labeled lines should certainly be present if Öhrwall was correct, or if there are pheromonelike tastes. But to the extent that tastes vary continuously, or even if there are very many tastes, as Henning contends, then the neural AFP model would be more appropriate. In fact, the psychophysical vindication of Öhrwall would nearly rule out AFP theory. (Smith's position, presented in Chapter 7 of this volume, combines both theories.) So the validity of these theories may be tested psychophysically.

Therefore, a clear topic exists that may be used to evaluate these theories; we are back to the psychophysical question of whether there are only four tastes. As discussed above, only a few experiments bear directly on this issue, and these tend to suggest that, in agreement with AFP theory, there are more than four tastes, perhaps many more, or even a continuum. This does not rule out the possibility that by some definitions (such as Smith's), or in some stimulus realms (such as sugars vs. nonsugars), LL analysis is relevant.

In addition to testability, the truth of a theory depends on its generality; how much data will it explain? The generality of these theories extends beyond the area of taste. It is unlikely that the nervous system would evolve totally separate coding systems for each sensory modality. I have found the AFP position very useful for sensory processes in general (Erickson, 1968), and for nonsensory processes as well (Erickson, 1974, 1982b). The resolution of this issue in taste will probably turn out to be part of more general considerations of neural organization. I prefer to believe that the nervous system is parsimonious in its principles, and that taste is not categorically different from the rest.

Typologies: What Are They, and Why Do We Need Them?

Are there neural groupings, as required by the LL position? First, it should be pointed out that although this theory requires four groups of neurons, each of which has a particular meaning for the organism, sweet, sour, salty, or bitter, it is not essential that each member of a group respond to stimuli in exactly the same way as every other member of that group (Erickson, 1977). Similarly, AFP theory makes no demands concerning neural grouping in a physiological sense. Therefore, physiological studies of neural grouping may not be expected to be definitive concerning these models. Even with this in mind, evidence of physiological groupings of neurons has been used as rough support for the LL position.

A serious problem here is that the search for such typologies often proceeds without definition of what a neuron type is. It is certain that if a clear definition is not given, *any* position on typologies may be correct. For example, are there types of people? This question cannot be answered "yes" or "no" unless the meaning of "type" is clarified: types by sex? nationality? height? personality? moral standards? We need a clear statement of the question before we can give an answer. (This is obviously a very broad problem area extending beyond taste psychophysics and neurophysiology, to other areas of sensory neurophysiology where the issues are not as simple as had been hoped—for example, visual X, Y, and W cells, and simple, complex, etc., cells—and to general areas of taxonomic analysis beyond these fields.)

In fact, there could be many definitions of typology that are useful. However, the typologies in each case might include different groupings of the neurons. They could be defined in terms of spontaneous activity, or fiber size and conduction velocity, or perhaps extent of dendritic field or shape of the soma. Some definitions could be unequivocal, such as clearly different areas of termination in the brain, and differences in neurotransmitters. There is nothing inherently wrong with any such method of groupings, but it would be unlikely as well as unnecessary for each method of grouping to produce the same groups. One must ask what the purpose of the typology is. For example, typologies of people by sex would not correspond with the typologies by nationality.

As examples of this in taste, if the topic is the coding of taste quality, it is problematic when methods of grouping not clearly relevant to this topic are used (e.g., Boudreau and Alev, 1973; Boudreau, 1974; Boudreau et al., 1975, 1982). For example, the LL position would not be threatened if "sweet-best" neurons contained the same neurotransmitter as "sour-best" neurons, or if they conducted at the same velocity. Also, it would not necessarily rule out the reality and role of "sweet-best" neurons if some methods of classification showed physiological or anatomical differences *within* the group; for example, if it is determined that "sweet-best" neurons included cell bodies of clearly different shapes, that need not nullify the "sweet-best" categorization. Depending on the method of classification, one is *assured* of "finding" or not finding groups. It is essential that the relevant criteria are defined before the experimental approach to the subject; otherwise, a "yes" or "no" answer to the question of groupings is meaningless.

Then how should we define neural typologies in taste? If we pursue the problem in terms of the coding of taste quality, the relevant data presumably consist of the way the neurons respond to qualitatively different stimuli, although other criteria have been used.

Most workers in the field would agree that certain data on the response of taste neurons to various stimuli are true. The area of agreement is that these neurons are arranged neither exactly evenly, as keys on a piano, nor into a few totally homogeneous groups, along a base line or through a multidimensional space (as in Figure 1, top left panel). The difficult question at this time is whether or not to use the term "types" for such an uneven distribution. The answer to this question depends on a clear definition of "type." Several possible definitions, of varying utility, follow.

"Homogeneous" Definitions

Photopigments can be used to illustrate one definition of "type." In this case, the meaning of type is that each member of the group is, in terms of the defining criteria, *exactly* the same as every other member of that group. A given photopigment does not vary; the absorption spectrum of rhodopsin is

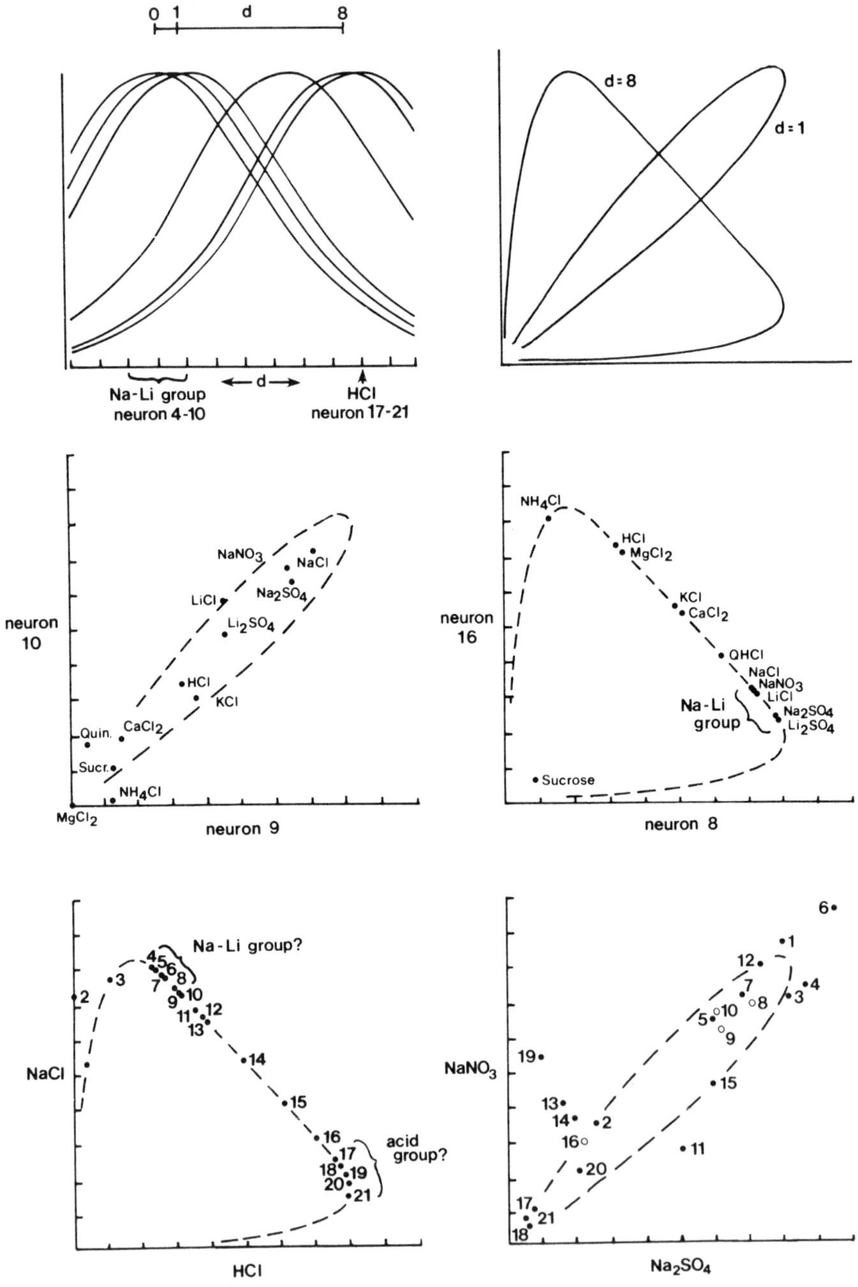

0 1 d 8
d=8
d=1
Na-Li group
neuron 4-10
d
HCl
neuron 17-21
neuron 10
neuron 9
NaNO3
NaCl
Na2SO4
LiCl
Li2SO4
HCl
KCl
Quin.
CaCl2
Sucr.
NH4Cl
MgCl2
neuron 16
neuron 8
NH4Cl
HCl
MgCl2
KCl
CaCl2
QHCl
NaCl
NaNO3
LiCl
Na2SO4
Li2SO4
Na-Li group
Sucrose
NaCl
HCl
Na-Li group?
acid group?
NaNO3
Na2SO4

presumably exactly the same in every receptor it inhabits (in a given species). I will call this a "homogeneous" definition of type. This definition could be roughly analogous to typing people by sex or nationality.

With this definition, taste neurons very probably are not arranged as types. If this were the case, the placements of the NRFs should fall exactly into a few groups, such as four. Any variance from such putative groups could only be due to experimental error. However, Woolston and Erickson (1979) found that there is too much variance or dispersion among the NRFs to conclude that they fell into homogeneous types.

"Polar" Definitions

"Homogeneous" is only one definition of type, however, and certainly an arbitrary one. The term "type" is often used in another very useful manner. For example, we speak of "easterners" and "westerners," and even of "New Yorkers." There are also the personality types of "introvert" and "extrovert." While our language makes good use of such terms, it is important to note that these terms do not imply that every individual in a group is identical to all others, i.e., that all "easterners" are the same. And the introversion-extroversion scale is not bipartite; these are just the poles between which people exist in a continuous array. (It is interesting, perhaps prophetic, that in his early studies of the responses of gustatory neurons, Pfaffmann held out little hope for typologies, commenting that neuron types in gustation might turn

---

Figure 1. Scatterplot analysis of relationships between neurons and between stimuli (derived from Erickson et al., 1965). Upper left: model gustatory neural response functions arranged along a base line of various taste stimuli; for demonstration purposes, schematic locations of various neurons and stimuli, to be illustrated below, are indicated (rat nucleus of tractus solitarius [NTS]; normalized data). Distance values (d) are arbitrary. Upper right: scatterplot forms that would be produced by stimuli or neurons at the given d values (NaCl and HCl would produce scatterplots of about the form d = 8; two Na or Li salts would produce scatterplots of about the form d = 1. Middle panels: scatterplots indicating that neurons 9 and 10 are similar (d = 1), and that neurons 8 and 16 are dissimilar (d = 8). Notice rough grouping of Na and Li salts in each scatterplot. Lower panels: scatterplots indicating that NaCl and HCl are dissimilar (d = 8), and that $NaNO_3$ and Na_2SO_4 are similar (d = 1). Locations of neurons used in middle panels are indicated with open circles; notice that neurons 8, 9, and 10 (used in the middle panels) fall within a rough Na-Li group in the lower panels. Neuron 16 (used in the right middle panel) falls near the rough "acid" group in the lower left panel.

out to be as diverse as personality types [1959].) These typologies might be termed "polar," with each of the primary poles constituting the extreme, limiting case. The corners of Henning's taste tetrahedron could be seen as such "primary" poles.

In this sense, gustatory NRFs could usefully be described in terms of the poles to which they were nearest, such as "salt-best," etc. Then the answer to the question of whether such neural typologies exist could only be "yes." In fact, there is, in a practical sense, no way for a neuron not to respond better to one of the four primaries than to another, just as any individual must be either an easterner *or* westerner, introvert *or* extrovert, etc. This would be true even if the NRFs were distributed along a base line (such as one of Henning's tetrahedral edges) in an absolutely even manner. This could be a useful shorthand language to help us talk about these neurons.

In summary, both psychophysically and neurally, unless a relevant and clear definition of type is given, the question of whether there are types cannot be resolved. With clear definitions, the use of the concept of typologies could be very beneficial.

Techniques for Defining Types

As a necessary part of the question of psychophysical or neuron typologies, mathematical techniques of analysis have been developed to supplement the earlier, more casual and intuitive classifications of neural activity. These mathematical techniques have included factor analysis, cluster analysis, multidimensional scaling (MDS), and others. These have each been discussed in detail elsewhere. The use of these methods is still in its early evolutionary stages. We have indulged in these methods for illustrating the organization of neurons (Erickson, 1963, 1967; Woolston and Erickson, 1979; Erickson et al., 1980), and Smith (Chapter 7 of this volume) presents some of the most elegant recent advances.

I would like to comment briefly on what these techniques are able to tell us, and what they cannot. It is important to understand that they do not give definitive answers accompanied with probability values of whether or not the groups exist. In all cases, after the analyses are complete, the groupings exist only in the eye of the beholder. If *a priori* definitions are given of what a group is, more definitive statements might possibly be made. For example, I have mentioned that Woolston and Erickson (1979) statistically showed that there was too much variance to accept neural groupings if the "homogeneous" definition was used. Although these analyses cannot decide the issue of typologies, they give very convenient illustrations of the relationships among the data.

Scatterplots. In the methods just mentioned, we necessarily get rather far from the data. Correlations or some other metric are derived, and then complex procedures are performed on these, such as factor analysis, cluster analysis, or MDS. On the other hand, we have used a technique (Erickson et al., 1965) that is a more direct representation of the data and may be one of the most powerful for showing in an undistorted way the relationships between stimuli or between neurons.

The method is simply to display in scatterplots the counts of numbers of evoked impulses (Erickson et al., 1965; Woolston and Erickson, 1979). Figure 1 is an illustration of this. By displaying the responses of a variety of neurons to a pair of stimuli, the relationships between the neurons may be seen. Such a plot will also show how similar those two stimuli are to each other (*d* values in the figure; see Erickson et al., 1965). In this figure, the form of the scatterplot shows that Na_2SO_4 and $NaNO_3$ are similar. The comparison of NaCl and HCl shows some clustering of neurons in the Na-Li region as well as in the HCl region; it also shows much dispersion among the neurons, as well as the fact that these stimuli are dissimilar.

On the other hand, as shown in this figure, if the responses of two neurons to a variety of stimuli are plotted, the relationships among the stimuli may be seen, as well as how similar those two neurons are. Rough clusterings of Na and Li salts are always seen.

It is very important to note that although we are using scatterplots, we are not using correlations. One reason is that the information concerning groupings and degrees of differences is in the *form* of the scatterplot, and the correlation does not give an explicit description of this form.

As with cluster analysis and MDS, the final analysis with scatterplot analysis is in the eye of the beholder; no P values are given. The inconvenience of this technique is that it only evaluates the data in terms of two neurons or two stimuli at one time, and one cannot feed the data into a computer and press a button for this analysis. Still, I think that it holds the promise of being both very close to the data and very powerful in finding underlying relationships among stimuli and neurons.

Some fairly clear trends are shown in this method. Some stimuli always form a loose group, such as the Na and Li salts, sweet stimuli, and to some extent acids. On the other hand, most stimuli, even within groups, appear to be quite individualistic, with definite but distant relations to other stimuli.

With the present method, perhaps more valid mappings of stimulus spaces may be devised. This is one of our current research directions.

For neurons, it is very difficult to see much evidence for groupings in the scatterplots, although there seems to be more grouping in the periphery than centrally. These results appear in data from our laboratory, as well as in those from other laboratories to the extent that we have been able to analyze them.

Neural mass differences. Another problem exists with the use of correlations. When used in cluster analysis or MDS, low correlations mean that the stimuli or neurons are far apart. Because of this they tend to provide much of the structure of the analyses. However, relatively ineffective stimuli and poorly responding neurons tend (even after normalization) to produce low correlations as a spurious result of producing weak responses. Thus they can seriously weaken such higher-order analyses. This may be corrected by using "neural mass differences" instead of correlations as measures of differences between stimuli or neurons, a method being developed by J. M. Gill II in my laboratory. The effect of using this measure with stimuli may be seen in Figure 2. It should be clear that I am not drawing any conclusions about groupings of neurons or stimuli; this cannot be done without definitions of the terms "group" or "type."

Conclusions

It should be clear from the foregoing that I do not claim that there are not neuron types or four tastes, because there are no clear and generally accepted definitions of what a type or a taste is. Nor do I think that I have proven that coding is by across-fiber patterns, or that labeled lines do not exist, because the bulk of the data has not been evaluated in terms of both theories, and unequivocal tests to differentiate between them are not common.

What I want to express, at least in the area of taste psychophysics and neural coding, is that our goals are rather fuzzy, and the methods of approaching these goals are not always clearly relevant. For example, it is not always clear whether the concept of four tastes is being implied. Is it being used in studies on "suppression"? Is it being used in studies of neural typologies? The answer to these questions is, frustratingly, sometimes "yes," sometimes "no," or "yes and no" or "maybe." Walking on a fence can be nerve-wracking and dangerous, but coming down on both sides can be painful.

The terms "sweet, sour, salty, and bitter," plus perhaps one or two more, will probably always be with us. It is hard to see how we could get along without them. But it will make an immense difference to our progress and

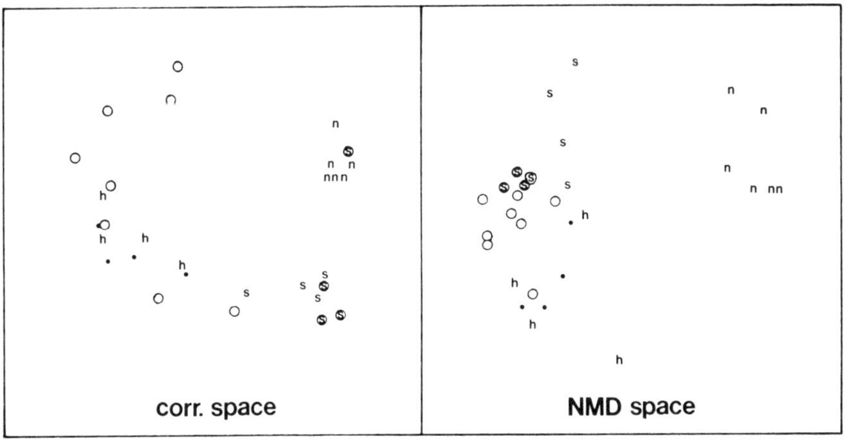

Figure 2. Demonstration of effects of using correlations with stimuli producing generally low responses; hamster PTA. Such stimuli (open circles) will typically produce low correlations with other stimuli, spuriously indicating high discriminability. Thus, these stimuli will be placed widely throughout the space (left panel), strongly determining its form. This problem may be reduced by using the absolute differences in numbers of impulses produced across all responding neurons by each pair of stimuli, neural mass differences (NMD). This collapses the effect of such stimuli as seen in the right panel. *n*, Na-Li salts; points, other salts; h, acids; s, sugars and other sweet stimuli.

understanding in this field if we can keep in mind the radical difference Öhrwall's four unique tastes and the continua of tastes *bounded* by Henning's primary four.

References

Bartoshuk, L. M. 1975. Taste mixtures: is mixture suppression related to compression? *Physiol. Behav.* 14:643–649.

———. 1978. Gustatory system. In *Handbook of Behavioral Neurobiology. I. Sensory Integration.* R. B. Masterton, ed. New York: Plenum Press, pp. 503–567.

Beebe-Center, J. G., M. S. Rogers, W. H. Atkinson, and D. N. O'Connel. 1959. Sweetness and saltiness of compound solutions of sucrose and NaCl as a function of concentration of solutes. *J. Exp. Psychol.* 4:231–234.

Boring, E. G. 1942. *Sensation and Perception in the History of Experimental Psychology.* New York: Appleton-Century-Crofts.

Boudreau, J. C. 1974. Neural encoding in cat geniculate ganglion tongue units. *Chem. Sens.* 1:41–51.

Boudreau, J. C., and N. Alev. 1973. Classification of chemoreceptive tongue units of the cat geniculate ganglion. *Brain Res.* 54:157–175.
Boudreau, J. C., W. Anderson, and J. Oravec. 1975. Chemical stimulus determinants of cat geniculate ganglion chemoreceptive group II unit discharge. *Chem. Sens.* 1:495–517.
Boudreau, J. C., J. J. Oravec, and N. K. Hoang. 1982. Taste systems of goat geniculate ganglion. *J. Neurophysiol.* 48:1226–1242.
Boynton, R. M., W. Schafer, and M. E. Neun. 1964. Hue-wave length relation measured by color-naming method for three retinal locations. *Science (Wash. DC).* 146:666–668.
Dethier, V. G. 1971. A surfeit of stimuli: a paucity of receptors. *Am. Sci.* 59:706–715.
———. 1973. Electrophysiological studies of gustation in *Lepidopterus larvae.* II. Taste spectra in relation to foodplant discrimination. *J. Comp. Physiol.* 82:103–134.
———. 1977. The taste of salt. *Am.* Sci. 65:744–751.
———. 1978. The role of chemosensory patterns in the discrimination of food plants. *Colloq. Int. Cent. Natl. Rech. Sci.* 265:103–114.
Dethier, V. G., and R. M. Crnjar. 1982. Candidate codes in the gustatory system of caterpillars. *J. Gen. Physiol.* 79:549–569.
DiLorenzo, P. M., S. W. Kiefer, A. G. Rice, and J. Garcia. 1982. Taste coding of ethyl alcohol: electrophysiological and behavioral data. *American Chemoreception Society Abstracts.*
Erickson, R. P. 1963. Sensory neural patterns and gustation. In *Olfaction and Taste I.* (Proceedings of the First International Symposium). Y. Zotterman, ed. Oxford: Pergamon Press Ltd., pp. 205–213.
———. 1967. Neural coding of taste quality. In *The Chemical Senses and Nutrition.* M. Kare, and O. Maller, eds. Baltimore, MD: The Johns Hopkins University Press, pp. 313–327.
———. 1968. Stimulus coding in topographic and nontopographic afferent modalities: On the significance of the activity of individual sensory neurons. *Psychol Rev.* 75:447–465.
———. 1974. Parallel "population" neural coding in feature extraction. In *The Neurosciences: Third Study Program.* F. O. Schmitt, and F. G. Worden, eds. Cambridge, MA: MIT Press, pp. 155–169.
———. 1977. The role of "primaries" in taste research. In *Olfaction and Taste* VI (Proceedings of the Sixth International Symposium). J. LeMagnen, and P. MacLeod, eds. London: IRL Press, pp. 369–376.
———. 1982a. Studies on the perception of taste: do primaries exist? *Physiol. and Behav.* 28:57–62.
———. 1982b. The "across-fiber pattern" theory: an organizing principle for molar neural function. *Contr. Sens. Physiol.* 6:79–110.
Erickson, R. P., and E. Covey. 1980. On the singularity of taste sensations: what is a taste primary? *Physiol. Behav.* 25:527–533.
Erickson, R. P., E. Covey, and G. S. Doetsch. 1980. Neuron and stimulus typologies in the rat gustatory system. *Brain Res.* 196:513–519.
Erickson, R. P., G. S. Doetsch, and D. A. Marshall. 1965. The gustatory neural response function. *J. Gen. Physiol.* 49:247–263.
Frank, M. 1973. An analysis of hamster afferent taste nerve response functions. *J. Gen. Physiol.* 61:588–618.
———. 1974. The classification of mammalian afferent taste nerve fibers. *Chem. Sens.* 1:53–60.
Frank, M., and C. Pfaffmann. 1969. Taste nerve fibers: a random distribution of sensitivities to four tastes. *Science (Wash. DC).* 164:1183–1185.
Gillan, D. J. 1982. Mixture suppression: the effect of spatial separation between sucrose and NaCl. *Percept. Psychophys.* 32:504–510.

———. 1983. Taste-taste, odor-odor, and taste-odor mixtures: greater suppression within than between modalities. *Percept. Psychophys.* 33:183–185.
Guttman, N. 1965. Effects of discrimination formation on generalization measured from a positive-rate baseline. *Stimulus Generalization.* D. J. Mostofsky, ed. Stanford: Stanford University Press, pp. 210–217.
Hellekant, G. 1975. Different types of sweet receptors in mammals. In *Olfaction and Taste* V (Proceedings of the Fifth International Symposium). D. A. Denton, and J. P. Coghlan, eds. New York: Academic Press, Inc., pp. 15–21.
Henning, H. 1916. Die Qualitätenreihe des Geschmacks. *Zeitschrift für Psychol.* 74:203–219.
Indow, T. 1969. An application of the *t* scale of taste: interaction among the four qualities of taste. *Percept. Psychophys.* 5:347–351.
Kamen, J. M., F. L. Pilgram, H. J. Gutman, and B. J. Kroll. 1961. Interaction of suprathreshold taste stimuli. *J. Exp. Psychol.* 62:348–356.
Kiesow, F. 1896. Beitrage zur physiologischen Psychologie des Geschmackssinnes. *Philosophische Studien.* 12:255–278.
———. 1898. Contribution à la psycho-physiologie de la cavité buccale. *Arch. ital. Biol.* 30:377–398.
Kroeze, J. H. A. 1982. The relationship between the side tastes of masking stimuli and masking in binary mixtures. *Chem. Sens.* 7:23–37.
Kuznicki, J. T., and N. Ashbaugh. 1979. Taste quality differences within the sweet and salty taste categories. *Sensory Processes.* 3:157–182.
———. 1982. Space and time separation of taste mixture components. *Chem. Sens.* 7:39–62.
Lawless, H. T. 1977. The pleasantness of mixtures in taste and olfaction. *Sensory Processes.* 1:227–237.
———. 1979. Evidence for neural inhibition in bitter-sweet taste mixtures. *J. Comp. Physiol. Psychol.* 93:538–547.
———. 1982. Misadventures in physiologizing: adaptation and taste mixology. *American Chemoreception Society Abstracts.*
MacKay-Sim, A., P. Shaman, and D. G. Moulton. 1982. Topographic coding of olfactory quality: odorant-specific patterns of epithelial responsivity in the salamander. *J. Neurophysiol. (Bethesda).* 48:584–596.
Marshall, D. A. 1968. A comparative study of neural coding in gustation. *Physiol. Behav.* 3:1–15.
McBurney, D. H. 1974. Are there primary tastes for man? *Chem. Sens. Flav.* 1:17–28.
———. 1978. Psychological dimensions and perceptual analyses of taste. In *Handbook for Perception, VIA. Tasting and Smelling.* E. C. Carterette and M. P. Friedman, eds. New York: Academic Press, Inc., pp. 125–155.
McBurney, D. H., and J. F. Gent. 1979. On the nature of taste qualities. *Psychol. Bull.* 86:151–167.
McCutcheon, B., and L. Brown. Response to NaCl taste in mixture with sucrose by sodium deficient rats. *Physiol. Behav.* in press.
Moskowitz, H. R. 1972. Perceptual changes in taste mixtures. *Percept. Psychophys.* 11:257–262.
Nowlis, G. H., and M. Frank. 1977. Qualities in hamster taste: behavioral and neural evidence. In *Olfaction and Taste VI* (Proceedings of the Sixth International Symposium). J. LeMagnen and P. MacLeod, eds. London: IRL Press, pp. 241–248.
Nowlis, G. H., M. E. Frank, and C. Pfaffmann. 1980. Specificity of acquired aversions to taste qualities in hamsters and rats. *J. Comp. Physiol. Psychol.* 94:932–942.
O'Connell, R. J. 1975. Olfactory receptor responses to sex pheromone components in the red-

banded leafroller moth. *J. Gen. Physiol.* 65:179–205.
———. 1981. The encoding of behaviorally important odorants by insect chemosensory receptor neurons. In *Perception of Behavioral Chemicals.* D. M. Norris, ed. Amsterdam: Elsevier/North Holland Biomedical Press, pp. 133–163.
O'Connell, R. J., and M. M. Mozell. 1969. Quantitative stimulation of frog olfactory receptors. *J. Neurophysiol. (Bethesda).* 32:51–63.
Öhrwall, H. 1891. Untersuchungen ueber den Geschmackssinn. *Skand. Arch. Physiol.* 2:1–69.
O'Mahoney, M., and L. Buteau. 1982. Taste mixtures: can the components be readily identified? *IRCS (Int. Res. Commun. Syst.) Med. Sci. Libr. Compend.* 10:109–110.
Pangborn, R. M. 1961. Taste interrelationships. II. Suprathreshold solutions of sucrose and citric acid. *J. Food Sci.* 26:648–655.
———. 1962. Taste interrelationships. III. Suprathreshold solutions of sucrose and sodium chloride. *J. Food Sci.* 26:495–500.
Pfaffmann, C. 1955. Gustatory nerve impulses in rat, cat and rabbit. *J. Neurophysiol. (Bethesda).* 18:429–440.
———. 1959. The afferent code for sensory quality. *Am. Psychol.* 14:226–232.
———. 1974. Specificity of the sweet receptors of the squirrel monkey. *Chem. Sens.* 1:61–67.
Pfaffmann, C., M. Frank, L. M. Bartoshuk, and T. C. Snell. 1976. Coding gustatory information in the squirrel money chorda tympani. *Prog. Psychobiol. Physiol. Psychol.* 6:1–27.
Schiffman, S. S., and R. P. Erickson. 1980. The issue of primary tastes versus a taste continuum. *Neurosci. Biobehav. Rev.* 4:109–117.
Schiffman, S. S., D. A. Reilly, and T. B. Clark. 1979. Qualitative differences among sweetners. *Physiol. Behav.* 23:1–9.
Scott, T. R. 1974. Behavioral support for a neural taste theory. *Physiol. Behav.* 12:413–417.
Scott, T. R., and R. P. Erickson. 1971. Synaptic processing of taste quality information in the thalamus of the rat. *J. Neurophysiol. (Bethesda).* 34:868–884.
Shimada, I., and K. Isono. 1978. The specific receptor site for aliphatic carboxylate anion in the labellar sugar receptor of the fleshfly. *J. Insect Physiol.* 24:807–811.
Shimada, I., A. Shiraishi, H. Kijima, and H. Morita. 1974. Separation of two receptor sites in a single labellar sugar receptor of the fieshfly by treatment with *p*-chloromercuribenzoate. *J. Insect Physiol.* 20:605–621.
Smith, D. V., R. L. Van Buskirk, J. B. Travers, and S. L. Bieber. 1983. Gustatory neuron types in the hamster brainstem. *J. Neurophysiol. (Bethesda).* 50:522–540.
Smith, J. C., T. W. Castonguay, D. F. Foster, and L. M. Bloom. 1980. A detailed analysis of glucose and saccharin drinking in the rat. *Physiol. Behav.* 24:173–176.
Smith, J. C., and D. F. Foster. 1980. Some determinants of intake of glucose + saccharin solutions. *Physiol. Behav.* 25:127–133.
Smith, J. C., D. F. Foster, and L. M. Bartoshuk. 1982. The synergistic properties of pairs of sweeteners. In *The Psychobiology of Human Food Selection.* L. M. Barker, ed. Westport, CT: AVI Publishing Co., pp. 123–138.
Valenstein. E. S., V. C. Cox, and J. W. Kakolewski. 1967. Polydipsia elicited by the synergistic actions of a saccharin and glucose solution. *Science (Wash. DC).* 157:552–554.
von Skramlik, E. 1926. *Handbuch der Physiologie der niederen Sinne. I. Die Physiologie des Geruchs- und Geschmackssine.* Leipzig: Georg Thieme-Verlag.
Woolston, D. C., and R. P. Erickson. 1979. Concept of neuron types in gustation in the rat. *J. Neurophysiol. (Bethesda).* 42:1390–1409.
Wundt, W. 1907. *Lectures on human and animal psychology.* New York: Macmillan Co.

7. Brainstem Processing of Gustatory Information

Introduction

Both peripheral and central gustatory neurons in a variety of mammalian species typically respond to more than one of the stimuli representing the four basic taste qualities (Pfaffmann, 1941, 1955; Fishman, 1957; Frank and Pfaffmann, 1969; Ogawa et al., 1968; Doetsch and Erickson, 1970; Erickson et al., 1965; Scott and Erickson, 1971; Smith and Travers, 1979; Travers and Smith, 1979; Van Buskirk and Smith, 1981). It was this lack of stimulus specificity in the responsiveness of individual chorda tympani fibers that first led Carl Pfaffmann (1941, 1955, 1959) to propose that taste quality might be coded by the pattern of activity across these broadly tuned afferents. This across-fiber pattern hypothesis was then further elaborated and given a quantitative basis by Erickson (1963, 1967, 1968, 1974, 1982; Erickson et al., 1965). Initially, the across-fiber pattern idea was accepted as the most likely mechanism for the neural coding of taste and has received support from a number of behavioral studies (Erickson, 1963; Scott, 1974; Smith et al., 1979). This theoretical view of quality coding made the multiple sensitivity of gustatory neurons an essential part of the neural code for quality, thus dealing with the ambiguity in the response of a single taste neuron to both qualitative and intensive parameters of the stimulus.

In the meantime, however, a number of findings appeared to support a more specific processing of gustatory quality. Psychophysical studies were suggesting the existence of four independent taste qualities (see McBurney, 1974). Some of these investigations suggested relatively specific mechanisms for the coding of saltiness (Smith and McBurney, 1969), sourness (McBurney et al., 1972), and sweetness (McBurney, 1972) as a result of cross-adaptation experiments. Taste-modifying substances, such as extracts of *Gymnema sylvestre* and miracle fruit, appeared to have specific effects on taste quality (see McBurney, 1974). These data and others (see McBurney 1974; McBurney and

---

David V. Smith, Department of Psychology
University of Wyoming, Laramie, Wyoming 82071

Gent, 1979) supported the notion of four (i.e., salty, sour, sweet, and bitter) independent, if not actually primary, taste qualities. Thus, to many investigators in this area there appeared to be a discrepancy between the psychophysical data, suggesting four independent qualities, and the neurophysiological data, which repeatedly demonstrated the multiple sensitivity of gustatory neurons. As a consequence, a great deal of controversy has arisen in recent years over both the nature of the taste qualities themselves (McBurney, 1974; Erickson, 1977; McBurney and Gent, 1979; Erickson and Covey, 1980; Schiffman and Erickson, 1980; Nowlis and Frank, 1981) and the mechanism of neural coding (Erickson, 1982; Erickson et al., 1980; Frank, 1973, 1974; Nowlis and Frank, 1977, 1981; Nowlis et al., 1980; Pfaffmann, 1974; Pfaffmann et al., 1976; Smith et al., 1979; Smith and Travers, 1979; Smith et al., 1983a, 1983b; Travers and Smith, 1979; Van Buskirk and Smith, 1981; Woolston and Erickson, 1979).

This controversy arose as a consequence of a proposal that seemed to resolve the discrepancy between the neural and psychophysical data on taste quality. This alternative view, which came to be known as the labeled-line hypothesis, began in Dr. Pfaffmann's laboratory with Frank's (1973) observations that hamster chorda tympani fibers could be classified into categories on the basis of their maximum sensitivities and that the neural response profiles within each class were similar and orderly. It was subsequently proposed that the coding of taste quality might simply be accomplished by activity in these separate neural "channels," one for each quality (Pfaffmann, 1974; Pfaffmann et al., 1976; Nowlis and Frank, 1977, 1981; Nowlis et al., 1980). Thus, "sweetness" would be coded by activity in sucrose-best neurons, "saltiness" by activity in NaCl-best neurons, etc. Such a proposal assumes, for example, that the activity in sucrose-best neurons that is elicited by nonsweet stimuli (such as NaCl or HCl) represents "noise" in the coding system. To the extent that these classes of neurons are narrowly tuned to one class of stimuli, the amount of noise generated by these side-band stimuli is relatively small. Such a mechanism would be possible in some insect systems, where the sensitivity of gustatory cells is relatively narrowly restricted to one class of stimuli (see Dethier, 1976). Even in the peripheral taste neurons of rats and hamsters, the response profiles have been shown to be relatively, although not completely, specific to one quality class (see Nowlis et al., 1980; Nowlis and Frank, 1981). Thus, even though Carl Pfaffmann (1965) has claimed that "*De gustibus non est disputandum*," these two theoretical approaches have led to considerable dispute in the taste literature in recent years. It was against this background of controversy that we began a series of studies of taste quality processing in the hamster brainstem (Smith and Travers, 1979; Smith et al., 1979; Travers and Smith, 1979; Smith, 1980; Van Buskirk and Smith, 1981; Smith et al., 1983a, 1983b).

Breadth of Responsiveness of Brainstem Neurons

Using the same stimuli as those used by Frank (1973) to characterize the responsiveness of hamster chorda tympani fibers, it was seen that the gustatory neurons in the hamster medulla (Travers and Smith, 1979) and pons (Van Buskirk and Smith, 1981) were systematically more broadly responsive to stimuli representing the four basic taste qualities. In fact, the sucrose-best class of neurons, which makes one of the best cases for a narrowly tuned "channel" in the periphery (see Frank, 1973, 1974; Nowlis and Frank, 1977, 1981), shows the greatest increase in its breadth of responsiveness from peripheral to medullary to pontine levels (Travers and Smith, 1979; Smith, 1980; Van Buskirk and Smith, 1981). The breadth of responsiveness of these cells was quantified by calculating the "entropy" of their response distributions to the four basic taste stimuli (Smith and Travers, 1979). The entropy (H), or breadth of responsiveness, is given by $H = K\Sigma p_i \log p_i$, where K is a scaling factor and p_i is the proportion of each response (to the four basic stimuli) to the total response to all four for a given neuron (see Smith and Travers, 1979). This measure reflects the diversity of a cell's responsiveness across these stimuli, such that a cell responding equally to all four stimuli has an entropy = 1.0 and a cell responding exclusively to only one stimulus has an entropy = 0. The distributions of the entropies of sucrose-best cells in the chorda tympani nerve (CT), nucleus tractus solitarius (NTS), and parabrachial nuclei (PbN) of the hamster are shown in Figure 1. What appears to be a relatively narrowly tuned class of gustatory fibers in the chorda tympani nerve becomes a rather broadly tuned class in the pons, raising questions about the viability of a labeled-line code at this level of the nervous system (see Smith and Travers, 1979; Travers and Smith, 1979; Smith, 1980; Van Buskirk and Smith, 1981). This question is raised because of the ambiguity that arises as a cell becomes responsive to more than one stimulus parameter (i.e., quality and intensity). To the extent that a cell's responses can be modulated by changes in both the quality and concentration of the stimulus, activity in a single neuron cannot reliably signal either parameter (Pfaffmann, 1955, 1959; Erickson, 1968, 1974, 1982; Perkel and Bullock, 1968).

In the meantime, there was some question about the existence of distinct classes of gustatory neurons (Woolston and Erickson, 1979; Erickson et al.,

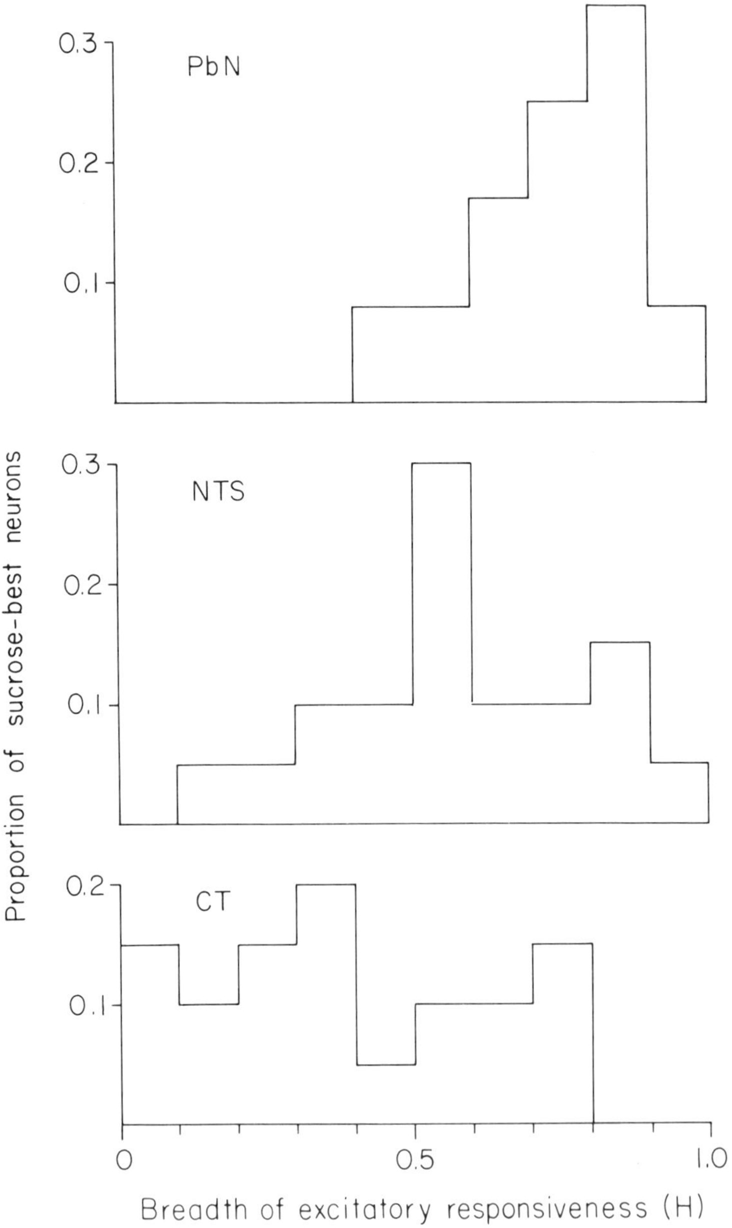
PbN
NTS
CT
0.3
0.2
0.1
0.3
0.2
0.1
0.2
0.1
0
0.5
1.0
Proportion of sucrose-best neurons
Breadth of excitatory responsiveness (H)

1980). Using statistical techniques borrowed from taxonomy (e.g., hierarchical cluster analysis), these investigators argued that the categorization of rat taste neurons into tight classes [e.g., sucrose-, NaCl-, HCl- or quinine (QHCl)-best] was an arbitrary subdivision of a population of neurons that varied more or less continuously in their response profiles. Thus, the notion that there were discrete classes of taste cells that could function as "channels" for coding taste quality in a labeled-line fashion was called into question. However, even though our own work had suggested a lack of specificity of hamster taste cells at brainstem levels, it was quite apparent from our familiarity with the response characteristics of these cells that their response profiles appeared, at least, to fall into discrete categories. Consequently, we decided to examine the response profiles of hamster brainstem neurons using multivariate statistics in an attempt to see how discrete these categories of taste neurons might be and, further, to see what role these classes of cells, if they existed, might play in the definition of the across-fiber patterns of activity elicited by gustatory stimuli. In other words, what is the empirical relationship between the labeled-line approach (as characterized by particular neuron classes) and the across-fiber pattern approach (as characterized by the response of the population of neurons to particular stimuli)? By examining the importance of each neuron type to the generation of the across-neuron patterns of activity, we hoped to show the interdependence of these two analytical approaches to gustatory quality coding, i.e., how they both describe, albeit from different perspectives, the same set of empirical phenomena.

Evidence for Neuron Groups

It has been common practice for investigators in taste to classify neurons into categories on the basis of their response profiles (Frank, 1973, 1974; Pfaffmann, 1974; Pfaffmann et al., 1976; Nowlis and Frank, 1977, 1981; Travers and Smith, 1979; Van Buskirk and Smith, 1981) or on the basis of other neurophysiological criteria (Boudreau and Alev, 1973; Boudreau, 1974).

Figure 1. Normalized distributions of the excitatory breadth of responsiveness (entropy) of sucrose-best neurons at three levels of the hamster gustatory system. Data from the chorda tympani (CT) were supplied by M. Frank. NTS, nucleus tractus solitarius; PbN, parabrachial nuclei. Neurons responding equivalently to all four stimuli have an entropy of 1; those responding exclusively to one stimulus have an entropy of 0. From Van Buskirk and Smith (1981).

Such a classification makes the assumption that these cells do not actually differ in a continuous fashion, making such a subdivision arbitrary (see Woolston and Erickson, 1979). To examine this question statistically, Woolston and Erickson (1979) investigated the relationships among the response profiles of cells in the nucleus tractus solitarius (NTS) of the rat using hierarchical cluster analysis and multidimensional scaling. These workers found no evidence in their data for distinctly separate classes of taste-sensitive neurons in the rat medulla or, in another study (Erickson et al., 1980), in the chorda tympani nerve of the rat.

Since the hamster gustatory system possesses a larger proportion of sugar-sensitive neurons than that of the rat, and since the sugar-sensitive cells make the best case for a narrowly tuned "labeled line" in the hamster's peripheral gustatory system, we felt that it would be appropriate to examine the question of neuron types in this species. Hierarchical cluster analysis provides a statistical tool for examining the grouping of elements within a multivariate set of data (see Everitt, 1980). This technique is commonly used in numerical approaches to taxonomy, where the relative similarity on a number of observed characteristics determines the degree of relationship among the members of a number of taxonomic classes. Thus, decisions about group membership can be based upon information, provided by the cluster analysis, about the similarity of each element to others within a group and dissimilarity to those in other groups. To the extent that there appear to be distinctions between groups of elements, one can discern the appropriateness of separating them into taxonomic classes. This approach provides an advantage over the simple practice of placing elements into different groups on the basis of their measurement on one or very few variables (see Woolston and Erickson, 1979; Smith et al., 1983a).

As data for this analysis, responses of 31 neurons in the parabrachial nuclei (PbN) of the hamster to an array of 18 stimulus compounds were arranged into an 18 × 31 multivariate matrix. This matrix was then examined using two multivariate statistical techniques, hierarchical cluster analysis and multidimensional scaling. For the hierarchical cluster analysis, we chose X^2 as the distance measure in the clustering process so that the relative similarity of the response profiles across the 18 stimuli would be the measure of difference, with no influence of the overall rate of impulse discharge (i.e., cells with similar profiles but differing by a large amount in firing rate would be considered to be similar by this criterion). The cluster program begins with as many clusters as there are neurons and then amalgamates the two neurons with the most similar response profiles (as defined by X^2) into a cluster. At succeeding steps, the two most similar neurons or clusters are joined to form a new cluster until a single cluster is obtained that contains all the neurons. As this amalgamation proceeds, the distance (X^2) between the neurons joined at each step is provided as an index of the relative similarity between the clusters joined. Small distances indicate strong similarity and large distances indicate less similarity. The series of intercluster distances is examined for the point at which the addition of a new cluster adds a great deal of dissimilarity. This point, where a sharp increase in the amalgamation distance occurs, provides an estimate of the number of clusters to be found in the data (see Everitt, 1980).

The results of a cluster analysis of the response profiles of neurons in the hamster PbN are shown graphically in Figure 2. This figure depicts the distance between clusters at successive stages of the clustering process so that one might appreciate the order in which the various neurons are entered into the amalgamation and the relative similarities between them at each stage. The neuron numbers and their best-stimulus (i.e., sucrose-, NaCl-, HCl-, or QHCl-best) designations are shown to the right of the figure. The two most similar neurons were 7 and 9, both sucrose-best, followed by 4 and 8, also sucrose-best, etc. At each successive stage of the clustering process, the next most similar neurons or clusters are joined together and the distance between clusters gradually increases. The cluster tree for this solution suggests three major clusters of cells, identified in the figure by S, H, and N. The distances between these three clusters are large in comparison to those within the clusters. Within the N group, one might identify two subgroups on the basis of the cluster distance between them. However, these three main clusters stand far apart within this solution and comprise, for the most part, cells with the same best-stimulus designation. That is, all but one of the cells in the S group are sucrose-best, all but three of the H group are HCl-best, and all but two of the N group are NaCl-best. Thus, on the basis of the similarity of their neural response profiles across a wide array of stimuli, these cells are classified in a manner very similar to that provided by their "best" response to one of the four basic stimuli (sucrose, NaCl, HCl, or QHCl). Further, the distance between the groups appears to be very much greater than the distances between cells within the groups, suggesting that placing these cells into classes on the basis of their response profiles is not purely an arbitrary exercise. Even though these pontine neurons are more broadly tuned than those at medullary or peripheral levels (Smith, 1980; Van Buskirk and Smith, 1981), the similarities in their response profiles strongly suggest three distinct classes of neurons (see also Smith et al., 1983a).

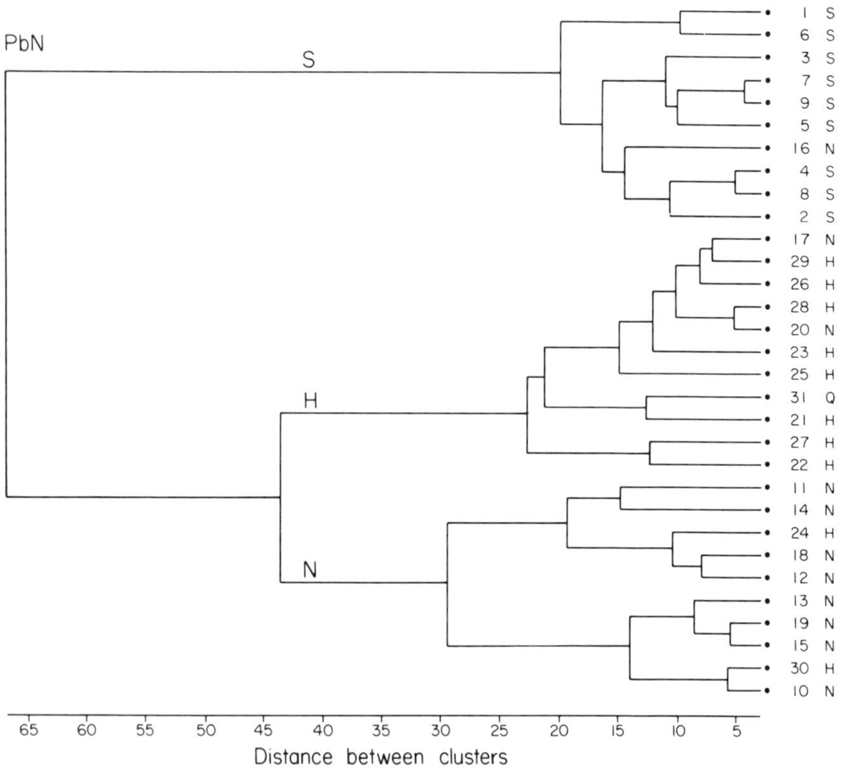

Figure 2. Cluster tree depicting the order of clustering for 31 PbN neurons. The neuron numbers and the best-stimulus classifications (S, N, or H) are shown on the right of the figure. The distance between the neurons or clusters joined at each step is shown along the abscissa. The three major clusters are indicated by S, H, and N. From Smith et al. (1983a).

So that one might appreciate the similarities in these response profiles as well as the breadth of responsiveness of these cells across this array of stimuli, the response profiles of the individual cells in these three groups are shown in Figures 3–5. The response profiles of cells in the S group are depicted in Figure 3, arranged in the order of their inclusion into the cluster solution. The appropriate portion of the cluster tree is shown to the right of the figure. All but one of these cells responded best to sucrose (of the four basic stimuli) and that one (neuron 16) actually responded better to fructose than to NaCl (which was the best of the four basic stimuli). By examining the response profiles and the cluster tree, one can appreciate the degree of similarity represented by the cluster distances shown on the right. Cells with extremely similar response profiles (such as neurons 9 and 5 and cells 4 and 8) are clustered together very early in the amalgamation process, whereas those with somewhat less similarity (e.g., neuron 2 or neuron 3) are brought into the cluster later on. All of these neurons in the S group are characterized by their relatively broad sensitivity across this wide range of stimuli, responding well to sodium salts, nonsodium salts, acids, and urea, in addition to the five sweet-tasting stimuli in the array (sucrose, fructose, D-glucose, Na-saccharin, and DL-alanine).

The response profiles of the H neurons to these 18 stimuli are shown in Figure 4, which also depicts the appropriate portion of the cluster tree. These cells are characterized by a broad sensitivity to the sodium salts and the nonsodium salts and acids, with some sensitivity to urea and QHCl, but relatively little to the sweet-tasting stimuli. Neurons 27 and 22 cluster somewhat differently from the others, primarily on the basis of their lack of sensitivity to both sweet-tasting and bitter-tasting (i.e., urea and QHCl) stimuli. Since the cluster analysis proceeded on the basis of the relative response profiles, which included the background spontaneous discharge as part of the response measure (because X^2 cannot be computed with negative numbers), neuron 31 appeared in the solution to be very similar to cell 21. With this obvious exception, all of these cells are seen to have similar, broadly sensitive response profiles.

For those cells in the N group, the individual response profiles are shown in Figure 5, along with the appropriate portion of the cluster tree. These cells are characterized by their responsiveness to the sodium salts and, to a lesser degree, the nonsodium salts and acids. The cells cluster into two subgroups before joining into one larger one. One of these subgroups (cells 11, 14, 24, 18, and 12) is characterized by a slight responsiveness to the sweet-tasting stimuli and to the organic acids, which is not characteristic of the other subgroup. Neuron 14, which is obviously different from the others because of its inhibitory responses, is included with these because the profiles were compared without subtracting spontaneous discharge rates (again because the X^2 criterion cannot deal with negative numbers). Overall, these sodium-sensitive neurons are much more similar to one another than to any cells in either the S- or H-neuron groups (see Figure 2).

The similarities among the response profiles were spatially mapped using multidimensional scaling (KYST). The input to KYST was a matrix of across-stimulus correlations (Pearson *r*) among all the neurons. Using these across-stimulus correlations as measures of similarity, the program then located the

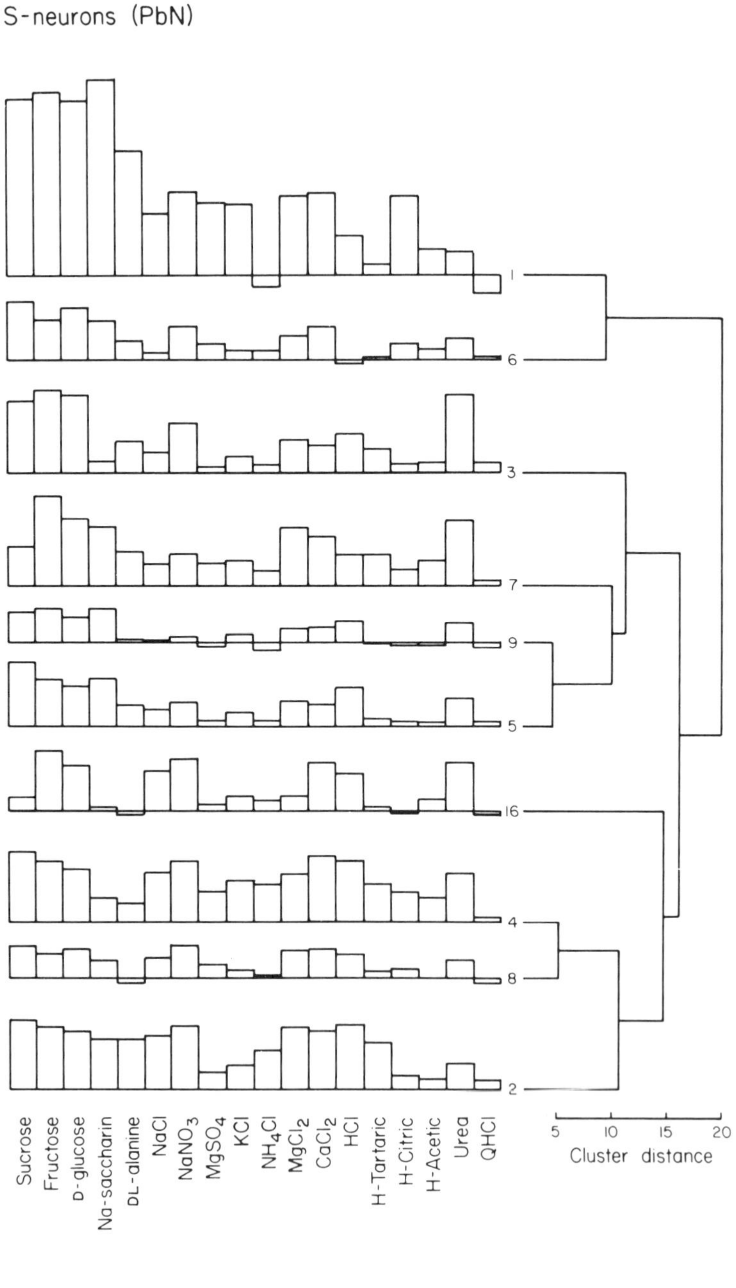
S-neurons (PbN)
1
6
3
7
9
5
16
4
8
2
Sucrose
Fructose
D-glucose
Na-saccharin
DL-alanine
NaCl
NaNO3
MgSO4
KCl
NH4Cl
MgCl2
CaCl2
HCl
H-Tartaric
H-Citric
H-Acetic
Urea
QHCl
5
10
15
20
Cluster distance

neurons in a multidimensional space in which similarity is represented by spatial proximity. The locations of these PbN neurons in a two-dimensional space are shown in Figure 6. To show the relationship between the hierarchical cluster solution and the multidimensional analysis of these data, the three-cluster stage of the cluster solution (dashed lines) is embedded in the two-dimensional space (see Shepard, 1980). These three clusters of neurons form three distinct groups in this two-dimensional space. The location of neuron 31 demonstrates a difference between these two multivariate techniques. Whereas the cluster solution grouped neuron 31 with the H neurons, this cell, which responded best to QHCl, was viewed quite differently by the multidimensional scaling solution, which placed it some distance from the other H neurons. A third dimension was necessary to accurately depict this neuron's relationships to the others (although there was only a slight further reduction in "stress" in a three-dimensional solution).

From these analyses it may be seen that the response profiles of taste-responsive neurons in the hamster pons are similar enough to warrant their classification into neural groups. Additional analyses have also shown that cells in the hamster medulla (Smith et al., 1983a) and chorda tympani nerve (Frank et al., in preparation) can be similarly classified. It is also apparent that the classification of these cells by these multivariate techniques is similar to that given by consideration of their "best" stimulus (S, N, H, or Q), since 80% of the cells in the PbN are categorized the same way by these separate approaches.

Neuron Types and the Across-Neuron Pattern

Work in our laboratory has suggested that taste neurons in the hamster brainstem are more broadly tuned than chorda tympani fibers (Travers and Smith, 1979; Smith, 1980; Van Buskirk and Smith, 1981), yet can still be grouped into relatively distinct classes on the basis of their response profiles (Smith et al., 1983a). Whether this neuron classification has any functional significance

---

Figure 3. Responses (impulses/5 s) of each of the S neurons in the PbN to each of the 18 stimuli. Neuron numbers and the appropriate portion of the cluster tree (showing cluster distances) are given on the right of the response profiles. Although the ordinate is not scaled, the maximum response in this figure (of neuron 1 to Na-saccharin) is 346 impulses/5 s. Responses are deviations from spontaneous rate. From Smith et al. (1983a).

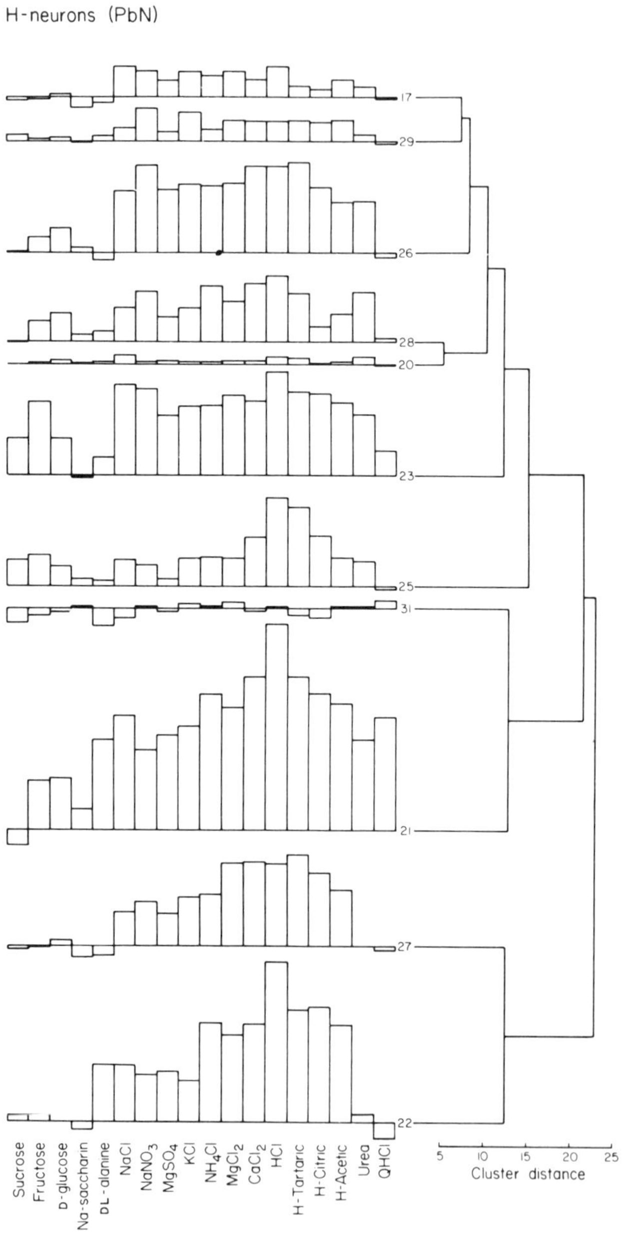
H-neurons (PbN)
17
29
26
28
20
23
25
31
21
27
22
Sucrose
Fructose
D-glucose
Na-saccharin
DL-alanine
NaCl
NaNO3
MgSO4
KCl
NH4Cl
MgCl2
CaCl2
HCl
H-Tartaric
H-Citric
H-Acetic
Urea
QHCl
5
10
15
20
25
Cluster distance

for coding taste quality is an obvious question. Therefore, we examined the roles played by these various neuron types in establishing and defining the across-neuron patterns of activity elicited by different gustatory stimuli. Similarities in the across-neuron patterns evoked by taste stimuli have traditionally been quantified with across-neuron correlations (Erickson et al., 1965) and the relationships among these patterns have been depicted using multidimensional scaling techniques (e.g., Perrotto and Scott, 1976; Woolston and Erickson, 1979). The following analyses address the question of the importance of the neuron groups in the PbN to the definition of the patterns elicited across these neurons by several taste stimuli.

The position of the S-neuron class in the across-neuron patterns elicited by five sweet-tasting stimuli (these stimuli taste sweet to humans and are classed as similar by hamsters in behavioral tests) is shown by the shaded area in Figure 7. In this figure, the neurons have been arranged along the abscissa according to their responsiveness to 0.1 M sucrose. It is quite clear that the most responsive neurons in the patterns belong to the S-neuron group. Thus, it is likely that the similarities among these across-neuron patterns (indicated by the correlation coefficients shown in the figure) would depend strongly on this group of neurons. A similar situation is shown for the acids in Figure 8. Again, the most responsive neurons, especially for the organic acids, belong to the H-neuron class. In this figure, the neurons have been arranged according to their responsiveness to 0.003 M tartaric acid. Several cells outside this group respond well to 0.003 M HCl (all of which are N neurons), suggesting that the organic acids are more specific stimuli than HCl for this class of cells.

Across-neuron correlations were computed for all possible pairs of the 18 stimuli, and this correlation matrix served as the input to a multidimensional scaling program. The relationships among all of these stimuli, as defined by their patterns of response across all of the PbN neurons in our sample, are depicted in the three-dimensional solution shown in Figure 9. In this solution, dimension I separates the five sweet-tasting stimuli from all the others, dimension II separates the sodium salts from the nonsodium salts and acids, and dimension III separates the two bitter-tasting stimuli (urea and QHCl) from the nonsodium salts and acids. Thus, these across-neuron patterns of activity

---

Figure 4. Responses (impulses/5 s) of each of the H neurons in the PbN to each of the 18 stimuli. Details and conventions are the same as those given for Figure 3, except that the maximum response (of neuron 21 to HCl) is 466 impulses/5 s. From Smith et al. (1983a).

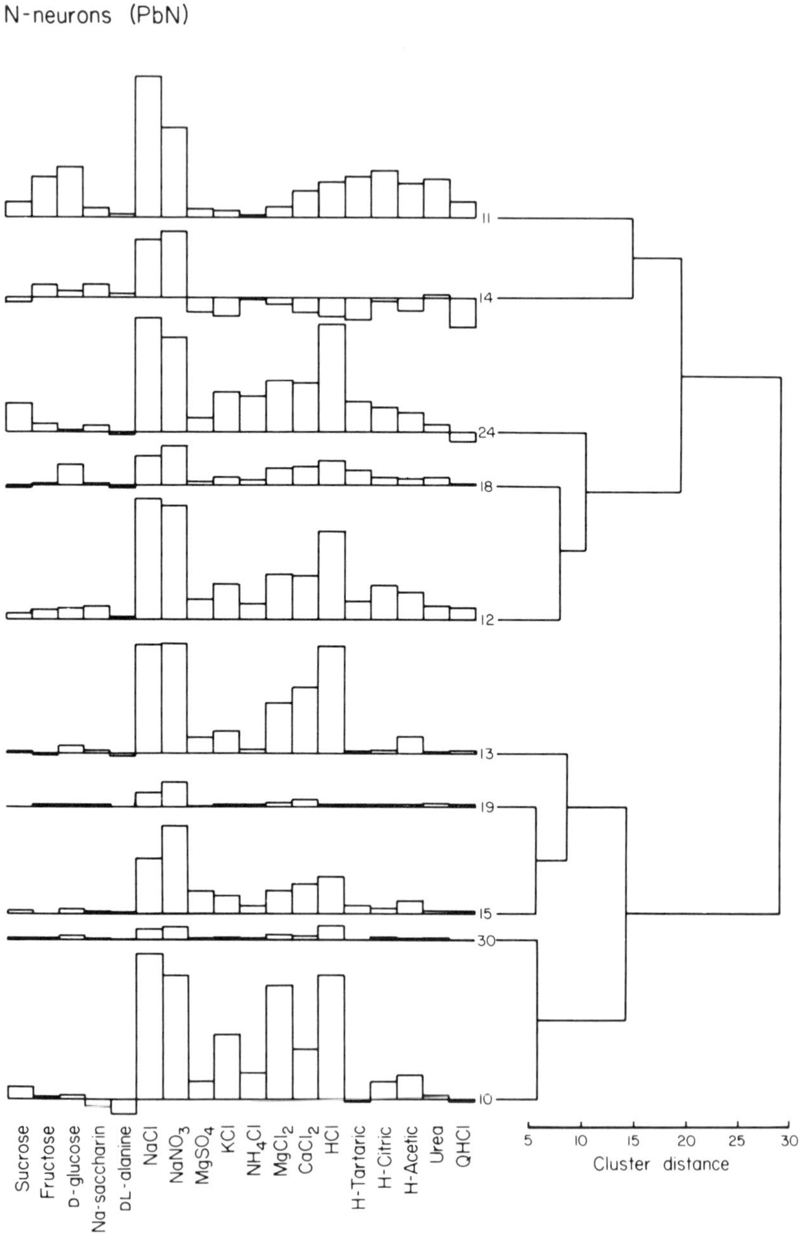
N-neurons (PbN)
11
14
24
18
12
13
19
15
30
10
Sucrose
Fructose
D-glucose
Na-saccharin
DL-alanine
NaCl
NaNO3
MgSO4
KCl
NH4Cl
MgCl2
CaCl2
HCl
H-Tartaric
H-Citric
H-Acetic
Urea
QHCl
5
10
15
20
25
30
Cluster distance

clearly separate groups of stimuli into behaviorally relevant categories (see Nowlis and Frank, 1977, 1981; Smith et al., 1979; Nowlis et al., 1980).

To determine the importance of the three neuron groups (S, H, and N neurons) in the definition of these stimulus relationships, this multidimensional scaling analysis was repeated for each neuron group alone (S neurons only, H neurons only, and N neurons only) and in the absence of each neuron group (without S neurons, without H neurons, and without N neurons). To simplify the presentation of these solutions, the third dimension (which separated QHCl and urea from the nonsodium salts and acids) is not shown. The two-dimensional solution for the stimulus relationships given by all the neurons is shown in Figure 10A. In this figure, the four groups of stimuli that are seen to be separated in the three-dimensional solution shown in Figure 9 are also clearly separated in two dimensions. The similarities in the across-neuron patterns evoked across the S neurons alone are depicted by the proximities of the stimuli in Figure 10B. All of the sweet-tasting stimuli evoke highly similar patterns across the S neurons, as shown by their tight grouping in the stimulus space. However, several other stimuli, which are not behaviorally similar to the sweet-tasting stimuli (see Nowlis and Frank, 1977, 1981; Smith et al., 1979; Nowlis et al., 1980), also evoke very similar patterns (e.g., citric acid, $MgSO_4$, and KCl). The bitter-tasting stimuli, on the other hand, evoke patterns across the S neurons that are very different from those elicited by the sweet-tasting stimuli. Thus, the S neurons can provide information on the similarities among the sweet-tasting stimuli but cannot by themselves distinguish the sweet compounds from several other, nonsweet stimuli. Similarly, the distinction between the sodium salts and the nonsodium salts and acids, which is so evident in Figures 9 and 10A, depends upon activity in both the H- and N-neuron classes. With only the H neurons, the nonsodium salts and acids remain tightly grouped in the stimulus space, but the sodium salts are also in this group (Figure 10C). This reflects an increase in the across-neuron correlation between the sodium salts and these other stimuli in the absence of the N neurons. Similarly, when only the N neurons are present, the sodium salts are not distinct from several of these other compounds (Figure 10D). Thus, both the H and N neurons are necessary for a neural distinction among these stimuli.

Figure 5. Responses (impulses/5 s) of each of the N neurons in the PbN to each of the 18 stimuli. Details and conventions are the same as those given for Figure 3, except that the maximum response (of neuron 10 to NaCl) is 276 impulses/5 sec. From Smith et al. (1983a).

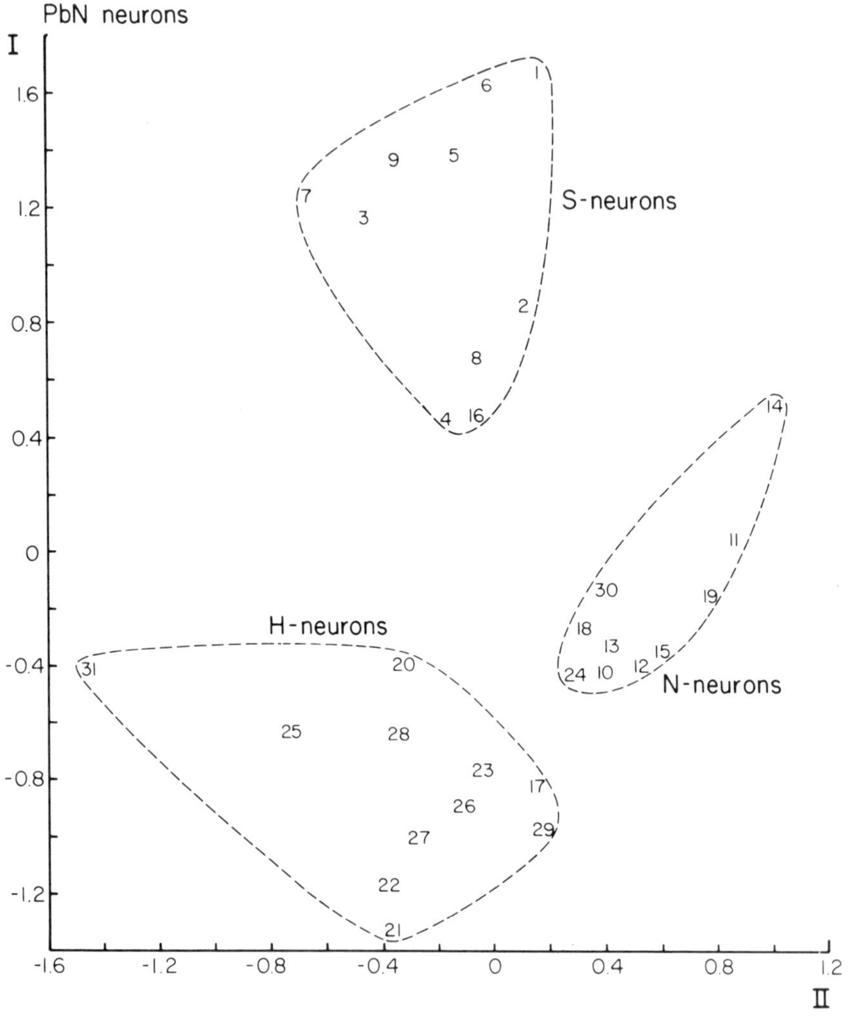

Figure 6. Locations of each of the 31 PbN neurons in a two-dimensional space obtained through multidimensional scaling (KYST). Neurons are identified by number, and the three clusters suggested by the hierarchical cluster solution are embedded in the two-dimensional space (dashed lines). The values of "stress" for 1, 2, 3, 4, or 5 dimensions were, respectively, 0.575, 0.103, 0.066, 0.042, and 0.032. From Smith et al. (1983a).

The results of similar analyses in the absence of each of the neuron groups in turn are shown in Figure 11. The two-dimensional solution based on the responses of all the neurons is repeated in Figure 11A. Without the contribution of the S neurons, the stimulus space changes to that shown in Figure 11B. In this figure, the sodium salts are still clearly separated from the nonsodium salts and acids, but the sweet-tasting stimuli are scattered throughout the space, reflecting a decrease in the similarity of the across-neuron patterns elicited by the sweet compounds. In fact, some of these patterns do not correlate at all without the contribution of the S neurons. Similarly, the absence of the H neurons results in a scattering of the nonsodium salts and acids throughout the stimulus space (Figure 11C), although the sweet stimuli and the sodium salts are still clearly separated from each other. Without the N neurons, the two sodium salts are still close together in the space (Figure 11D), but not at all separate from the nonsodium salts and acids. Thus, each of the neuron groups in the PbN plays a distinct role in the definition of the across-neuron patterns of activity elicited by these various stimuli. The similarities among the patterns evoked by like-tasting stimuli (e.g., the sweet compounds) depend upon activity in a particular neuron group (e.g., the S neurons). This is true for the sweet-tasting stimuli and for the nonsodium salts and acids, but not for the two sodium salts. Further, the distinctions in the patterns evoked by dissimilar-tasting stimuli (e.g., sucrose and citric acid) depend upon the activity in more than one neuron group (e.g., the S and H neurons). This is true for the distinctions between the sweet-tasting stimuli, the nonsodium salts and acids, and the sodium salts. All three neuron groups must be present in order for the PbN to sort out these three groups of stimuli on the basis of their across-neuron patterns.

Discussion

The present analyses were designed to demonstrate the relationship between these two very different approaches to the handling of neural data in taste. Some investigators have tended to approach the question of neural coding by categorizing neurons into functionally meaningful classes (Boudreau and Alev, 1973; Frank, 1973, 1974; Boudreau, 1974; Pfaffmann et al., 1976; Nowlis and Frank, 1977, 1981; Smith et al., 1979; Travers and Smith, 1979; Nowlis et al., 1980; Van Buskirk and Smith, 1981). Others have stressed the population approach to understanding neural coding in this system (Erickson et al., 1965; Erickson, 1968, 1974; Doetsch and Erickson, 1970; Scott and Erickson, 1970; Perrotto and Scott, 1976; Woolson and Erickson, 1979).

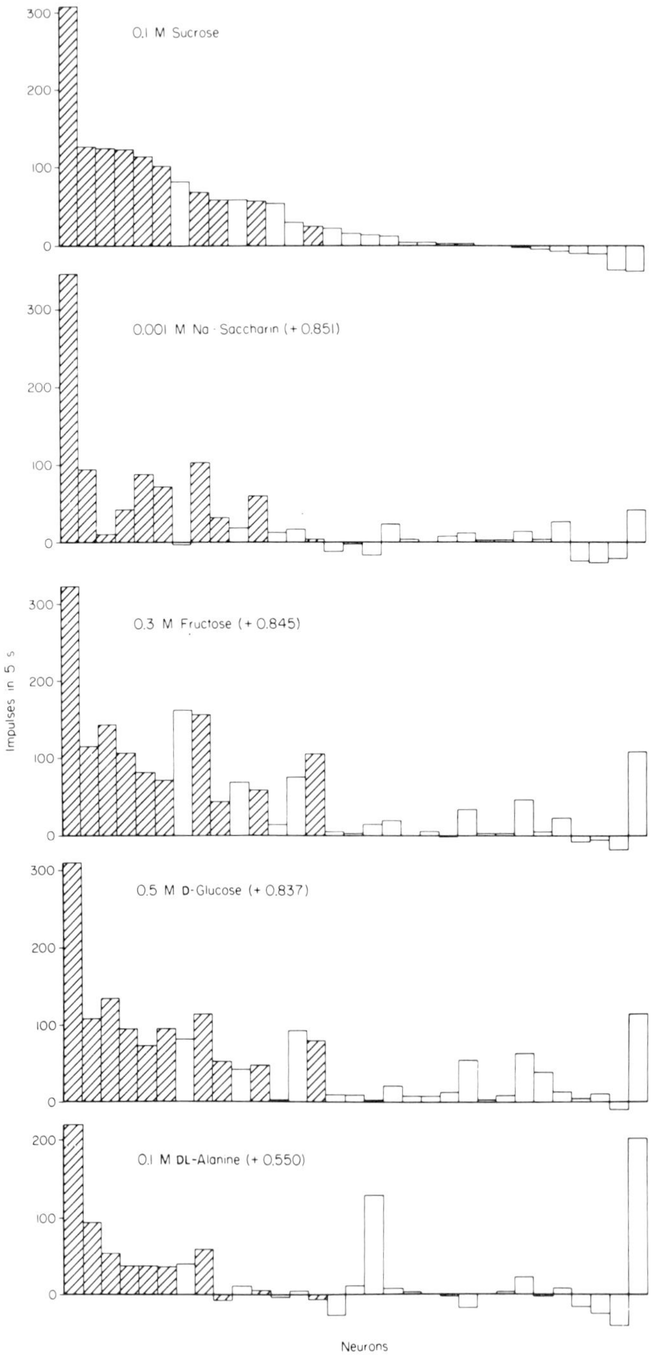
0.1 M Sucrose
0.001 M Na·Saccharin (+0.851)
0.3 M Fructose (+0.845)
0.5 M D-Glucose (+0.837)
0.1 M DL-Alanine (+0.550)
Impulses in 5 s
Neurons
300
200
100
0

Some investigators in this area have examined the responses of gustatory neurons with an eye to both analytical approaches (Frank, 1973; Pfaffmann et al., 1976; Smith et al., 1979; Travers and Smith, 1979; Van Buskirk and Smith, 1981). The results of the present analyses demonstrate the interdependence of these two approaches to the understanding of gustatory neural function. Although the theoretical implications of a labeled-line code and an across-fiber pattern code are quite distinct, a neuron-group approach and a population approach to the analyses of the neural data in this system are thoroughly interwoven.

In its purest form (see Erickson, 1968, 1974), the across-fiber pattern hypothesis of taste quality coding places a great deal of emphasis on the pattern of activity and does not distinguish separate roles among taste neurons. On the other hand, the labeled-line hypothesis (see Pfaffmann, 1974; Pfaffmann et al., 1976; Nowlis and Frank, 1977, 1981; Nowlis et al., 1980) suggests that activity in particular neurons per se is the code for gustatory quality. The present series of studies demonstrates that the across-fiber patterns are dominated by the responses of particular classes of neurons. A particular neuron group is necessary to establish the similarities among the patterns generated by a particular group of stimuli (e.g., S neurons for the sweet-tasting stimuli). However, no one neuron type alone is capable of providing information that can distinguish the across-neuron patterns evoked by dissimilar-tasting compounds. More than one neuron type must contribute to the pattern in order for the patterns evoked by unlike stimuli to be distinct. Thus, the viability of the across-neuron pattern as a code for taste quality must rest on the additional specification that these various neuron types are playing an important role in the definition of the patterns. One might conclude, therefore, that the S-neuron group is critical for the coding of sweetness, or that the H-neuron group is critical for the coding of sourness, regardless of one's theoretical bias regarding these two coding hypotheses. That is, the same cells are important for coding a particular quality, whether they are viewed from a labeled-line or an across-fiber pattern perspective.

Even though the same neurons may be important in both theoretical ap-

---

Figure 7. Across-neuron patterns for the sweet-tasting stimuli. The neurons are arranged in order of their responsiveness to 0.1 M sucrose in each case. The S-neuron group is shaded in all the patterns. The correlations of each pattern to that evoked by sucrose are shown above each pattern. Spontaneous rate for each neuron is shown as zero on the ordinate. From Smith et al. (1983b).

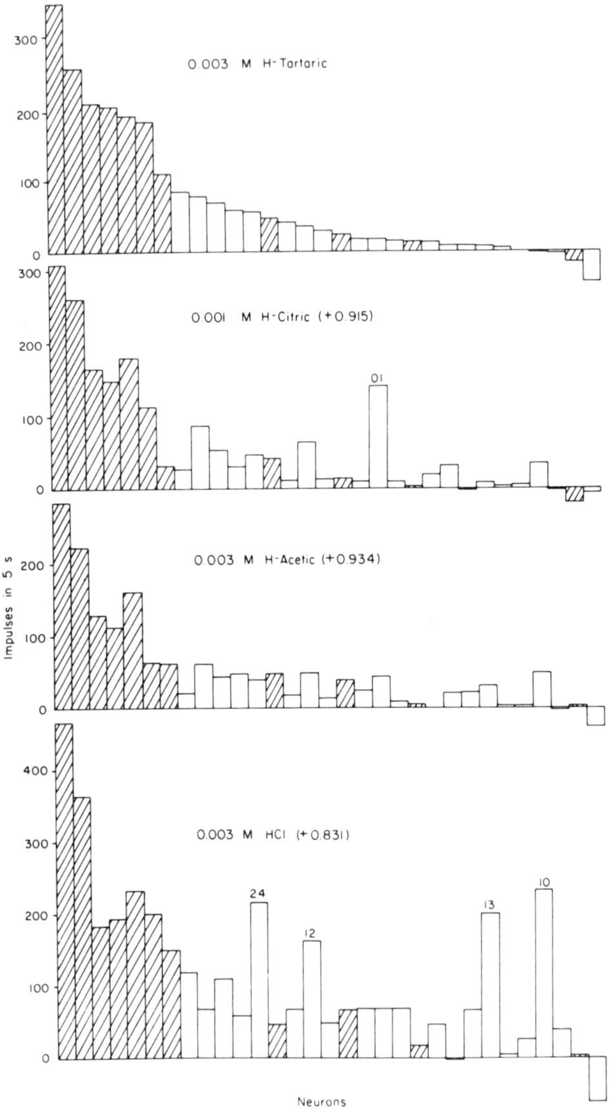

Figure 8. Across-neuron patterns for the acids. The neurons are arranged in order of their responsiveness to 0.003 M tartaric acid. The H-neuron group is shaded in each pattern and the correlation of each pattern to that for tartaric acid is shown. Spontaneous rate for each neuron is shown as zero on the ordinate. From Smith et al. (1983b).

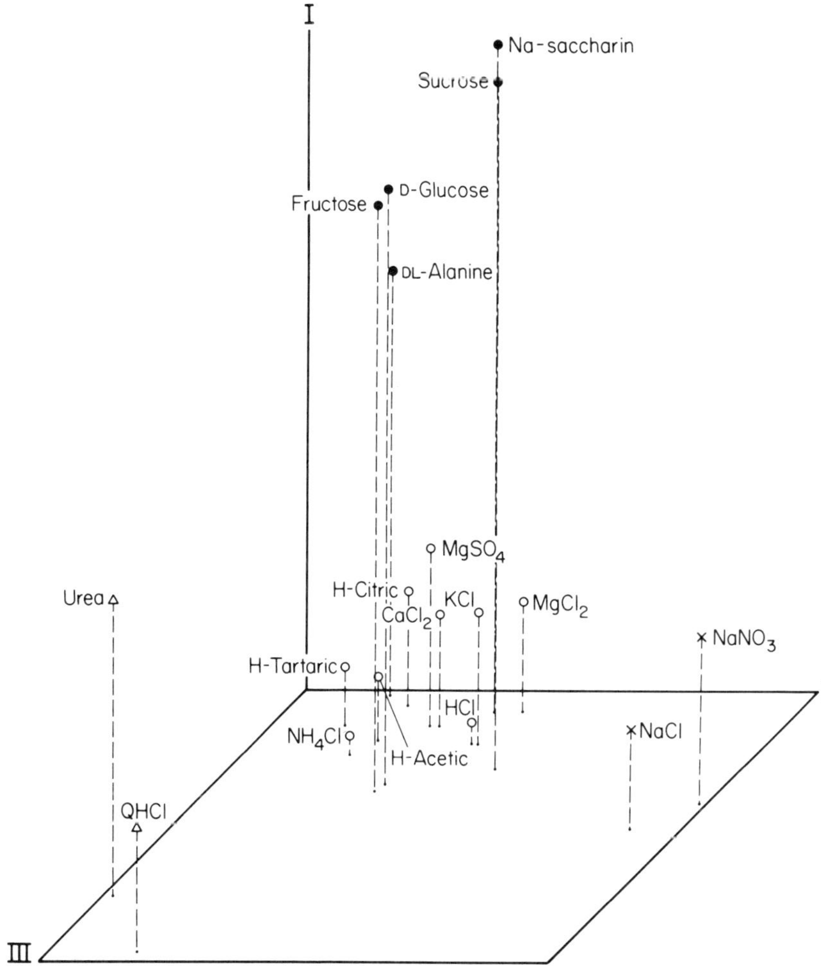

Figure 9. Three-dimensional space showing the relative similarities of the 18 stimuli, obtained from multidimensional scaling (KYST). All 31 neurons were used to generate the across-neuron correlations represented by this stimulus space. Four groups of stimuli are indicated by different symbols. The values of "stress" for 1, 2, 3, 4, or 5 dimensions were, respectively, 0.226, 0.065, 0.031, 0.015, and 0.010. From Smith et al. (1983b).

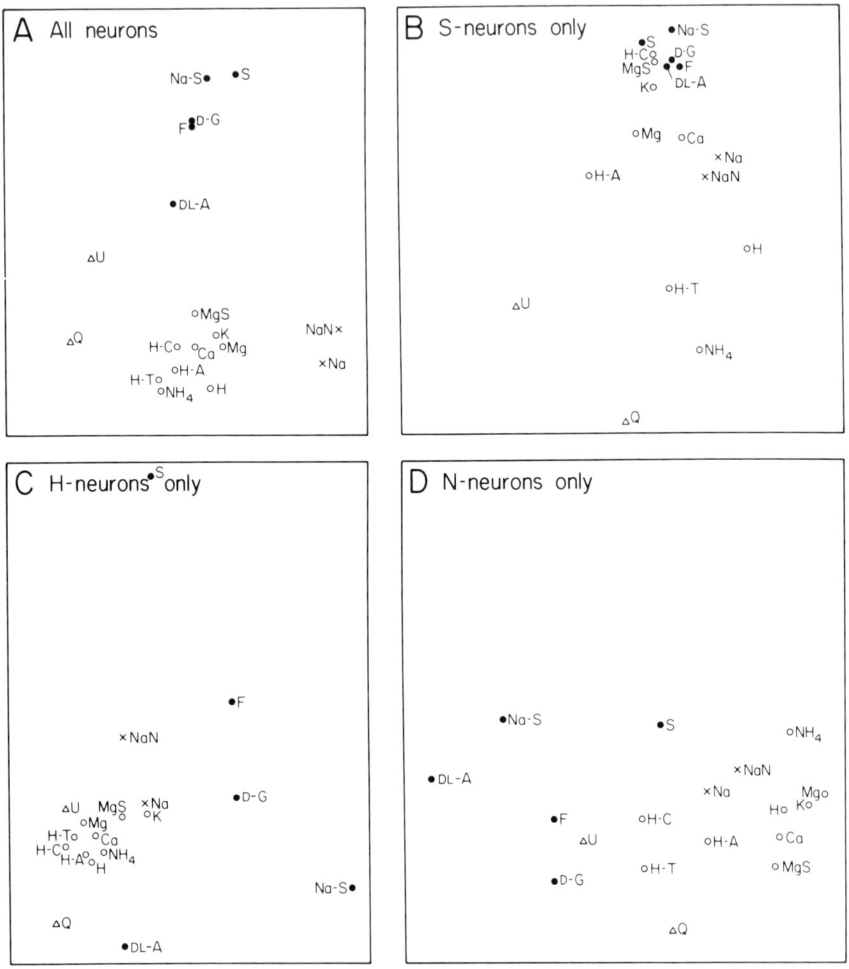

Figure 10. Two-dimensional representations of the stimulus relationships obtained from multidimensional scaling. (A) All 31 neurons were used to generate the across-neuron correlations represented by this space. (B) Across-neuron correlations were calculated across the S neurons only. The values of "stress" for 1, 2, 3, 4, or 5 dimensions were, respectively, 0.137, 0.053, 0.035, 0.014, and 0.010. (C) Across-neuron correlations were calculated across the H neurons only. The values of "stress" for 1, 2, 3, 4, or 5 dimensions were, respectively, 0.199, 0.060, 0.027, 0.012, and 0.009. (D) Across-neuron correlations were calculated across the N neurons only. The values of "stress" for 1, 2, 3, 4, or 5 dimensions were, respectively, 0.162, 0.084, 0.045, 0.018,

proaches, the implication for the role of individual neurons in coding taste quality is quite different in the labeled-line and across-fiber pattern hypotheses. A labeled-line code implies that the information content of a message is contained in its source, i.e., the neuron in which it is occurring (see Perkel and Bullock, 1968; Dethier, 1976). Activity in such a sensory channel will be interpreted as a particular message regardless of how it is evoked. Thus, in taste, each neuron type, or "labeled line," would contribute a particular quality to the overall taste of a stimulus. If a stimulus evokes activity in both the "sweet" and "salty" labeled lines, for example, it would, according to this hypothesis, taste both sweet and salty. To the extent that these labeled lines are specifically tuned to one class of stimuli, such an hypothesis is not unreasonable (as in the gustatory receptors of the blowfly; see Dethier, 1976). However, we are faced with the situation in the mammalian gustatory system that the neurons are not completely specific, even in the peripheral nerves (Frank, 1973; Pfaffmann et al., 1976; Nowlis et al., 1980). It was Carl Pfaffmann (1941, 1955, 1959) who first described this lack of specificity and who proposed that a comparison of activity across gustatory neurons could be the code for taste quality. Since taste neurons respond with increasing impulse frequency to changes in both stimulus concentration and quality, they cannot code either parameter with complete certainty. Further, as we have shown repeatedly, cells in the hamster brainstem are even more broadly tuned than those in the chorda tympani nerve (Smith et al., 1979; Travers and Smith, 1979; Smith, 1980; Van Buskirk and Smith, 1981). Thus, we have to try to understand the function of these very broadly tuned neuron "types" in the brainstem. On the one hand, the existence of these neuron groups suggests that they may be functionally significant, as in a labeled-line system, but on the other hand, their lack of specificity to one of the classic taste stimuli evokes the familiar arguments for an across-fiber pattern code. To the extent that these neuron groups, which would be termed "labeled lines" by some investigators (e.g., Pfaffmann, 1974; Pfaffmann et al., 1976; Nowlis and Frank, 1977, 1981; Nowlis et al., 1980), are critical in establishing and defining the "across-neuron patterns" that others (e.g., Erickson et al., 1965; Erickson,

---

and 0.010. S, sucrose; F, fructose; D-G, D-glucose; Na-S, Na-saccharin; DL-A, DL-alanine; Na, NaCl; NaN, $NaNO_3$; NH_4, NH_4Cl; K, KCl; MgS, $MgSO_4$; Ca, $CaCl_2$; Mg, $MgCl_2$; H, HCl; H-T, tartaric acid; H-C, citric acid; H-A, acetic acid; U, urea; Q, quinine hydrochloride. Four groups of stimuli are indicated by different symbols. From Smith et al. (1983b).

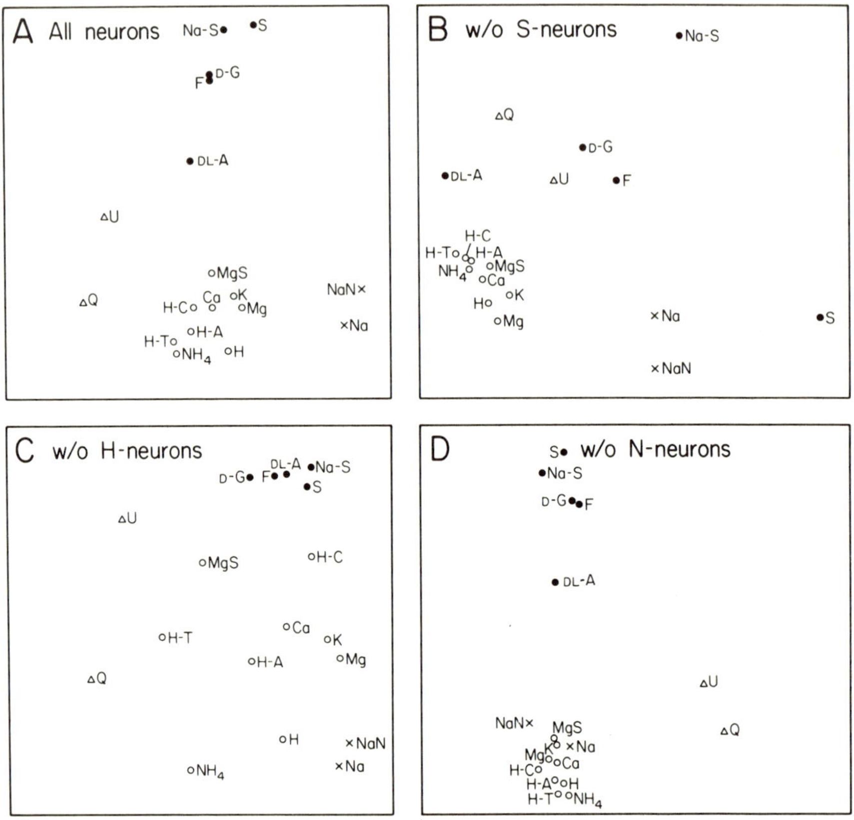

Figure 11. Two-dimensional representation of the stimulus relationships obtained from multidimensional scaling. (A) All 31 neurons were used to generate the across-neuron correlations represented by this space. (B) Across-neuron correlations were calculated without the responses of the S neurons. The values of "stress" for 1, 2, 3, 4, or 5 dimensions were, respectively, 0.251, 0.098, 0.050, 0.020, and 0.014. (C) Across-neuron correlations were calculated without the responses of the H neurons. The values of "stress" for 1, 2, 3, 4, or 5 dimensions were, respectively, 0.270, 0.076, 0.033, 0.018, and 0.010. (D) Across-neuron correlations were calculated without the responses of the N neurons. The values of "stress" for 1, 2, 3, 4, or 5 dimensions were, respectively, 0.158, 0.046, 0.018, 0.010, and 0.010. Abbreviations are the same as in Figure 10. From Smith et al. (1983b).

1968, 1974, 1982) would argue are the code for taste quality, perhaps little is to be gained by further arguments in this area. Whether one chooses to accept a labeled-line or an across-fiber pattern interpretation of neural coding depends, not upon conclusive data, but upon whether one believes that activity in a single neuron type can represent, in and of itself, a single taste quality. Unless this point proves to be experimentally testable, the distinction between these two theoretical approaches is primarily a philosophical one. "*Ergo, post omnes has disputationes, de gustibus non diutius sit disputandum.*"

Acknowledgments

This research was supported in part by National Institute of Neurological and Communicative Disorders and Stroke Grant NS-10211 and Research Career Development Award NS-00168. Thanks are due to my colleagues, Drs. Steven L. Bieber, Joseph B. Travers, and Richard L. Van Buskirk, for their invaluable contributions to the work reported here, and to Dr. Marion Frank for supplying data on chorda tympani fibers, to which we have made several comparisons in our work on brainstem neurons, and for her significant input to our discussions about these issues.

References

Boudreau, J. C. 1974. Neural encoding in cat geniculate ganglion tongue units. *Chem. Sens. Flav.* 1:41–51.

Boudreau, J. C., and N. Alev. 1973. Classification of chemoreceptive tongue units of the cat geniculate ganglion. *Brain Res.* 54:157–175.

Dethier, V. G. 1976. *The Hungry Fly: A Physiological Study of the Behavior Associated with Feeding.* Cambridge: Harvard University Press.

Doetsch, G. S., and R. P. Erickson. 1970. Synaptic processing of taste quality information in the nucleus tractus solitarius of the rat. *J. Neurophysiol.* (*Bethesda*). 33:490–507.

Erickson, R. P. 1963. Sensory neural patterns and gustation. In *Olfaction and Taste* (Proceedings of the First International Symposium). Y. Zotterman, ed. Oxford: Pergamon Press Ltd., pp. 205–213.

———. 1967. Neural coding of taste quality. In *The Chemical Senses and Nutrition* M. R. Kare and O. Maller, eds. Baltimore: Johns Hopkins Press, pp. 313–328.

———. 1968. Stimulus coding in topographic and nontopographic afferent modalities: on the significance of the activity of individual sensory neurons. *Psychol. Rev.* 75:447–465.

———. 1974. Parallel "population" neural coding in feature extraction. In *The Neurosciences: Third Study Program.* F. O. Schmitt and F. G. Worden, eds. Cambridge: MIT Press, pp. 155–170.

———. 1977. The role of "primaries" in taste research. In *Olfaction and Taste VI* (Proceedings of the Sixth International Symposium). J. LeMagnen and P. MacLeod, eds. London: IRL Press, pp. 368–376.

———. 1982. The across-fiber pattern theory: an organizing principle for molar neural function. In *Contributions to Sensory Physiology*, vol. 6. W. D. Neff, ed. New York: Academic Press, Inc. pp. 79–110.

Erickson, R. P., and E. Covey. 1980. On the singularity of taste sensations: what is a taste primary? *Physiol. Behav.* 25:527–533.

Erickson, R. P., E. Covey, and G. S. Doetsch. 1980. Neuron and stimulus typologies in the rat gustatory system. *Brain Res.* 196:513–519.

Erickson, R. P., G. S. Doetsch, and D. A. Marshall. 1965. The gustatory neural response function. *J. Gen. Physiol.* 49:247–263.

Everitt, B. 1980. *Cluster Analysis*. New York: Halsted Press.

Fishman, I. Y. 1957. Single fiber gustatory impulses in rat and hamster. *J. Cell. Comp. Physiol.* 49:319–334.

Frank, M. 1973. An analysis of hamster afferent taste nerve response functions. *J. Gen. Physiol.* 61:588–618.

———. 1974. The classification of mammalian taste nerve fibers. *Chem. Sens. Flav.* 1:53–60.

Frank, M., and C. Pfaffmann. 1969. Taste nerve fibers: a random distribution of sensitivities to four tastes. *Science (Wash. DC)*. 164:1183–1185.

McBurney, D. H. 1972. Gustatory cross-adaptation between sweet-tasting compounds. *Percept. Psychophys.* 11:225–227.

———. 1974. Are there primary tastes for man? *Chem. Sens. Flav.* 1:17–28.

McBurney, D. H., and J. F. Gent. 1979. On the nature of taste qualities. *Psychol. Bull.* 86:151–167.

McBurney, D. H., D. V. Smith, and T. R. Shick. 1972. Gustatory cross-adaptation: sourness and bitterness. *Percept. Psychophys.* 11:228–232.

Nowlis, G. H., and M. Frank. 1977. Qualities in hamster taste: behavioral and neural evidence. In *Olfaction and Taste VI*. (Proceedings of the Sixth International Symposium). J. LeMagnen and P. MacLeod, eds. London: IRL Press, pp. 241–248.

———. 1981. Quality coding in gustatory systems of rats and hamsters. In *Perception of Behavioral Chemicals*. D. M. Norris, ed. Amsterdam: Elsevier/North-Holland Biomedical Press, pp. 59–80.

Nowlis, G. H., M. E. Frank, and C. Pfaffmann. 1980. Specificity of acquired aversions to taste qualities in hamsters and rats. *J. Comp. Physiol. Psychol.* 94:932–942.

Ogawa, H., M. Sato, and S. Yamashita. 1968. Multiple sensitivity of chorda tympani fibres of the rat and hamster to gustatory and thermal stimuli. *J. Physiol. (Lond.)*. 199:223–240.

Perkel, D. H., and T. H. Bullock. 1968. Neural coding. *Neurosci. Res. Progr. Bull.* 6:221–348.

Perrotto, R. S., and T. R. Scott. 1976. Gustatory neural coding in the pons. *Brain Res.* 110:283–300.

Pfaffmann, C. 1941. Gustatory afferent impulses. *J. Cell. Comp. Physiol.* 17:243–258.

———. 1955. Gustatory nerve impulses in rat, cat and rabbit. *J. Neurophysiol. (Bethesda)*. 18:429–440.

———. 1959. The afferent code for sensory quality. *Am. Psychol.* 14:226–232.

———. 1965. De gustibus. *Am. Psychol.* 20:21–33.

———. 1974. Specificity to the sweet receptors of the squirrel monkey. *Chem. Sens. Flav.* 1:61–67.

Pfaffmann, C., M. Frank, L. M. Bartoshuk, and T. C. Snell. 1976. Coding gustatory information in the squirrel monkey chorda tympani. *Prog. Psychobiol. Physiol. Psychol.* 6:1–27.

Schiffman, S. S., and R. P. Erickson, 1980. The issue of primary tastes versus a taste continuum. *Neurosci. Biobehav. Rev.* 4:109–117.

Scott, T. R. 1974. Behavioral support for a neural taste theory. *Physiol. Behav.* 12:413–417

Scott, T. R., and R. P. Erickson. 1971. Synaptic processing of taste-quality information in thalamus of the rat. *J. Neurophysiol.* (*Bethesda*). 34:868–884.

Shepard, R. N. 1980. Multidimensional scaling, tree-fitting, and clustering. *Science (Wash. DC).* 210:390–398.

Smith, D. V. 1980. Processing gustatory information by hamster brainstem neurons. In *Olfaction and Taste VII* (Proceedings of the Seventh International Symposium). H. van der Starre, ed. London: IRL Press, pp. 267–270.

Smith, D. V., and D. H. McBurney. 1969. Gustatory cross-adaptation: does a single mechanism code the salty taste? *J. Exp. Psychol.* 80:101–105.

Smith, D. V., and J. B. Travers, 1979. A metric for the breadth of tuning of gustatory neurons. *Chem. Sens. Flav.* 4:215–229.

Smith, D. V., J. B. Travers, and R. L. Van Buskirk. 1979. Brainstem correlates of gustatory similarity in the hamster. *Brain Res. Bull.* 4:359–372.

Smith, D. V., R. L. Van Buskirk, J. B. Travers, and S. L. Bieber. 1983a. Gustatory neuron types in the hamster brainstem. *J. Neurophysiol. (Bethesda).* 50:522–540.

———. 1983b. Coding of taste stimuli by hamster brainstem neurons. *J. Neurophysiol. (Bethesda).* 50:541–558.

Travers, J. B., and D. V. Smith. 1979. Gustatory sensitivities in neurons of the hamster nucleus tractus solitarius. *Sensory Processes.* 3:1–26.

Van Buskirk, R. L., and D. V. Smith. 1981. Taste sensitivity of hamster parabrachial pontine neurons. *J. Neurophysiol. (Bethesda).* 45:144–171.

Woolston, D. C., and R. P. Erickson. 1979. Concept of neuron types in gustation in the rat. *J. Neurophysiol. (Bethesda).* 42:1390–1409.

Mechanisms of Behavioral Responses to Taste

8.
Time as a Factor in Gustation: Temporal Patterns of Taste Stimulation and Response

There is water in the river, but it always flows and does not remain the same. At the edges of the river, bubbles form and disappear again and again; they do not remain the same. People and homes in this world are like them. (Kamo-no-Chōmei, 1212)

Correct and timely decisions are the necessary outcome of successful tasting. To permit this adaptive result, the motor components of gustation (Halpern, 1977, 1983) and their associated sensory processes must have closely coupled temporal properties. Ideally, a single lick, sip, or bite would provide a sufficient chemosensory sample for such decisions. The organism could then continue its interaction with a chemosensory environment by repeating these discrete samplings of external environments or by sustained intraoral manipulation. In the case of ingested substances, the intraoral mastication and/or transport (Halpern, 1977) may be terminated by a series of swallows, or by expulsion of the ingesta from the mouth. If the decision is to reject, it should be possible to initiate this behavior before any swallowing has occurred. Conversely, if repeated samplings are made, the gustatory system ought to retain sufficient responsiveness to provide behaviorally useful evaluation of the *n*th sample. These putative design standards for tasting are compatible with the proposal that "the mouth is an especially acute haptic organ . . . which functions to lay hold of the environment" (Gibson, 1967). "This activity is not only one that prepares the food for swallowing. It is also exploratory and stimulus-producing" (Gibson, 1966). This chapter will evaluate the extent to which mammalian tasting actually fits this description.

Decision during Licking: The Laboratory Rat

Gustatory behavior provides both the *raison d'être* and the boundary condi-

---

Bruce P. Halpern, Department of Psychology and Section of Neurobiology and Behavior, Cornell University, Ithaca, New York 14853 tions of gustatory physiology. Of the several ingestive gustatory behaviors, licking/lapping of liquids has been most extensively studied (Halpern, 1977). Licking/lapping by the laboratory rat is particularly useful for evaluating models of gustatory function, since both the behavioral and the neurophysiological aspects of taste in this animal have received close attention. The detailed geometry and parameters of licking by this animal change somewhat as a function of ease of access to the drinking tube and of the liquid being drunk, as well as of the cumulative duration of licking. Lapping, in which only liquid is contacted by the tongue, has characteristics that differ somewhat from licking. However, differences in access do not produce large modifications in all licking/lapping measures. Thus, interlick intervals (ILI), i.e., time from the beginning of a contact between the tongue and the drinking tube to the start of the next contact (= start-start time), generally are 140-175 ms in duration during the first 5-10 s of licking, while comparable intervals for lapping are 149-235 ms in duration. A volume of 5-7 µl is removed from the drinking tube by each of the licks, while each lap removes from the liquid pool and delivers to the mouth 8-10 µl (Marowitz and Halpern, 1973; Halpern, 1975, 1977; Table I).

Quality Coding

Quality coding time for gustation, i.e., "the time period needed for the quality of a tasted material to be coded in the impulse patterns of the nerves emerging from the tongue and oral cavity" (Halpern and Tapper, 1971), has been estimated for the laboratory rat from the temporal parameters of licking. Both the ILI itself (Scott, 1974) and the components (microstructure) of each lick cycle (Halpern and Tapper, 1971; Halpern, 1975, 1977) have been used as the basis for such estimates. These measures allow evaluation of maximum possible quality coding time for a given set of external sensory factors and internal conditions. The reasoning is that the necessary time period for coding cannot be greater than the time interval from beginning of contact between the tongue and the liquid (usually assumed to be synonymous with contact with the drinking tube) to the end of the licking burst. Termination of licking is taken as a behavioral index of a decision to stop licking. However, termination of licking per se is ambiguous. It can be considered to be the end of the last contact with the liquid (or its logical equivalent, the number of licks divided by the typical [Scott, 1974] or measured licking rate), or as the end of the last contact with the liquid *plus* the time interval (stop-start time) that would normally elapse until the beginning of the next lick-contact. The difference between these two measures of licking termination, the stop-start time or lick

Table I. *Rat Median Drinking Parameters*

Environment[1]	*Dur*[2]	*Int*[3]	*ILI*[4]	*Rate*	*Total pauses*[5]	*Volume* µl/*lick*
Lick: 1st 10 Licks	65	80	145	7.8	0	6.43
1st 20 Licks	72	75	149	7.2	0	
Lap: 1st 10 Laps	72	71	149	7.4	0	8.5
1st 20 Laps	70	66	150	6.9	0	

Median drinking parameters for six adult (296 g) male Holtzman rats drinking distilled water after 18-h water deprivation (water *ad libitum* from noon until 4 p.m., then water removed; tested the next morning. 12 hr light/dark, with lights on at 7:45 a.m.; temperature 23°–24°C). Times are in milliseconds. Duration (Dur) and intervals (Int) are corrected for drinkometer on, and off, latencies.

1. Ingestion environment. Lick is the high restriction licking environment described in Marowitz and Halpern (1973) and in Halpern (1975). Lap is a 9 × 9 mm (inside dimensions) square Pyrex glass container (2 mm wall thickness), open at the top, which is 12 mm above the metal floor of a 16 cm wide × 28 cm long enclosure in which the rat is tested. The water level in this container is maintained such that the bottom of the water miniscus is 6 mm below the open top by automatic filling from the bottom. The front wall of the lapping container, which extends 1 cm higher than the other walls, is 3 cm from the front (long axis) of the enclosure; the container is 2 cm from a side wall of the enclosure. A 12-cm-long clear plastic internal wall, 5 cm from and parallel to the front of the enclosure, constrains the rats to drink while positioned on the long axis of the enclosure.
2. Duration of contact between tongue and liquid at each lick or lap (start-stop time).
3. Time between the end of one contact and the start of the next (stop-start time).
4. Interlick interval or interlap interval (start-start time).
5. Time periods > 125 ms between the end of one contact and the start of the next.

interval, is 80-100 ms long. Such a difference will be significant for quality coding time calculations only if the total time estimate is relatively short. In either case, termination of licking is recognized by the occurrence of a pause (see Tables I and II). A pause is a time period between the start of licks (= start-start time) that is appreciably greater than the median (or mean) ILI. Criteria that have been used for a pause include: (*a*) an ILI either ≥200 ms, or twice the median ILI, whichever is greater (Halpern and Tapper, 1971); and (*b*) a lick interval (stop-start time) > 125 ms (Halpern, 1975; Tables I and II). This is approximately equivalent to an ILI > 180 ms.

Table 11. *Rat Median Licking Parameters*

Liquid[1]	*Dur*[2] *ms*	*Int*[3]	*ILI*[4]	*Rate Licks/s*	*Licks*[5] *to 1st pause*	*Total*[6] *number pauses ms*	*Total*[6] *pause duration*	*Time to 1st pause*[5]
	ms					*ms*		
H_2O	73	76	143	7.5	52	0	0	7,561
Sucrose (500 mM)	56	73	129	9.2	6	2	1,221	899
H_2O	65	80	145	7.8	57	0	0	8,390
NaCl (500 mM)	56	73	133	8.0	4	2	900	657

Median licking parameters over first 10 licks for seven adult (325 g) male Holtzman rats maintained on a 20 hr water deprivation schedule. Testing in a high-restriction environment was done for each liquid after 17–19 hr of deprivation. Durations (Dur) and intervals (Int) are corrected for "on" and "off" latencies of the electronic drinkometer. See Marowitz and Halpern (1973) and Halpern (1975) for high-restriction environment.

1. First H_2O testing sessions shown was 5 d before the sucrose session; second, 1 d before NaCl session. Six H_2O testing sessions and 30 d separated the sucrose and the NaCl tests. All testing liquids contained 0.05% F.D. & C. Red no. 40.
2. Duration of contact between tongue and drinking tube (start-stop time).
3. Interval between end of one lick contact and beginning of the next (stop-start time).
4. Interlick interval between the start of contact for one lick, and the start of the next lick contact (start-start time).
5. A pause is a lick interval (stop-start time) > 125 ms. Time to first pause was obtained by multiplying the median number of licks to the first pause by the median ILI, and then adding 125 ms. The first 70 licks were measured for each rat; the number of licks to the pause, and the tme to the pause, are based on all 70 licks.
6. First 10 licks only.

Highly controlled behavioral situations are required for satisfactory estimates of maximum possible gustatory quality coding time. A drinking tube arrangement (Halpern, 1973, pp. 344–345; Halpern, 1977, pp. 62–71) is often used for such studies. When combined with barriers that permit only the tongue to contact the drinking tube (see Halpern, 1975, for a photograph), this highly restricted drinking situation limits the chemosensory stimulation component of each lick to the 6 µl removed from the drinking tube by the tongue and subsequently delivered into the mouth (Baron and Halpern, 1978). Consequently, taste-dependent decisions that lead to the termination of a licking burst under these conditions must depend for their quality encoding sensory input upon such relatively small volumes. Not only is the amount of liquid per lick small in this drinking situation, but the time between the start of licks, the ILI, is short. During the first 2 s of such a licking burst, ILI are 131-149 ms (Halpern, 1975; Table I). The brief durations of the ILI and its components (duration, interval) are important because they establish the accuracy of behavioral measurements for estimates of maximum gustatory taste quality coding time. This permits a conservative argument that if licking stopped within 2 s of the beginning of a burst, gustatory decision times were shorter than the number of licks emitted (plus one?) multiplied by 149 ms. Thus, if a licking burst ended after 10 licks, both quality encoding and a subsequent decision to terminate licking would have required < 1,490 ms (or, if the $n + 1$ licks rule is used, 1,639 ms).

Minimum number of licks by a laboratory rat to a behavioral taste rejection decision were measured in two laboratories in the early 1970s. In 1971, Halpern and Tapper used a conditioned aversion technique to determine that a single lick of 500 mM NaCl was sufficient to permit recognition of and behavioral reaction to this solution. The logical analysis presented earlier in this chapter would permit a calculation that the total gustatory recognition and response time could not exceed 298 ms, with ILI taken as 149 ms, and the $n + 1$ licks rule used. The sensory basis for the behavior was identified as recognition of NaCl, rather than a perhaps simpler discrimination between H_2O and all other liquids. This characterization of the basis for the behavior was possible because comparably trained rats, presented with 500 mM sucrose after training instead of 500 mM NaCl, showed a median of seven licks over a 915-ms licking-burst time, before a pause in licking occurred. It was also noted that "since central analysis, decision-making, and motor activity take up a significant portion of the total time needed to stop the licking behavior . . . it is plausible that the quality of the taste was coded in the peripheral afferent input in a much shorter period after onset of the stimulus." The stimulus solutions used in this study, H_2O, 500 mM sucrose, and 300 or 500 mM NaCl, were selected on the basis of behavioral taste quality categorization measures (Tapper and Halpern, 1968) to be very different from each other for the laboratory rat. This was done because the goal was to identify the absolute minimum taste quality coding time. If this time is assessed in number of ILIs for a single-lick behavioral reaction, it cannot fall below one interval plus the time period up to calculated contact time of a second, unobserved lick.

For quality coding times that are estimated to be a few hundred milliseconds or less, adding an extra ILI to the duration of the observed number of licks would represent from 50% (as in the case of the Halpern and Tapper single lick data) to 15% (for a 1,000-ms estimate) of the total time estimated. The lick cycle, which comprises the complete ILI (start-start time), is easily divided into two major components: contact duration (start-stop time) and lick interval, i.e., the time between the end of contact of one lick and the beginning of the contact of the next lick (stop-start time). During the first 2 s of licking, lick intervals (stop-start time) are 73-80 ms long (Halpern and Tapper, 1971; Halpern, 1975; Table I). (All observations on lick contact duration and lick interval are corrected for the on and off latencies of the drinkometer used, as described by Halpern [1977], p. 65.) Therefore, a conservative estimate of brief taste quality coding time would be secured by adding to the sum of the ILIs (start-start times = lick cycle durations) of the number of observed licks the duration of the next (unobserved) lick interval (= stop-start time). For the limiting case, in which licking stops after one counted lick, the quality coding time would be < 229 ms (149 ms ILI + 80 ms lick interval).

A longer period of sensory input might be expected to be required for discriminations between relatively similar taste stimuli; this was explored by Scott (1974; 1981). A conditioned aversion procedure and a behavioral situation similar to that used by Halpern and Tapper (1971) was used. Four taste stimulus pairs that had been selected to range from "very distinct" ($MgCl_4$-sucrose) through "somewhat distinct" (KCl - sucrose) to "intermediate similarity" (KCl-HCl) or "similarity" (HNO_3- HCl) provided data on maximum possible quality coding time for behavioral discriminations based upon taste (Table III). The number of licks and the time required for the discriminations clearly does increase as the degree of similarity between stimulus pairs increases. However, even for HNO_3 and HCl, a pair Scott considered similar (based upon neurophysiological criteria), maximum possible quality coding time is conservatively estimated to be no more than 773 ms, with four to five licks required.

The motivational conditions produced by a combination of radiation-induced conditioned gustatory aversion and water deprivation might yield taste quality coding times much shorter than those that would occur under less-demanding conditions. If this were the case, then the Halpern and Tapper (1971) and the Scott (1974) data just reviewed, while indicating encoding and subsequent behavioral decision/discrimination times that are within the capacity of the laboratory rat, would represent only the extreme end of a continuum. To evaluate this possibility, an appreciably milder situation was used: seven male Holtzman rats were adapted to a 20-hr water deprivation schedule

Table III. *Rat Quality Coding Times and Number of Licks*

Stimulus pair	*Maximum quality coding time, ms*	*Calculated number of licks*[1]	*Similarity*[2]
100 mM $MgCl_2$, 1,000 mM sucrose	280	1.3[3]	0.15
300 mM KCl, 1,000 mM sucrose	391	2	0.61
300 mM KCl, 30 mM HCl	633	3.5	0.74
30 mM HNO_3, 30 mM HCl	773	4.4	0.94

Maximum possible quality coding times, and number of licks required for a behavioral discrimination, for four taste stimulus pairs of increasing similarity (modified from Scott, 1974).

1. Number of licks: Scott's behavioral discrimination time for the stimulus pair, i.e., the quality coding times tabulated above minus 80 ms, divided by 158 ms, which is Scott's average interlick interval.
2. Correlation coefficients between single unit responses to each stimulus pair, recorded in the rat nucleus of the fasciculus solitarius (NFS); see Table I of Scott (1974).
3. Observed mean was 1.23 licks.

over a 5-mo period, with 4 hr of unlimited access to water from a drinking tube every afternoon in each animal's home cage. Food ad libitum (Purina Laboratory Chow) was always available in the home cage; a 12-h light-dark cycle was used (on at 7:45 a.m.), and temperature ranged between 23° and 24°C. Testing was done in the morning in a high-restriction licking environment (Halpern, 1975), with each water-deprived rat permitted access to liquid (distilled water, sometimes containing 0.05% Red no. 40 dye during training) for 3 min. Over a 5-mo period, during which the rats had a total of 24 training sessions, licking rates became high and stable, with no pauses (stop-start intervals > 125 ms) during the first 50–70 licks (= approximately 16 s of licking without a pause). A new stimulus liquid was then introduced for one drinking session. After six additional sessions with distilled water, distributed over 30 d, a second novel stimulus liquid was introduced for one session. The first new liquid was 500 mM sucrose; the second, 500 mM NaCl. Both of these liquids will be drunk at a high rate if they are not novel and if the rats have no accumulated water deficit. As Table II indicates, introducing a novel taste stimulus leads to a temporary termination of licking within 1 s of the beginning of licking. For 500 mM sucrose, six licks, requiring a total of 774 ms, gave a maximum quality coding time of 854 ms; for 500 mM NaCl, four licks, occurring over 532 ms before a pause in licking, give a maximum possible taste quality coding time (and subsequent decision to stop licking) of 612 ms. Thus, novelty alone is sufficient to produce quality coding estimates comparable to those reported by Halpern and Tapper and by Scott. The apparent differences between the number of licks to the first pause and the times preceding pauses (time to 1st pause) for sucrose and NaCl may be related to both the greater motivational power of sucrose and to sensory differences.

Decision Time vs. Quality Coding Time

The discussion has thus far focused upon estimates of taste quality coding time. A behavioral endpoint, specifically a temporary (pause) or more-or-less permanent cessation of licking for a particular stimulus liquid, was used as the basis for this estimate. This outcome, of course, requires not only that sufficient sensory information be encoded to allow an appropriate decision, but also that the decision be made and then expressed as a change in the ongoing ingestive behavior. Whatever gustatory encoding models one adopts (e.g., Pfaffmann, 1978; Erickson and Covey, 1980; Nowlis and Frank, 1981; Scott, 1981; Yamamoto et al., 1981), completion of quality coding must precede a neurophysiological decision to alter a licking burst. This neural decision event, in turn, would necessarily occur before measurable changes in motor activity. However, these distinctions will not permit a division of the total behavioral measure into separate times attributable to coding, decision, and output, unless the beginning and end of one or more of these processes can itself be measured. A more modest goal would be an attempt to identify a time at which peripheral quality coding could begin, and a time by which changes in motor output and/or the beginning of the next lick could be detected. Latency of sensory input would be relevant to the start of quality coding, while either electrophysiological or more overt signs of tongue or jaw activity would signal the temporal region that could follow a decision.

Gustatory neural latency to stimulation of the tongue with solutions of salts or inorganic acids, measured from the laboratory rat in a peripheral nerve (chorda tympani) innervating lingual fungiform papillae taste buds, is typically 26–45 ms (Faull and Halpern, 1972; Marowitz and Halpern, 1977; Kelling and Halpern, 1983a, footnote 12). This latency represents a time period from the moment of contact between a stimulus liquid and the tongue to the beginning of a gustatory neural response to the stimulus liquid. The estimates of total gustatory quality coding and decision time that have been done so far in this chapter have all started at the beginning of the first contact between the tongue and the liquid in the drinking tube. Therefore, these estimates of total quality coding and decision time have all included, as part of the calculated time, the gustatory neural latency. Since no information on the chemosensory characteristics of the taste stimulus appears in the sensory system during the latency period, this time should not be included in calculations of maximum possible quality coding and decision time. Consequently, it is appropriate to subtract at least the shortest latency, 26 ms, from the previously calculated total gustatory coding and decision times. Since the observed range of quality coding and decision time had been from 229 ms for single-lick recognition and rejection of 500 mM NaCl to 773 ms for a discrimination between HNO_3 and HCl, uniform subtraction of a 26-ms gustatory neural afferent latency for rat chorda tympani nerve will yield a corrected range. For very discriminable stimuli and optimal conditions, no more than 203 ms is required from the beginning of peripheral sensory input during the first lick contact to the elimination of further licking. Stimuli that are relatively similar can require as much as 747 ms from the start of afferent taste responses to a behavioral discrimination signaled by the end of licking. Of course, neither of these times exhausts the potential range. For example, neural afferent latencies to salts are probably shorter than those to sucrose (e.g., Halpern and Tapper, 1971, footnote 15; Kelling and Halpern, 1983a, footnote 12) by 30 ms to perhaps as much as 300 ms. Such long latencies, if correct, suggest that very little peripheral gustatory input may be needed for some taste discriminations by the laboratory rat. On the other hand, there are no doubt gustatory discriminations that would require many seconds of sampling. Nonetheless, it seems clear that a complete gustatory-afferent-input to behavioral-output sequence can occur in the laboratory rat in $\sim$ 200 ms, whereas 500–800 ms is adequate time for such a sequence when the taste stimuli are not strikingly different.

The decision, motor output, and behavioral components of the taste-quality-coding/behavioral-discrimination complex can only be incompletely and hesitatingly disentangled at present. More specifically, the observable movements of jaw and tongue will be considered in some detail to identify a time during a total lick cycle when such movements begin, and the bioelectric signs of the underlying muscular activity, i.e., electromyograph (EMG), will be used to validate the measures of movement. Data from these sources will serve to shorten the total time available for gustatory quality coding and a subsequent decision to change ongoing licking behavior. However, the moment in time when quality coding ends and the decision process begins will not be elucidated. It may be that little meaningful distinction between these two actually exists, if the notions of parallel and distributed processing and multiple feedback loops are applicable to gustation.

A lick by a laboratory rat begins with a lowering of the jaw, and quickly progresses to a stereotyped tongue-protrusion pattern (Marowitz and Halpern, 1973; Halpern, 1977, p. 68; Figure 1). About 40 ms elapse from the beginning of observable jaw motion to contact with the drinking tube. It follows that EMG in muscles that open the jaw should be detectable at a somewhat greater time interval before tongue contact than the gross movement of the jaw itself. This is indeed the case. Yamamoto and his colleagues (Yamamoto et al., 1982a; Yamamoto, 1983) find that EMG activity in a jaw-opener muscle, the digastric (Halpern, 1977, p. 4), begins about 60 ms before the tongue contacts the drinking tube during licking by a laboratory rat. The measures of jaw and tongue movement during licking, and the jaw opener EMG that precedes them, might permit the contention that the decision to make or not make a lick is made 40–60 ms before the tongue contacts the drinking tube during a lick cycle. Since the total time from the beginning of afferent input to contact with the drinking tube at the beginning of the *next* lick has already been estimated to be no more than 203 ms under optimum conditions of discriminability, the process of encoding and decision would have only about 140–160 ms available (203 ms total time minus the 40 ms needed for the jaw and tongue-protrusion motions of the next lick = 163 ms quality coding time; 203 ms minus the 60 ms that elapse from the first sign of electrophysiological activity in a jaw opener muscle to the next tongue contact = 143 ms). This seems to be rather brief, perhaps, but it is 30–40 ms *longer* than the time required for maximum chorda tympani nerve phasic response magnitude to be reached when gustatory afferent responses are recorded from this nerve in the anesthetized rat while NaCl solutions with a volume (5 µl) and a flow duration (50 ms) similar to that of a lick are applied to the anterior portion of the tongue (Marowitz and Halpern, 1977; Figure 2).

Temporal Characteristics of Gustatory Neural Responses in the Rat

The quality coding time estimates developed in this chapter provide an upper limit on the maximum possible necessary time period of afferent neural activity for taste quality to be adequately coded. For inorganic salts and sucrose, conservative estimates of the time period ranged from 203 to 828 ms after the start of the afferent chemosensory response (i.e., with a latency of 26 ms subtracted). What are the properties of gustatory afferent neural responses during this time period, roughly the first 1,000 ms of the response?

A series of rapid changes in peripheral gustatory afferent responses occur

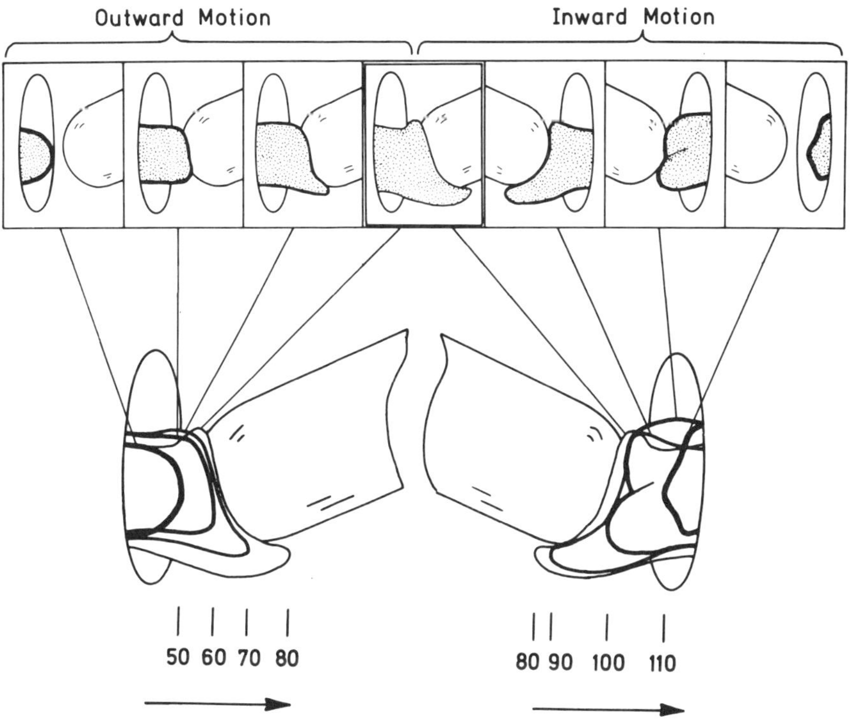

Figure 1. Licking sequence by a laboratory rat when drinking in a high restriction environment (Marowitz and Halpern, 1973; Halpern, 1975). Prepared from 200 frames/s motion pictures made during the 2nd lick of a prolonged (> 70 licks without pause [i.e., no intervals > 125 ms]) burst of licks of distilled water by a rat deprived of water for 18 h, maintained on a 20-h water deprivation schedule. General conditions as in Table II. The illustrated lick had a contact duration (start-stop time) of 56 ms and a lick interval (stop-start time) of 73 ms, which gives an interlick interval (start-start time) to the next (3rd) lick of 129 ms and an instantaneous rate of 7.75 licks/s. Top: Successive positions of the tongue at 10-ms separations in time (every other frame of the motion picture) during outward motion, i.e., protrusion (left-hand drawings), and retraction, i.e., inward motion (right-hand drawings) of the tongue. Bottom: Overlays of successive tongue positions at 10-ms intervals during outward motion (left) and inward motion (right) of the tongue. The most protruded position in the left-hand sequence is the same as the least retracted position in the right-hand sequence. Numbers are in milliseconds; the arrows indicate the general direction of motion with reference to the cylindrical drinking tube and the circular opening through which the tongue extends.

during the first 1,000 ms after stimulation of the tongue with NaCl, LiCl, KCl, quinine hydrochloride, or HCl. Both multiunit (e.g., Marowitz and Halpern, 1977; Figure 2) and single unit (Ichioka and Hayashi, 1974; Mistretta, 1972; Nagai and Ueda, 1981; Ogawa et al., 1973, 1974; Pritchard and Scott, 1982) observations indicate that maximum response to such stimuli occurs during the first 1,000 ms of the response. More specifically, maximum response magnitude generally occurs within the first 200 ms. Stated somewhat differently, response to salts such as NaCl, KCl, or LiCl, acids like HCl, and the alkaloid salt quinine hydrochloride all usually have a prominent initial burst of activity (phase component). The peak of this phasic burst is near or before the calculated maximum taste quality coding time for NaCl. The magnitude of the response, as well as the rise rate, change systematically with stimulus concentration (Faull and Halpern, 1972; Marowitz and Halpern, 1977). Other chemicals that produce similar phasic patterns in peripheral taste nerves of the laboratory rat include certain amino acids, such as L-cysteine hydrochloride and L-lyseine hydrochloride (Pritchard and Scott, 1982).

Other stimulus solutions have less dramatic phasic responses and somewhat longer response latencies. Sucrose is included in this group, as well as saccharin and such amino acids as L-arginine, L-histidine, and L-threonine (Ogawa, et al., 1973, 1974; Nagai and Ueda, 1981; Pritchard and Scott, 1982). Although less of the total response to a sustained stimulus is concentrated in the first few hundred milliseconds than would be the case for NaCl, much activity is seen in the first 1,000 ms. Thus there seems to be considerable information potentially available early in the peripheral gustatory afferent response, with either a monotonic decline in response magnitude after the phasic burst (Ogawa et al., 1973; Marowitz and Halpern, 1977), found for salts and acids, or rhythmic changes in response rate (Mistretta, 1972; Ogawa et al., 1974; Nagai and Ueda, 1981) observed with sucrose, saccharin, and quinine hydrochloride.

Maximum taste quality coding times under 1,000 ms are compatible with afferent neural responses of the types thus far described. However, other response patterns also occur. Thus, responses to amino acids such as glycine and DL-alanine are often much slower than those to NaCl (Halpern, 1963). A single unit analysis of rat chorda tympani nerve responses to glycine, L-proline, and L-alanine found a wide range of latencies, with some requiring many seconds (Pritchard and Scott, 1982). This could mean that prolonged sampling of such stimuli would be required before a decision could be reached or a discrimination made. However, some units had latencies of < 1 s to these amino acids, and showed considerable sustained response to them during the

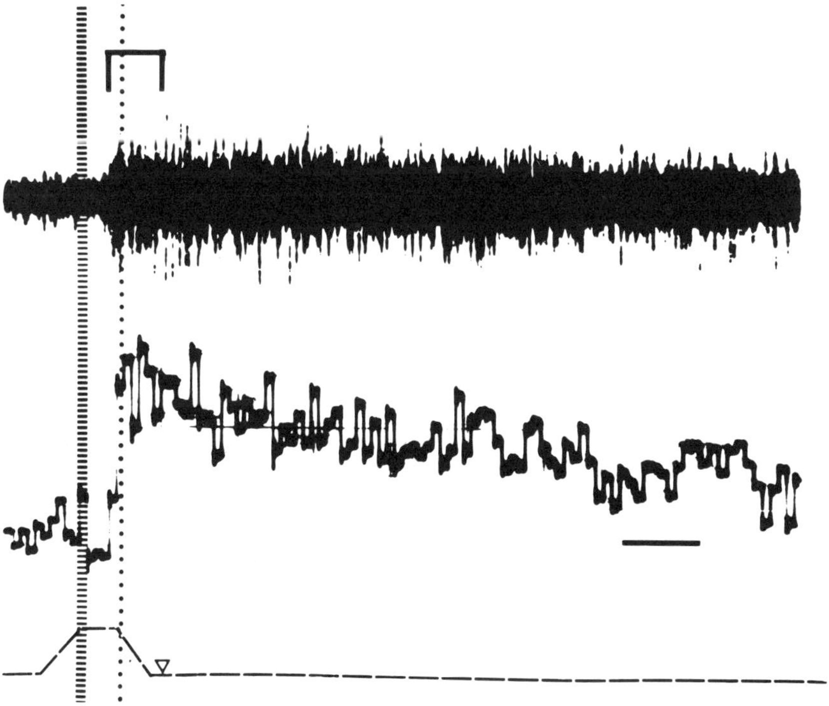

Figure 2. Relationship between gustatory afferent neural response and lick cycle in the laboratory rat. Stimulus presentation or lick contact start is indicated by the dashed vertical line; the dotted vertical line indicates end of stimulus flow (neural recording) or end of lick contact (lick cycle). Horizontal calibration line is 100 ms. Top and middle traces, chorda tympani nerve multiunit response (top) and digitally controlled summator-processed output to chorda tympani nerve response (Marowitz and Halpern, 1977). Chorda tympani response is to a single presentation of 5 μl of 200 mM NaCl flowing for 55 ms over a 6-mm^2 area of anterolateral tongue. Bottom trace, schematic of lick cycle. Broken line represents tongue position relative to drinking tube. Lowest horizontal position indicates that the tongue is inside mouth; highest, that the tongue is in contact with the drinking tube and stimulus liquid. A single lick is shown during which the tongue protrudes from the mouth, moves outward toward and contacts the drinking tube (left-hand diagonal); remains in contact with the drinking tube for the lick contact duration (horizontal segment between diagonals); and then moves inward back into the mouth (right-hand diagonal). The open triangle marks the time when outward tongue movement for a second lick would be expected to start. Inverted U at top, left-hand vertical element marks beginning of gustatory afferent response; right-hand vertical element, beginning of tongue protrusion movement for a second lick.

first 1,000 ms. It follows that a prediction of long quality coding times, while tempting, is not the only plausible outcome. Some behavioral data support this caution: rats trained in a radiation-induced conditioned gustatory aversion procedure (Tapper and Halpern, 1968) reject 500 mM DL-alanine after five licks (D. N. Tapper and B. P. Halpern, unpublished observations, cited in Halpern and Tapper, 1971). This is similar to the number of licks that preceded a pause (interval > 125 ms between the end of one lick contact and the beginning of the next lick) when 500 mM sucrose was introduced to rats accustomed to only distilled water (Table II). The actual quality coding time for the 500 mM DL-alanine consisted of a 600 ms burst of licking, plus 80 ms representing the interval from the end of the last lick to the potential moment of contact of the next, unobserved lick. The total, 680 ms, is similar to that found by Scott (1974) for a KCl-HCl discrimination (Table III), which was considered to involve liquids of intermediate similarity.

The response patterns seen in peripheral gustatory nerves are not preserved as central nervous system processing progresses. In particular, the phasic component becomes less distinct, especially in thalamus and cerebral cortex, and firing rates tend to increase in general in the hindbrain (nucleus of the fasciculus solitarius), and then to decrease at subsequent brainstem, diencephalic, and forebrain loci (e.g., Scott, 1981; Yamamoto et al., 1981; Yamamoto, 1983). Some long response latencies occur, while others are in the 60 ms range. Changes between peripheral input and central postsynaptic responses are of course to be expected. They are the usual finding in sensory systems. Understanding the nature of these changes, and the properties of processing, will require use of appropriate stimuli, and, quite probably, relevant behavioral situations.

Summary: Decisions during Licking in the Laboratory Rat

The considerable behavioral and neural data on licking in the laboratory rat permit a reasonably detailed analysis of the chemosensory basis for this behavior. The goal is to estimate the required duration of sensory taste responses from the receptors of the mouth for taste-dependent behavior. If the taste-dependent behavior involves a quality categorization, i.e., a recognition that the solution being licked is, for the rat, different in its taste properties from one or more comparison solutions, then a quality coding time can be measured. Behavioral measures necessarily overestimate this time since they are the outcome of not only a quality coding process but also neural decisions, motor output sequences, and development of sufficient muscle tension to produce overt changes in licking.

Licking itself can be measured at several increasingly detailed levels. The most molar measure is the number of licks made before licking stops. If that number is > 1, then the time interval between the start of successive licks, i.e., the ILI, must also be known. Alternatively, the total elapsed time from the beginning of contact by the first lick to the end of contact by the last lick could be determined. Either of these measures would estimate the total time between the first contact between tongue and liquid and the end of licking behavior. They would differ in that the elapsed time measure, extending from the start of contact with the liquid by lick 1 to stopping of contact during inward tongue motion by the final lick, would be shorter than the cumulative ILIs. The latter would include the time from the end of contact by the final lick to the expected start of contact by the next unobserved lick.

The observed taste quality coding time estimates, arrived at using very conservative behavioral measures of the total possible time, ranged from 229 ms for recognition of 500 mM NaCl, to 854 ms for a response to a novel 500 mM sucrose solution. These estimates can be refined by correcting them for the latency of the afferent gustatory response and for the time that elapses from the beginning of bioelectric activity (EMG) in jaw muscles at the start of the tongue protrusion to the contact of the tongue with the liquid being licked. After such corrections, minimum taste quality coding and decision time for NaCl is 143 ms. This time is still long compared with the initial components of the phasic gustatory afferent neural responses.

Decision during Sipping: Human Taste-dependent Decisions

Drinking by humans is quite different from licking by rats. Adult humans are primarily suction drinkers (Halpern, 1977, pp. 74–78). The basic unit in human liquid ingestion behavior is the sip. A sip sequence generally includes a smooth and stereotyped movement of a drinking vessel into contact with the lower lip, a period of contact between the drinker and the liquid in the drinking vessel, removal of the drinking vessel from the lower lip, and one or more swallows (Figure 3). The precontact approach phase of an initial sip has a duration of ~ 1,500–2,000 ms (Halpern, 1975; Table IV). During this movement of the drinking vessel toward the mouth, the angle of the drinking vessel is gradually changed such that when the leading edge of the drinking vessel reaches the lower lip, the liquid in the drinking vessel is at or near the top of the leading edge. Consequently, contact with the liquid can begin shortly after contact with the drinking vessel. The duration of contact with the liquid

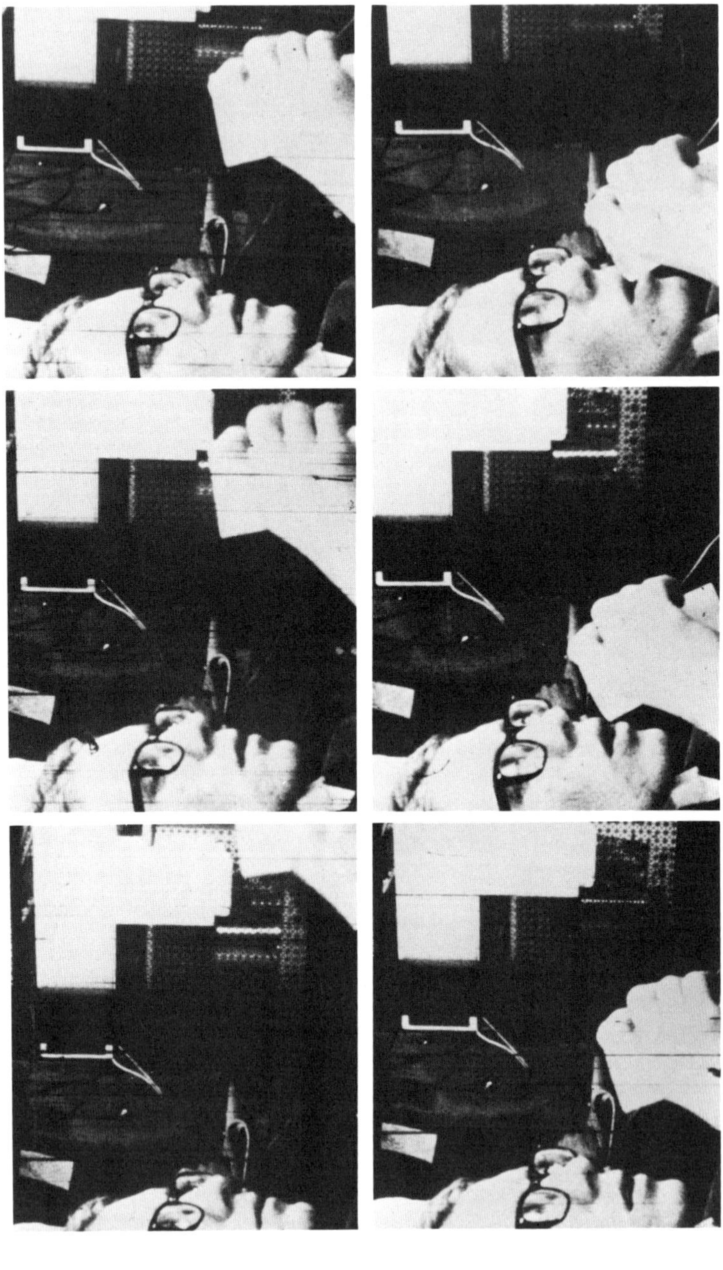

during a single sip has a mean of ~ 1,000 ms (Halpern, 1975; Halpern and Nichols, 1975; Table IV, Figures 5 and 6). Ingestion can occur only during this period of contact. When a single sip is made, swallowing usually follows the end of contact with the liquid, rather than beginning during the contact/ingestion component of the sip sequence. The volume ingested during a single sip can vary widely among indivuduals, but means are 12–14 ml per sip (Table IV).

Since the overall duration of each human sip sequence is relatively long, effects of the chemosensory characteristics of the liquid in the drinking vessel might be measurable during a single sequence, as well as in the length of the time interval between the start of successive sips. Behavioral measures of sip contact duration and sip volume showed no sizable mean differences when a cherry-flavored beverage (Cherry Kool-Aid), a cherry-flavored beverage containing 100 mM NaCl, a 309 mM sucrose solution, or water was drunk during a single sip (Halpern and Nichols, 1975; Halpern, 1975; Table IV). The absence of taste-dependent differences in sip contact duration or volume during a single sip could indicate that the chemosensory differences between the four solutions were too small to modify human responses, at least for laboratory measurement when a single sip is to be made. Alternatively, it

---

Figure 3. Human single-sip drinking during EMG recording experiment. Participant was a 19-yr-old, 63.5 kg male drinking and swallowing single sips of liquid at 23 ± 1°C from 295.7-ml capacity, frosted-glass drinking vessels. All liquids were colorless. Cherry Kool-Aid, distilled water, and cherry Kool-Aid containing 100 mM NaCl were presented in random order. 174 ml of liquid was presented on each trial. Individual frames from a motion picture sequence filmed at 40 frames/s are shown. The drinking glass contained cherry Kool-Aid. This was the 11th sip of liquid; it was preceded by three sips of cherry Kool-Aid, four sips of water, three sips of cherry Kool-Aid in 100 mM NaCl, and three control trials (see figure 4). The frame sequence is left-to-right in the top row (frames 1–3) and left-to-right in the bottom row (frames 4–6). Note miniature electorde on left upper lip (one of a bipolar pair) for EMG (orbicularis oris electromyograph) measurement, laryngophone (throat microphone) which detects swallow sounds (not visible in frame 1), and wire leading from drinking glass to drinkometer circuit for measurement of contact onset and duration (visible in frames 3, 5, and 6). Frame 1 is 800 ms after the drinking glass was lifted from a table at which the subject sat; 2, 925 ms; 3, 1,025 ms; 4, 1,225 ms; 5, 1,475 ms; 6, 1,575 ms. Contact between the subject and the liquid in the glass was detected at 1,750 ms after the glass was lifted from the table. During this one sip, 28.8 ml of cherry Kool-Aid was drunk by the subject.

Table IV. *Human Drinking Parameters*

	Volume per sip		*Time from lifting drinking glass to:*			*Time from start of contact to maximum EMG*	*Contact duration*
	Mean (SD)	*Max/Min*	*Contact start*	*Contact end*	*First swallow*		
Liquid drunk	*ml*			*ms*		*ms*	*ms*
H_2O	12.75 (6.97)	29/4	1,926 (388)	2,890 (553)	3,540 (362)	550	964
308 mM sucrose	13.32 (6.94)	30.8/5.6	1,971 (391)	2,944 (209)	3,456 (342)	500	973
Cherry Kool-Aid in 308 mM sucrose[1]	13.61 (6.82)	30.9/5.4	1,975 (492)	3,026 (738)	3,288 (1,098)	700	1,051

Mean and standard deviation (in parentheses) drinking parameters for single sips of three liquids, with 174 ml of each presented 6 to 8 times each in random order to each of 12 adult (mean age, 30.1 yr) subjects. The liquids were either distilled water or made up with distilled water, were presented at 23 ± 1°C, and had the same red color (absorbance ~ 500 nm) produced by adding F.D. & C. Red no. 40 to the water and the sucrose. Eleven of the subjects were female.

1. Significantly different EMG magnitude from H_2O and sucrose, starting at 500 ms after contact ($P < 0.001$, F-test).

could be that sip contact duration and volume are insensitive measures of human taste effects. A physiological measure, upper lip surface EMG from the orbicularis oris sphincter muscle of the mouth, was also taken (Figures 4–6). The EMG did differ as a function of the liquid in the drinking vessel. However, these differences were not present in the EMG record made while the drinking vessel approached the mouth, or in the first few hundred milliseconds after contact with the liquid started (Table IV; Figure 6). Statistically significant ($P < 0.001$, F test) differences were observed late in the sip-contact period, between 500 and 1,051 ms after the start of contact.

The long delay between the start of contact with a liquid during a sip and an

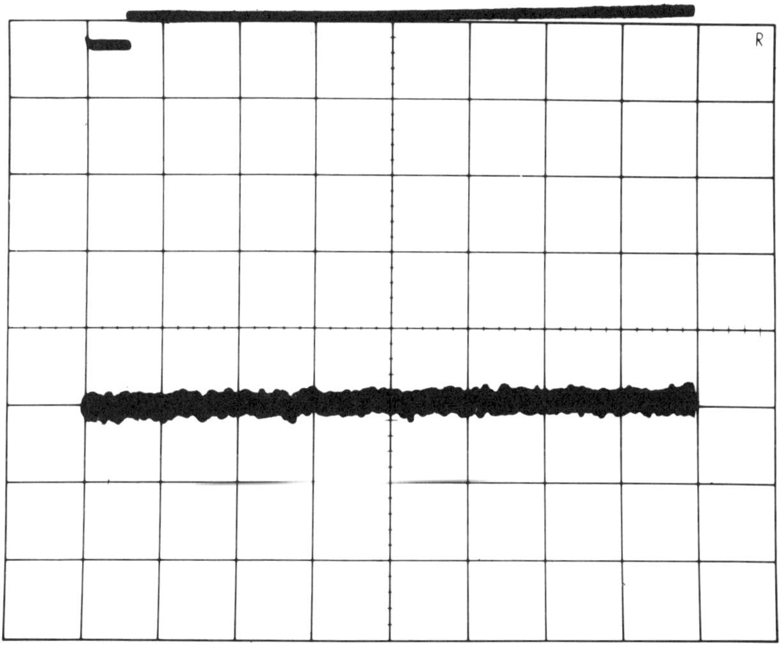

Figure 4. Control trial EMG record. The EMG (lower trace) was recorded with bipolar electrodes from the subject's upper lip while he waited for the "ready" signal. On this control trial, no "ready" signal was given. Instead, a silent switch was pressed at the predetermined time at which a "ready" signal would normally be given. The switch closure produced a voltage change (upper trace) equivalent to the voltage that is switched when a drinking glass is lifted at the beginning of an experimental trial. Time, 1 s/division.

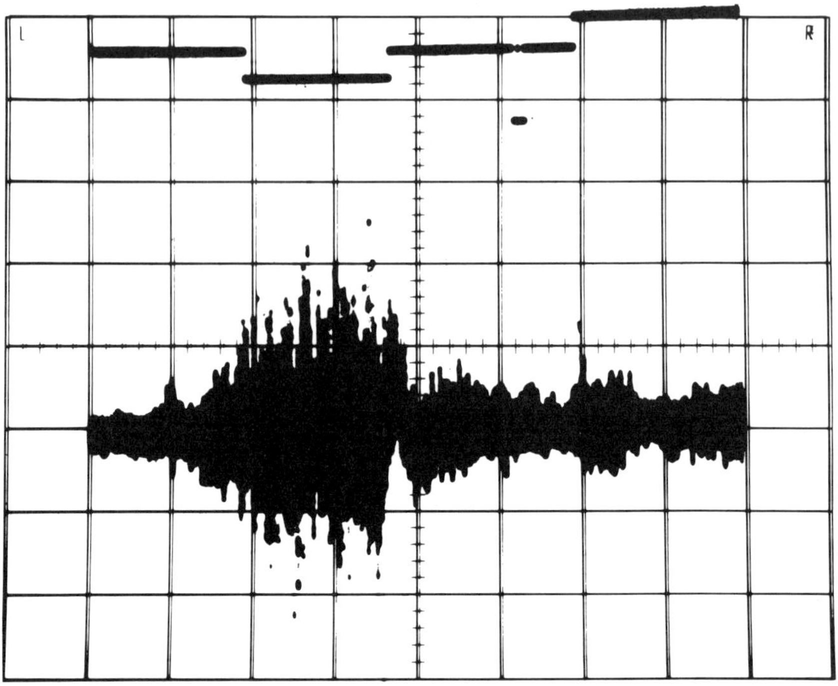

Figure 5. Single-sip human drinking pattern. A 27-yr-old female drank one sip of cherry Kool-Aid made up in an aqueous solution of 308 mM sucrose and 100 mM NaCl. A volume of 9.2 ml was drunk. This was the 11th sip of the experiment, in which H_2O, cherry Kool-Aid made up in 308 mM sucrose, and cherry Kool-Aid made up in 308 mM sucrose and 100 mM NaCl were randomized with each other and with control trials (on which no "ready" signal, to indicate that a single sip should be taken, was spoken; (Figure 4). All liquids had the same red color (Table IV). 174 ml of liquid was offered in a 296-ml, clear plastic drinking glass on each trial. Bottom trace: EMG recorded from upper lip using surface electrodes (Halpern and Nichols, 1975; Halpern, 1975; Figure 3). EMG calibration: One division, 200 mV (vertical); 1,000 ms, horizontal. Sweep started when the subject lifted the drinking glass to begin the sip sequence. Top trace: sip sequence event indicator. Trace started when the drinking glass was lifted from its support on the table at which the subject was seated. The trace continued at this drinking glass-lifted position until contact between the liquid in the drinking glass and the subject occurred. For this sip, the precontact time was 1,847 ms. At contact, the event indicator moved down to a lower position, where it remained during contact. Contact duration on this sip was 1,766 ms. When contact ended, the indicator moved upward to the position it had occupied before contact started but after the drinking glass had been lifted by the subject. Swallows, detected by a throat micro-

effect of the particular liquid on the upper lip EMG could indicate that human reaction times to taste stimuli are relatively long, of the order of 500 ms or more. This is indeed the case, as a number of studies have demonstrated (e.g., Kelling and Halpern, 1983a, b; Lester and Halpern, 1979; Yamamoto and Kawamura, 1981; Yamamoto et al., 1982b). However, the cause of this relatively long human gustatory reaction time, which is more than twice the behaviorally measured minimum taste quality coding time for the laboratory rat (see first section of this chapter), as well as more than twice the time required for human simple visual or auditory reaction times (Kelling and Halpern, 1983a, footnote 3), remains unresolved. Some possible explanations which might be offered, and sometimes have been, include very long peripheral afferent neural taste response latencies in humans, and/or a necessity for prolonged gustatory stimulation in humans before any behaviorally meaningful taste quality coding can occur.

Both of these proposed explanations for the relatively long human gustatory reaction times are rejected by experimental data. Firstly, it has been demonstrated that long stimulus durations are not required to produce typical human taste reaction times. Stimulus durations of 200 ms (Lester and Halpern, 1979) or less (Kelling and Halpern, 1983a) yield gustatory reaction times to NaCl or sodium saccharin that are at least as short as those to the same stimulus solutions, but with long durations. Similar observations have also been made for HCl and $MgSO_4$ (Kelling and Halpern, 1983b). Secondly, brief taste stimuli, e.g., from 100 ms durations (Kelling and Halpern, 1983a) down to 50 ms durations (Kelling and Halpern, 1983b) are sufficient to permit judgments of taste quality and intensity by humans. At 100 msec durations and above, the quality judgments given as answers to the question "What was the taste?" differed little from those to stimuli with durations of 1,000–2,000

---

phone functioning as a laryngophone (Figure 3), caused the event indicator to move to its lowest position for the duration of the swallow sound. The first swallow during this sequence occurred at 1,513 ms after contact ended; the second, at 1,582 ms after contact ended between the subject and the liquid in the drinking glass. When the drinking glass was returned to its support on the table in front of the subject, the event indicator trace moved to its highest position. Total length of this single sip sequence, from lifting of the drinking glass to return of the drinking glass, was 5,766 ms. Event trace calibration: one division, 1,000 ms. The EMG and all sequence event voltages were recorded with an FM instrumentation tape recorder operated at 7.5 in/s, and subsequently A/D converted and averaged.

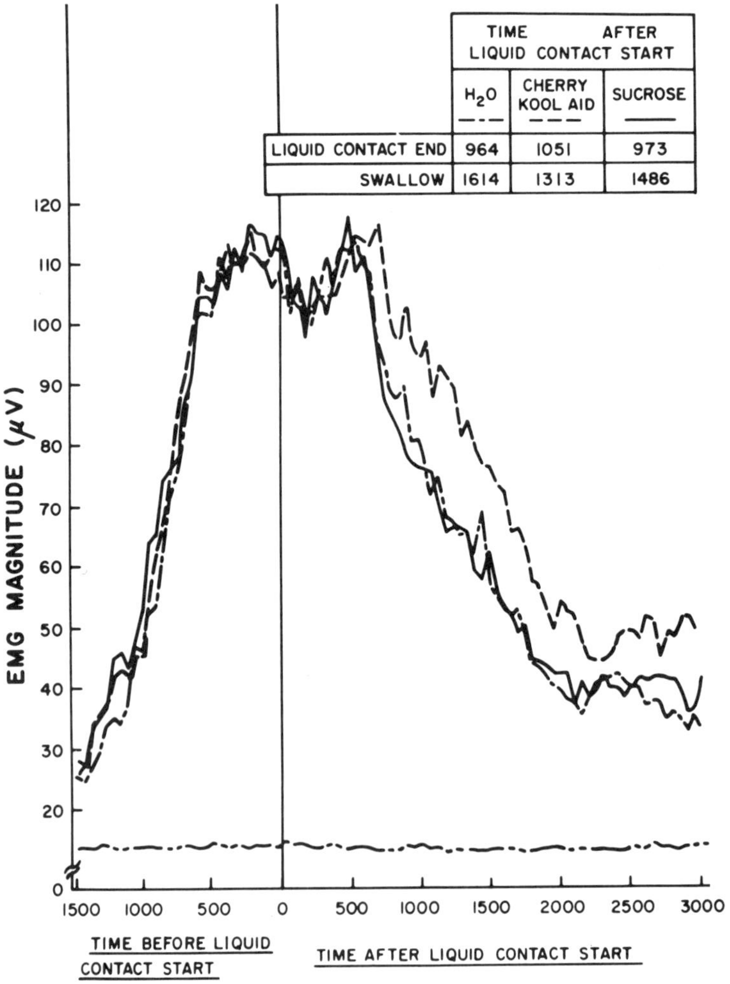

Figure 6. Averaged (within and between subjects) upper-lip EMG recorded from 12 subjects (Table IV) during single sips of red dyed H_2O, cherry Kool-Aid made up in 308 mM sucrose, or 308 mM sucrose. Recording procedures are described in Figure 5. The EMG for cherry Kool-Aid is significantly different ($P < 0.001$) from the EMG for H_2O or sucrose starting at 500 ms after contact. The bottom line (dots and dashes) is the average of control trials (Figure 4). The inserted table identifies the line types used in the graph for each liquid's averaged EMG response, and gives mean times both from start of contact between the liquid in the drinking glass and the subject to the end of contact with the liquid and to the first swallow. All times are given in milliseconds.

ms (intensity *was* dependent upon stimulus duration, with 1,000–2,000 ms duration stimuli judged to be twice as intense as 80–100 ms duration stimuli). Finally, peripheral gustatory afferent neural responses recorded from human chorda tympani nerves (Zotterman, 1967) have relatively rapid rise times (Kelling and Halpern, 1983a, footnote 14) and do not appear to be obviously slower than responses from the comparable gustatory nerve in rats (B. Oakley, cited in Lester and Halpern, 1979).

In *general*, it *appears* that central nervous system characteristics, rather than peripheral gustatory afferent taste quality coding dissimilarities, will *probably* account for the *apparently* slower taste reaction times in humans than in the laboratory rat. It may be that the tongue or jaw movement should be measured in humans, as it is in the rat, in order to secure a more relevant taste reaction-time behavior. Another possibility is that no motor output change is possible for a considerable period after taste quality coding is completed, so that the danger of tongue damage during biting and mastication is reduced. Alternatively, perhaps human taste quality coding requires a prolonged period of central nervous system processing, but can operate on a rather brief peripheral afferent input (Kelling and Halpern, 1983a). If the latter explanation is correct, then human taste processing and decisions are perhaps as complex and difficult as is processing in the auditory and visual domains.

Natural drinking: are the sips of a multisip series independent samples? Under normal conditions, more than one sip may be taken before the drinker interrupts the series. Are there sizable changes in taste responses between the first and second sips, the second and third, and so on, due perhaps to an accumulation of ingesta in the mouth and cumulative sorption to chemoreceptor membranes? Conversely, it might be that the one or more swallows that usually accompany sips (Figures 5 and 6, Table IV) result in a sufficiently thorough wipe of the mouth to return the oral cavity and the chemoreceptor arrays to the salivary base line that preceded the first sip (Halpern, 1977). The time period between successive sips will of course be a major factor. Initial experiments suggested that intersip intervals of 1.5–4 s occur early in a sipping series (J. Stremlin and B. P. Halpern, unpublished data, cited in Halpern and Meiselman, 1980). Consequently, it might be possible to simulate sipping by alternately presenting a stimulus solution and a liquid intended to model some aspects of the effects of the swallow.

An attempt to measure the psychophysical effects of a series of simulated sips and swallows has been reported (Halpern and Meiselman, 1980). The procedure was to alternate 1-, 2-, or 3-s flows of taste stimulus solution with equal duration flows of either distilled water or an "artificial saliva" containing Na, K, Ca, Cl, and HCO_3 ions at approximately salivary levels. Magnitude estimates of total taste intensity, made during 70 s of this "simulated drinking" of NaCl or sodium saccharin solutions, showed no decrease in intensity over time. In contrast, alternate pulses of stimulus solution only showed significant adaptation from the first magnitude estimate measurement, as early as 7.4 s after alternating flow started. In both cases, liquids flowed at 5 ml/s over the anterior portion of the tongue. These data could be interpreted to mean that little or no taste adaptation would occur during a series of normally spaced sips, but would be expected with continuous stimulation.

A number of differences exist between the Halpern and Meiselman (1980) "simulation of human drinking" and actual multiple sip ingestion. These differences include successive involvement of many or all of the oral chemoreceptor arrays during normal sips rather than the limited anterior fungiform population that was studied in the simulation, a smaller removal liquid volume and probably lower flow rate during swallowing than during the distilled water or artificial saliva flow of the simulation, and a greater opportunity for ingested liquid to be sorbed and/or mechanically trapped during natural drinking. Consequently, absence of adapation during alternating flows of a stimulus liquid and a lower concentration removal liquid does not equivocally demonstrate that taste adaptation will be minimal if not absent during normal drinking.

Each of the differences between the simulated drinking study and natural drinking could be explored separately to assess their importance. A more direct approach would be to measure adaptation during natural drinking. An initial attempt toward this has been made. First a preliminary study was done to obtain more data on the temporal pattern of normal drinking. For this study, six subjects were provided with 174 ml of cherry Kool-Aid in a drinking glass, plus a beaker with additional liquid, and were asked to drink as much cherry Kool-Aid as they wanted. Mouth-to-liquid contact measures, upper lip EMG, swallowing, etc., were recorded, using techniques similar to those from which Figures 3–6 and Table IV were obtained. It was found that contact duration was ~ 1,300 ± 611 ms (median, ± standard error of the median) during the first sip, and 2,000–2,500 ms during later sips. Median intervals between the start of sip contacts were ~ 1,800 ms between the 1st and 2nd sips; 3,100 ms between the 2nd and 3rd, 4,100 ms between the 3rd and 4th, and 9,200 ms between the 4th and 5th (J. Stremlin and B. P. Halpern, unpublished observations); standard errors of the median were 20–50% of these durations. A linear equation was fitted to the intersip contact times: $t = 2.1x$ -

0.8, where t is the time interval in seconds between two sips and x is the ordinal position of the first member of a pair of sips. Thus, this equation would call for an interval of 1.3 s between the 1st and 2nd sips of a series; 18.1 s between the 9th and 10th sips.

The main experiment, itself preliminary in nature, used this intersip time series. Seven subjects who had passed a screening test for reliability of taste judgments (Lester and Halpern, 1979) were presented with an array of eight drinking glasses, each containing 174 ml of a transparent liquid. After two pretesting trials, during which a modulus was assigned to a liquid in a separate glass, subjects were seated before the eight-glass array and asked to take one sip from the first glass upon hearing a tone; one sip from the second glass upon hearing the tone again, and so on for all eight glasses. They were also told that they were to make a magnitude estimate of total taste intensity of each sip when asked to do so. Measures of sip contact time, volume, and duration, as well as time of swallows, but *not* EMG, were made. All drinking glasses and the two glasses used to assign the magnitude estimate modulus contained an aqueous solution of 180 mM sucrose in 10 mM citric acid, prepared just before each experiment. Tones were prerecorded. The time series of intertone (and therefore intended intersip) intervals was 9.7, 11.8, 13.9, 16, 18.1 19.2, and 22.3 s. (In each case, the interval is the one the equation would call for later in the sequence. However, it was believed that the four shorter intervals called for by the equation, 1.3, 3.4, 5.4, and 7.6 s, were too brief to permit a sip, a magnitude estimate, smooth return of the glass to the table, and smooth drinking from the next glass.) Magnitude estimates were requested 500 ms after the first swallow of each sip. Three eight-sip sessions were done for each subject, with the first considered a practice session. Data from the second session are in Figure 7. We found (D. Barsky and B. P. Halpern, unpublished observations) that magnitude was highest during the first sip, and declined during later sips (Figure 7). The data were fit by an exponential function: $\% \; Max = 89.765e^{-0.014T}$, where % *Max* is percent of maximum judged total taste magnitude and T is time, in seconds, from the first sip. An equation of similar form was fit by Gent (1979) to taste adaptation data. Our preliminary data suggest that considerable taste adaptation may occur during a series of sips. However, even after some 111 s of drinking, judged intensity is still ~ 17% of the initial maximum.

Before this observation of partial adapation during natural drinking can be accepted with confidence, a number of control procedures are required. Assuming that it is essentially correct, its generality must then be tested, its parameters established, and the crucial differences between this procedure

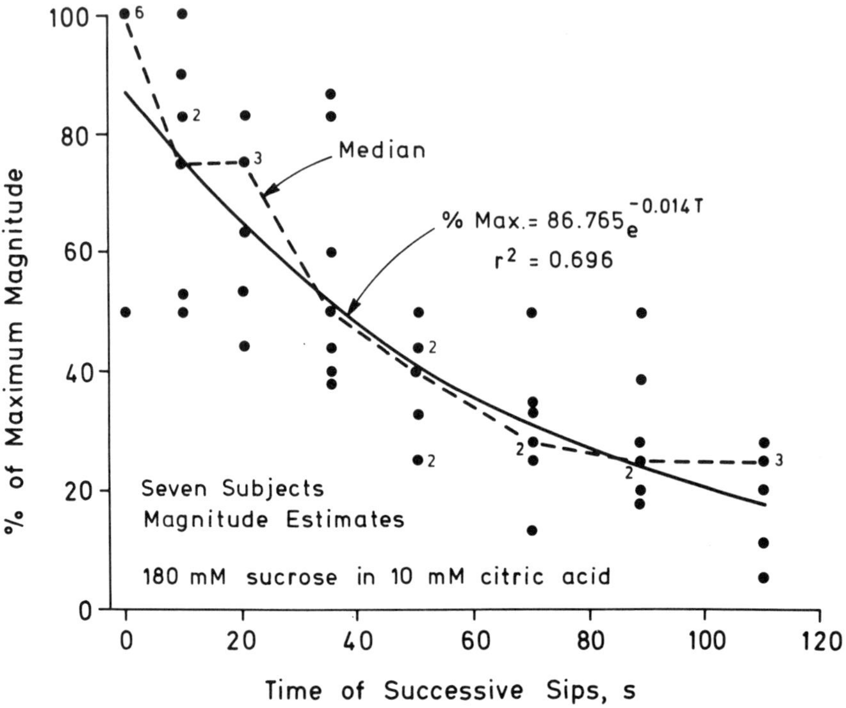

Figure 7. Taste intensity during multiple sip drinking. Seven subjects each took eight successive sips from eight separate drinking glasses, each glass containing 174 ml of clear liquid. All drinking glasses, including the standard solution glasses (see below), contained 180 mM sucrose in 10 mM citric acid. Before the multiple sip drinking of each session began, subjects were presented with a glass containing 174 ml of liquid, and were told that they would be judging and writing down the total taste intensity of the liquid in the drinking glass, that the liquid was a standard solution, that the taste intensity of the liquid in the glass had been assigned an intensity of 10, and then asked to sip the liquid, hold the sip until they felt that they could determine the taste intensity, and then swallow the liquid. After 2 min, this procedure was repeated. Next the multiple sip procedure was explained. "You will hear a tone that will signal you to sip from the next glass in sequence, to the left of the previous glass you sipped from . . . Sip as you normally do, do not allow the liquid to linger in your mouth an abnormally long period of time. After I have seen that you have swallowed, I will say *Now*, signaling you to make and write down your magnitude estimate of the total taste intensity of the liquid you sipped. Remember to compare it to the standard solution of 10." Tones occurred at 0, 9.7, 21.5, 35.4, 51.4, 69.5, 88.7, and 111.0 from the beginning of the multiple sip series. Magnitude estimates were requested ("Now") 0.5 s after the first

References

Baron, D., and B. P. Halpern. 1978. Liquid lost during drinking. *Physiol. Behav.* 21:663–666.

Erickson, R. P., and F. Covey. 1980. On the singularity of taste sensations: what is a primary taste? *Physiol. Behav.* 25:527–533.

Faull, J. R., and B. P. Halpern. 1972. Taste stimuli: time course of peripheral nerve response and theoretical models. *Science* (*Wash. DC*). 178:73–75.

Gent, J. F. 1979. An exponential model for adaptation in taste. *Sensory Processes.* 3:303–316.

Gibson, J. J. 1966. *The Senses Considered as Perceptual Systems.* Boston: Houghton Mifflin Co., p. 138.

———. 1967. The mouth as an organ for laying hold on the environment. In *Symposium on Oral Sensation and Perception.* J. F. Bosma, ed. Springfield, IL: Charles C. Thomas Publisher, pp. 111–136.

Halpern, B. P. 1963. Chemical coding in taste-temporal patterns. In *Olfaction and Taste* (Proceedings of the First International Symposium). Y. Zotterman, ed. Oxford: Pergamon Press Ltd., pp. 275–284.

———. 1973. The use of vertebrate laboratory animals in research on taste. In *Methods of Animal Experimentation, Vol. IV.* W. I. Gay, ed. New York: Academic Press, Inc., pp. 225–362.

———. 1975. Temporal patterns of liquid intake and gustatory neural responses. In *Olfaction and Taste V* (Proceedings of the Fifth International Symposium). D. Denton and P. Coghland, eds. New York: Academic Press, Inc., pp. 47–52.

———. 1977. Functional anatomy of the mouth and tongue of mammals. In *Drinking Behavior: Oral Stimulation, Reinforcement, and Preference.* J. A. W. M. Weijnen and J. Mendelson, eds. New York: Plenum Publishing Corp., pp. 1–92.

———. 1983. Tasting and smelling as active, exploratory sensory processes. *Am. J. Otolaryngol.* 4:246–249.

Halpern, B. P., and H. L. Meiselman. 1980. Taste psychophysics based on a simulation of human drinking. *Chem. Sens.* 5:279–294.

Halpern, B. P., and T. L. Nichols. 1975. Human drinking: orbicularis oris EMG patterns and liquid contact durations. *Physiologist.* 18:238 (Abstr.).

Halpern, B. P., and D. N. Tapper. 1971. Taste stimuli: quality coding time. *Science* (*Wash. DC*). 171:1256–1258.

Ichioka, M., and H. Hayashi. 1974. Spatio-temporal nerve impulse patterns in rat chorda tympani fibres in correlation with four primary taste qualities. *Proc. Jpn. Acad. Sci.* 50:392–395.

Kamo-no-Chōmei. 1212. *The Ten Foot Square Hut (Hōjōki).* Japan. The translation is by the faculty and graduate students of the Osaka University Dental School, Department of Oral Physiology; it was prepared while B.P.H. was a Visiting Professor (NIH-Fogarty Senior International Fellow) in the department. For a published complete translation, see: A. L. Sadler, trans. 1972. *The Ten Foot Square Hut and Tales of the Heike: Being Two Thirteenth-Century Japanese Classics, the "Hōjōki" and Selections from the "Heike Monogatari".* Tokyo: Charles E. Tuttle Co. Inc.

Kelling, S. T., and B. P. Halpern. 1983a. Taste flashes: reaction times, intensity and quality. *Science* (*Wash. DC*). 219:412–414.

———. 1983b. Effects of stimulus duration and concentration on taste perception of $MgSO_4$ and HCl. Program of the 5th Annual Meeting of the Association for Chemoreception Sciences (Abstr.).

Lester B., and B. P. Halpern. 1979. Effect of stimulus presentation duration on gustatory reaction time. *Physiol. Behav.* 22:319–324.
Marowitz, L. A., and B. P. Halpern. 1973. The effects of environmental constraints upon licking patterns. *Physiol. Behav.* 11:259–263.
———. 1977. Gustatory neural responses of the chorda tympani to lick duration stimuli. *Chem. Sens.* 2:121–133.
Mistretta, C. M. 1972. A quantitative analysis of rat chorda tympani fiber discharge patterns. In *Olfaction and Taste IV* (Proceedings of the Fourth International Symposium). D. Schneider, ed. Stuttgart: Wissenschaftliche Verlagsgesellschaft MBH, pp. 294–300.
Nagai, T., and K. Ueda. 1981. Stochastic properties of gustatory impulse discharges in rat chorda tympani fibers. *J. Neurophysiol.* (*Bethesda*). 45:574–592.
Nowlis, G. H. and M. Frank. 1981. Quality coding in gustatory systems of rats and hamsters. In *Perception of Behavioral Chemicals.* D. M. Norris, ed. Amsterdam: Elsevier, pp. 59–80.
Ogawa, H., M. Sato, and S. Yamashita. 1973. Variability in impulse discharges in rat chorda tympani fibers in response to repeated gustatory stimulation. *Physiol. Behav.* 11:469–479.
Ogawa, H., S. Yamashita, and M. Sato. 1974. Variation in gustatory nerve fiber discharge pattern with change in stimulus concentration and quality. *J. Neurophysiol.* (*Bethesda*). 37:443–457.
Pfaffmann, C. 1978. The vertebrate phylogeny, neural code, and integrative processes of taste. In *Handbook of Perception*, vol. VIA. *Tasting and Smelling.* E. C. Carterette and M. P. Friedman, eds. New York: Academic Press, Inc., pp. 51–123.
Pritchard, T. C., and T. R. Scott. 1982. Amino acids as taste stimuli. II. Quality coding. *Brain Res.* 253:93–104.
Scott, T. R. 1974. Behavioral support for a neural taste theory. *Physiol. Behav.* 12:413–417.
———. 1981. Brain stem and forebrain involvement in the gustatory neural code. In *Brain Mechanisms of Sensation.* Y. Katsuki, R. Norgren, and M. Sato, eds. New York: John Wiley & Sons Inc., pp. 177–196.
Tapper, D. N., and B. P. Halpern. 1968. Taste stimuli: a behavioral categorization. *Science* (*Wash. DC*). 161:708–710.
Yamamoto, T. 1983. Neural mechanisms of taste function. *Front. Oral Physiol.* 4:102–130.
Yamamoto, T., T. Kato, R. Matsuo, N. Araie, S. Azuma, and Y. Kawamura. 1982b. Gustatory reaction time under variable stimulus parameters in human adults. *Physiol. Behav.* 29:79–84.
Yamamoto, T., and Y. Kawamura. 1981. Gustatory reaction time in human adults. *Physiol. Behav.* 26:715–719.
Yamamoto, T., R. Matsuo, T. Fujiwara, and Y. Kawamura. 1982a. EMG activities of masticatory muscles during licking in rats. *Physiol. Behav.* 29:905–913.
Yamamoto, T., N. Yuyama, and Y. Kawamura. 1981. Central processing of taste perception. In *Brain Mechanisms of Sensation.* Y. Katsuki, R. Norgren, and M. Sato, eds. New York: John Wiley & Sons, pp. 197–207.
Zotterman, Y. 1967. The neural mechanism of taste. *Prog. Brain Res.* 23:139–154.

9. Taste Mixtures: An Analysis of Synthesis

In the following discussion, we plan to consider taste mixtures within these two contexts: (1) the search for psychophysical laws (motivated by the fact that our normal taste world is one of mixtures rather than simple tastes) and (2) the importance of the nature of mixture phenomena to the classification of taste as analytic or synthetic.

Kiesow vs. Öhrwall

Those who cannot remember the past are condemned to repeat it. (Santayana)

The modern controversies over taste mixtures are reminiscent of a battle between two giants of the nineteenth-century taste world: Friedrich Kiesow and Hjalmar Öhrwall. Kiesow did his Ph.D. research with Wundt, who established, in Leipzig, the first laboratory of experimental psychology. Öhrwall studied with the color physiologist Holmgren. Both Wundt and Holmgren studied with Hermann Helmholtz (responsible for distinguishing between modality and quality) who had in turn studied with Johannes Müller, the father of the Doctrine of Specific Nerve Energies. Kiesow and Öhrwall through their academic lineages were obviously in the mainline tradition of sensory science in the late nineteenth century.

Both of these men accepted the same basic tastes: sweet, salty, sour, and bitter. Contenders like alkaline and metallic were ultimately eliminated as actually stimulating more (e.g., touch) than the sense of taste (von Skramlik, 1926). However, Kiesow and Öhrwall did not see eye to eye about the relationships among the four tastes. Both invoked the teaching of Helmholtz that stimuli from different modalities cannot be ordered along a continuum, whereas qualities within a modality can be so ordered. However, they differed on how to apply these teachings to taste.

L. M. Bartoshuk and J. F. Gent
John B. Pierce Foundation Laboratory, New Haven, Connecticut 06519; and The University of Connecticut, Farmington, Connecticut 06032

Öhrwall (1891): Taste Is Analytic

Öhrwall wrote:

> I support the view that there are no different kinds of the four taste sensations, sweet, sour, salty, and bitter. If this view is right then it is obvious that there is even less a continuous transition from one of these kinds of tastes to any of the others through a series of qualitatively different sensations as is the case with different colors or with tones of different frequencies. In this respect taste sensations differ essentially from visual and auditory sensations.
>
> Perhaps one can express this difference of the taste sensations with visual sensations by saying that the spectrum of the taste sense is discontinuous, consisting of a few widely separated lines which cannot be put into a determined order (perhaps even owing to defects in transitions). The simple taste sensations cannot, as colors can, be mixed into new sensations which cannot be analyzed into their simple constituents. If one mixed with one another two or more colors in the series which consists of a ring of the spectrum through purple, one obtains, as is known, either white or a color lying between the colors of the components plus white and it is impossible to analyze this mixture with the eye into its simple constituents. However, a mixture of salty and sour, for example, will not produce bitter or a new taste sensation (assuming, naturally, that a new chemical combination does not result from the mixture). When one mixes, a new taste never results and still less can one maintain something that corrresponds to white—that quality is missing. The mixture tastes either salty as well as sour or only salty or sour, etc. A solution, for example, of tartaric acid, quinine sulphate, sugar and NaCl in appropriate proportions tastes simultaneously sweet, sour, salty, and bitter.

Öhrwall went on to dispute earlier, somewhat anecdotal accounts of taste contrast (tasting one substance was said to enhance the taste of another substance in a manner analogous to colored after-images) and compensation (tastes were said to cancel one another in mixtures, analogous to the mixing of opponent colors).

> If there were, as has often been declared, such contrast and compensation phenomena among the various taste categories in the same way as between the various colors, then this would indicate that nearby tastes are combined with one another and would form an important argument against considering them as belonging to different senses.

Responding to this, Kiesow set out to search for examples of contrast and compensation that could not be dismissed as easily as the older observations cited by Öhrwall. He produced the first systematic studies of these phenomena in taste.

Kiesow (1894, 1896): Taste Is Synthetic

Contrast. Kiesow stimulated localized tongue areas either with a dropper or with a camel's-hair brush. For "successive contrast" experiments he used the front of the tongue and for "simultaneous contrast" experiments he used the tongue edges (stimulating homologous areas). Kiesow found examples of both kinds of contrast. In his successive contrast design, the pairs NaCl-sucrose, HCl-sucrose, and NaCl-HCl mutually contrasted with one another, but other pairs failed to do so. In the simultaneous contrast design, the NaCl-HCl pair did not show impressive contrast, but the other pairs that mutually contrasted with one another in the successive design, did so in the simultaneous design as well (1894).

Compensation. Kiesow compared mixed to unmixed solutions to assess changes in intensity and quality. Note that he (like all those who have followed him in mixture research) treated taste "mixing" like "adding" taste solutes. For example, a mixture of 1.0 M sucrose (342 g of sucrose in water to make 1 liter of solution) and 1.0 M NaCl (54 g of NaCl in water to make 1 liter of solution) is made by dissolving 342 g of sucrose and 54 g of NaCl in water to make 1 liter of solution. Kiesow searched diligently for cases in which two taste qualities would cancel one another in a mixture because he believed that such cases would be analogous to the cancellation of color produced by mixing opponent colors. He found one such case: a mixture of weak NaCl and sucrose. The mixture tasted neither salty nor sweet, but not tasteless either. Kiesow called it *fade*, which translates to "flat," and he considered it to be a quality distinct from salty and sweet.

In other mixtures, although cancellation of both qualities did not occur, both qualities were usually reduced in intensity. The component qualities were recognizable but Kiesow believed that they fused (*verschmeldet*) into a new quality which, for lack of better terminology, was named after the simple tastes to which it was similar (e.g., bittersweet).

Öhrwall (1901): Taste Is Still Analytic

Öhrwall rejected Kiesow's contrast and compensation in what may be the longest footnote in the taste literature (more than 750 words, Öhrwall, 1901, pp. 257–259). He noted that Kiesow's contrast demonstrations were weak, showed many individual differences, and the contrast was not always in the correct quality (e.g., HCl made NaCl sweeter). Öhrwall concluded that Kiesow's observations could not be considered truly analogous to color contrast.

Öhrwall's footnote went on to dismiss compensation as a taste phenomenon on the grounds that one hard-to-find case of mutual cancellation of qualities (i.e., weak NaCl and sucrose) in a mixture hardly qualified as evidence that taste qualities mix in the same way that colors do. Further, he noted that the loss of both weak qualities could be viewed as simply the reduction in intensity of both and not really as the production of a new quality Kiesow's general observation that qualities typically were reduced in intensity in mixtures was dismissed by Öhrwall as not tied specifically enough to taste. He argued that sensitivity can be reduced in a variety of ways in sensory processes (e.g., even across modalities). Öhrwall did not comment on Kiesow's contention that taste qualities fuse into a new quality in mixtures even though the components are still recognizable.

Search for Mixture Laws

What Happened to "Contrast" and "Compensation?"

"Successive contrast" has disappeared from the taste vocabulary because we now think of the phenomenon as "adaptation." When the tongue is exposed to one solution (i.e., adaptation) and then a second solution is tasted, we say that cross-enhancement occurs if the second solution is stronger than normal or that cross-adaptation occurs if the second solution is weaker than normal. According to this terminology, Kiesow found cross-enhancement. However, modern work on the "water taste" suggests a different interpretation of his data (McBurney and Bartoshuk, 1972, 1973). We now know that adaptation to a variety of substances makes water take on a taste. For example, adaptation to NaCl makes water taste sour or bitter (Bartoshuk et al., 1964); adaptation to urea makes water taste salty (McBurney, 1969); and adaptation to acids tends to make water taste sweet (Bartoshuk, 1968; McBurney, 1969). The apparent cross-enhancement in Kiesow's study was probably complicated by water taste artifacts. For example, Kiesow noted that if one first tastes an acid, a sugar solution tastes sweeter than normal. However, the adaptation to sour may actually be causing the water in which the sugar is dissolved to taste sweet. The "enhanced" sweet taste of the sugar solution is actually made up of the sum of the sweetnesses of the sugar and the water solvent.

Simultaneous contrast differs considerably from successive contrast. The two halves of the tongue project ipsilaterlly until the thalamus (Norgren and Leonard, 1973), so neural responses to taste stimuli on opposite edges of the tongue cannot interact until they reach that structure in the central nervous system. As long as taste stimuli are kept separate on the tongue, changes in their tastes can be attributed to the central nervous system. Thus the peripheral separation of the tongue halves allows one to do "central mixture experiments." That is, if one taste solution affects the taste of another on the other side of the tongue, then we can conclude that the interaction occurred in the CNS. It is important to note that the reverse is not true. Failure to find suppression when stimuli are spatially separated does not mean that mixture suppression does not have a central locus. Kiesow's early studies on simultaneous contrast are comparable to more recent studies of central "mixtures."

"Compensation," like "contrast," has disappeared from our taste vocabulary. In the years after Kiesow's work, those studying taste mixtures adopted a more empirical, less theoretical orientation. The emphasis was on a search for laws that would permit the prediction of the tastes of mixtures from the tastes of components rather than on the implications of the nature of taste mixtures for our view of taste qualities.

Post-Kiesow Mixture Research: Empirical Search for Mixture Laws

Kiesow's 1896 study set the stage for a variety of experiments to follow, many coming from psychophysicists working within food technology. The focus was often practical; can an undesirable taste be removed by adding something else? In addition, mixtures of substances with similar tastes, primarily sweet substances, became very important as the developing food industry sought to learn how to predict the effects of sweetener substitutions. However, as studies proliferated, instead of getting closure on the problems, contradictions multiplied across studies that all appeared to be competently done. In the discussion that follows, we will try to highlight some of the ideas that now help organize these early data.

Mixtures of substances with similar tastes. The substitution of one sweetener for another is commercially important as well as interesting in its own right. One of the most important early treatments of the problem was a monograph by Cameron (1947).

Cameron knew that the sweetness of a mixture of sweeteners was not the simple sum of the sweetnesses of the component substances. In fact, the sweetness of a mixture was often greater than that sum, leading to the idea that certain sweeteners can "synergize" with one another. Cameron's rule for predicting the sweetness of a mixture of two different sweeteners rested on expressing each of the two in terms of a standard sweetener, glucose. He did this by empirically matching a standard glucose series to a variety of other sweeteners. Once he could express any sweet substance at any concentration in terms of its "equivalent" glucose concentration (i.e., the one that was equally sweet), then he could predict the sweetness of a mixture by adding the glucose equivalents. His prediction: that the mixture would have the sweetness of the glucose sum. This method was not perfect but it was remarkably accurate under some circumstances.

Neurophysiological data and modern psychophysical data now show what Cameron really did and why it worked. In Cameron's day, psychophysicists had to rely on matching studies to evaluate suprathreshold concentrations. Thus they could find the concentrations of two substances that produced equivalent perceived intensities but they had no way of knowing the mathematical form of the function that relates concentration to perceived intensity. Pfaffmann and Hagstrom (Pfaffmann, 1959) measured this function from the chorda tympani of the rat. They found that the neural response was not always a linear function of concentration. In more recent years, the direct scaling methods developed by S. S. Stevens (1969) and his colleagues have produced psychophysical functions describing the growth of perceived sweetness with concentration by simply asking subjects to estimate the magnitudes of their sensations.

As Pfaffmann argued (1959), once we can actually see the underlying functions for sweetness, we can analyze the problem Cameron faced. Consider how the sweetness of a substance adds to itself. This can be shown by looking at the properties of psychophysical functions. Figure 1 shows three psychophysical functions. Consider function B. Concentration 1.0 produces a sweetness of 1.0. If we double that concentration (2.0), the perceived sweetness will double as well. In this case the sweetness of the "mixture" would be the simple sum of the sweetnesses of the components. However, if we depart from the linear function (B) and do the same thing with functions A and C, the results will be very different. For function A, doubling concentration less than doubles the perceived intensity and for function C, more than doubles the perceived intensity. Power functions with exponents < 1, like A, are said to show compression. Those with exponents > 1, like C, are said to show expansion.

These functions explain why some sweet mixtures seem to show "synergism." Suppose two sweet substances have functions like C in Figure 1. The sum of concentration 1.0 of each would produce a sweetness of 2.8. The simple sum of the sweetnesses of the components would be 2.0. Thus apparent synergism is actually additivity along a function like that shown by C. Additivity along A would seem to produce "suppression" when compared with simple additivity.

If all sweeteners really stimulated a single kind of receptor site so that they all had the same underlying sweetness function, then we could easily predict addition from the shape of that function. This is essentially what Cameron was trying to do. By expressing all sweeteners in terms of glucose equivalents, then adding the equivalents and noting the sweetness that that concentration of glucose would have, Cameron was acting as if all the sweeteners had a common underlying sweetness function (that of glucose) and he added them up along that function.

Of course, we know now that all sweeteners do not have the same functions (Stevens, 1969; Moskowitz, 1973) but the functions of many are very similar. The differences in functions may result because all sweets do not seem to stimulate a common receptor site (Andersson et al., 1950; Pfaffmann, 1969; Pfaffmann et al., 1976; McBurney, 1972; Schiffman et al., 1979; Van der Wel, 1972; Van der Wel and Arvidson, 1978; Faurion et al., 1980; Schiffman et al., 1981; Jakinovich, 1982; Gent and Bartoshuk, 1983; Lawless and Stevens, 1983), or because some sweets have taste qualities other than sweet that alter the sweetness function through the kind of mixture interactions to be discussed below, or because different sweets may have identical "sweet units" on the molecule but have properties that change the availability of those sweet units (e.g., cooling diminishes the sweetness of relatively dilute sucrose; see Bartoshuk et al., 1982). In any case, the variation in the shapes of psychophysical functions for sweeteners means that the simple additivity rule of Cameron should have failed at least part of the time as, indeed, it did. What is the perfect solution? It depends on our determining how sweets add when they are stimulating different receptor sites, and determining how to correct for the presence of nonsweet qualities (in other words, how to predict the mixture interactions between sweet and nonsweet tastes).

Incidentally, the psychophysical functions for the sweetness of sugars

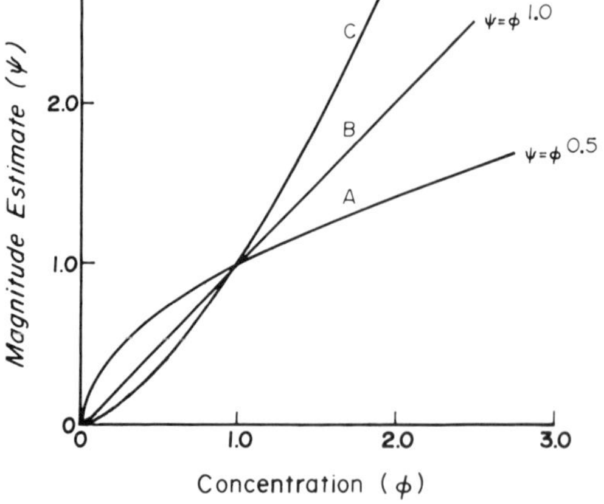

Figure 1. Three hypothetical psychophysical functions (power functions) plotted in linear coordinates. Function A illustrates compression (successive concentration increments are less and less intense). Function C illustrates expansion (successive concentration increments are more and more intense.

change shape if the method of stimulation varies (Meiselman, 1971; Bartoshuk and Cleveland, 1977). This allows us to demonstrate the dependence of sweetness mixing on the shape of the psychophysical functions of the components. Figures 2 and 3 show two mixture experiments. The stimuli were the same in both experiments but the method of stimulation was not. The data in Figure 2 were collected by flowing the sugar solutions (warmed to body temperature) over the tongue. For the data in Figure 3, the subjects sipped sugar solutions (at room temperature) and then spit them into a sink. Both experiments were designed to permit direct comparisons between the psychophysical functions of the unmixed components and a mixture function constructed in the following way. Concentrations of the sugars were selected to produce approximately equal intensities. These four sugar solutions were then used to construct the six possible two-component mixtures, the four possible three-component mixtures, and the one possible four-component mixture. The average responses to the mixtures of one, two, three, and four components could then be compared with responses to the unmixed components and with two, three, and four times these original concentrations. For both Figures 2 and 3, the psychophysical functions for the individual sugars (shown by the left four functions) look very much like the mixture function (shown on the

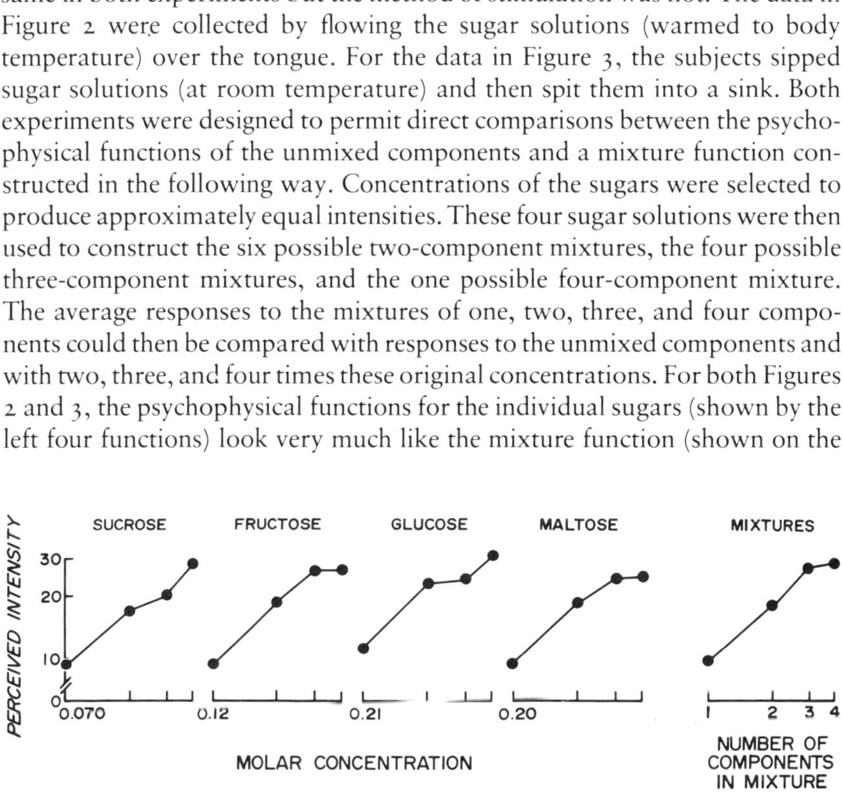

Figure 2. The four functions on the left show the perceived intensity (measured by magnitude estimation) of four sugars as a function of concentration. Stimuli (warmed to body temperature) were flowed over the extended tongue. The label at the left of the abscissa for each function gives the molar concentration of the weakest stimulus; the other stimuli were two, three, and four times as concentrated. The function on the right gives the average perceived intensity of the four unmixed components, the six two-component mixtures, the four three-component mixtures, and the one four-component mixture. The slopes of the best-fitting straight lines through these functions are: sucrose, 0.75; fructose, 0.78; glucose, 0.69; maltose, 0.70; and mixtures, 0.80 (from Bartoshuk and Cleveland, 1977).

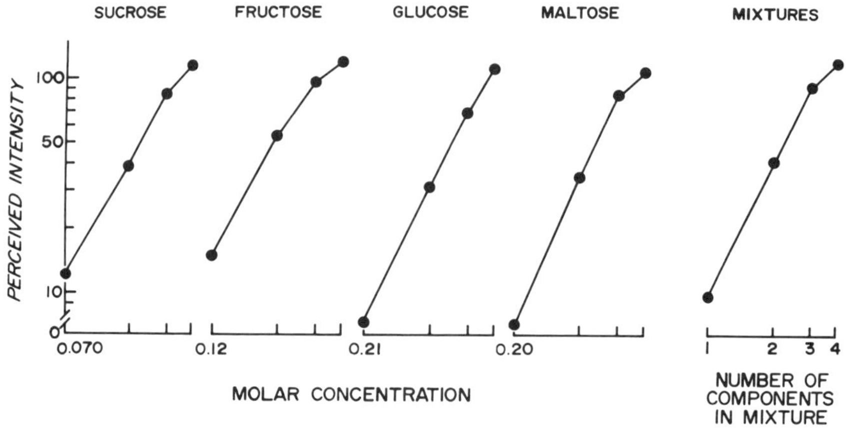

Figure 3. The stimuli were the same as for Figure 2 except that they were at room temperature and were tasted with the sip and spit procedure. The slopes are: sucrose, 1.67; fructose, 1.54; glucose, 1.97; maltose, 2.01; and mixtures, 1.87 (from Bartoshuk and Cleveland, 1977).

right). This means that the sugars acted much as if they were interchangeable stimuli and added along a common function. The most striking finding is shown by comparing Figures 2 and 3. In Figure 2, the psychophysical functions have slopes < 1 (which means that they show compression) and the mixtures show less than simple additivity (e.g., the mixture of all four sugars was about 0.75 of the sum of the components). In Figure 3, the psychophysical functions have slopes > 1 (which means that they show expansion) and the mixtures show greater than simple additivity or "synergism" (e.g., the mixture of all four sugars was about three times the sum of the components).

Does true synergism ever occur in taste? The "synergism" reported in mixtures of sweeteners (Stone and Oliver, 1969; Yamaguchi et al, 1970; Moskowitz, 1974; Moskowitz and Dubose, 1977) and mixtures of acids (Moskowitz, 1974) seems to be explained by addition along an expanding function. However, there is a case of true synergism in taste: the mixture of monosodium glutamate and the ribonucleotide disodium guanylate (Rifkin and Bartoshuk, 1980). At certain concentrations, mixtures of these substances produce a taste intensity greater than the prediction obtained by adding along a common function (and, incidentally, also greater than the simple sum of the component tastes).

Mixtures of substances with different tastes. We can easily imagine two chemically different substances that are identical to the taste system because they share a structure that is an effective taste stimulus. Such stimuli would be expected to produce psychophysical functions of the same shape and they would, of course, add along that common function. However, there is no obvious reason why mixtures of substances with different tastes should be related to the shapes of their psychophysical functions. In 1975, one of us (L. M. Bartoshuk) proposed, nonetheless, that such a relation seemed to show up in a study of mixtures of NaCl, sucrose, HCl, and quinine hydrochloride (QHCl). The stimuli were tasted with a McBurney Gustometer (e.g., McBurney and Pfaffmann, 1963; Smith and McBurney, 1969; McBurney, 1972), which flows solutions that have been warmed to body temperature over the extended tongue. The substance that showed the most compression when scaled, QHCl, showed successively more suppression as additional qualities were added to it. The substance that showed the least compression, HCl, showed virtually no suppression in mixtures. By manipulating the way solutions are tasted (e.g., sip and spit vs. dorsal flow), we can change the slopes (i.e., the amount of compression) in the psychophysical functions. If the compression-suppression theory were correct, then such changes would change the amount of mixture suppression. For one mixture, sucrose and QHCl, this happened (Bartoshuk, 1978, 1979). The degree of compression predicted the degree of suppression. Unfortunately for the theory, the other mixtures were much less cooperative. For example, suppression of the bitterness of QHCl in mixtures of NaCl and QHCl was much more constant and showed no relation to the psychophysical functions for QHCl (Bartoshuk, 1979, 1980). Yet the fact that any of the mixture interactions changed as the functions changed suggests that the original question as formulated by Kiesow may have no answer. Kiesow set out to determine the laws of two-component mixtures. That is, given the qualities of the components, how will the components be altered in the mixture? Perhaps knowledge of the qualities is not sufficient. In the case of sucrose-QHCl mixtures, it seems we must know their psychophysical functions as well.

Agreements across early studies. Some impressive agreement across studies deserves comment. Several investigatores (Fabian and Blum, 1943; Beebe-Center et al., 1959; Pangborn, 1960, 1962; Indow, 1969) have found enhanced sweetness in mixtures of weak NaCl and sucrose. Kroeze (1982) showed that this was related to the sweet taste of NaCl (first noted by Renquist, 1919). Kroeze generalized this and pointed out that mixture suppression experiments are often complicated by the fact that many simple stimuli have complex tastes. Subjects add these "side tastes" into the taste mixture such that when the side taste of one component is the same as the other component, then that other component appears to be suppressed less or even enhanced.

In mixtures of sucrose and quinine (Kiesow, 1896; Indow, 1969; Bartoshuk, 1975, 1978, 1979; Lawless, 1977, 1979; Bartoshuk and Seibyl, 1982), both components are usually suppressed. As discussed above, the shapes of the psychophysical functions for the sucrose and quinine appear to predict the suppression for the components (Bartoshuk, 1978, 1979; Lawless, 1979) and the shape of the functions in turn depends on the conditions of stimulation. We suspect that the agreement across studies may reflect the fact that the methods of stimulation produced compression functions.

Acid plus sucrose is another mixture for which there is agreement and a possible explanation of the agreement. Sourness is suppressed (Kiesow, 1896; Fabian and Blum, 1943; Pangborn, 1960, 1961; Kamen et al., 1961). Kuznicki and McCutcheon (1979) stimulated single papillae and found that some of this suppression occurs at the receptor site. They proposed that this may be the result of competition between the sucrose and the acid for the H receptor sites.

The suppression of bitterness in NaCl-quinine mixtures (Kiesow, 1896; Indow, 1969; Bartoshuk, 1975) may also have a peripheral explanation (see below).

Neural Locus of Mixture Interactions: Central or Peripheral?

Psychophysical experiments can provide information relevant to the neural mechanisms underlying sensory phenomena. The work discussed below provides examples of the way in which psychophysical studies can be clarified by and in turn can aid physiological investigations.

Spatially Separated Stimuli

There are anatomical and neurophysiological data relevant to spatially separated stimuli. In small areas, individual neurons branch and can innervate several receptors. This pattern of branching varies on different tongue loci (Wang and Frank, cited by Pfaffmann, 1970; Miller, 1976) and across different species (e.g., compare Wang and Frank with Oakley, 1975). Oakley (1975) measured the distance between fungiform papillae connected to a single neuron by probing the tongue with droplets while recording from the neuron. The largest separation found was 18 mm, the median being 2 mm.

Two mechanisms for lateral interactions have been proposed. Rapuzzi and Casella (1965) demonstrated that the frog glossopharyngeal nerve fibers divide into several branches that innervate various papillae. When several papillae are stimulated, action potentials travel up the branches and summate to produce a larger response than would have resulted from stimulation of a single papilla. However, when action potentials pass the branching points, antidromic impulses fire back out to the periphery and produce inhibition at the receptors (Taglietti et al., 1969; Filin and Esakov, 1968).

The second mechanism for lateral interactions was demonstrated in the rat by Miller ([1971, 1972] cited in Beidler, 1969). He recorded while stimulating a single papilla with NaCl. Stimulating the surrounding papillae with NaCl increased the response but stimulating the surrounding papillae with K benzoate decreased the response. He suggested that the inhibition might result from spread of hyperpolarization produced by the benzoate anion.

As we discussed above, the two halves of the tongue are innervated separately until the sensory projections reach the thalamus (Norgren and Leonard, 1973). Thus interactions between areas on opposite sides of the midline must occur in the CNS.

Contrast across stimuli with different taste qualities. A variety of early and modern psychophysical studies have found evidence of interactions when the tongue was stimulated with spatially separated stimuli (Kiesow, 1894; Bujas, 1934, 1937; Collings et al., 1976; Lawless, 1978, 1980; Gillan, 1982).

Lawless (1978, 1980) was the first to use direct scaling methods to demonstrate that the two halves of the tongue must be connected via the CNS. He showed that when sucrose was placed on one side of the tongue and quinine on the other, both the sweetness and bitterness were reduced. A similar experiment with sucrose and NaCl (Gillan, 1982) did not show an interaction when the stimuli were on opposite sides of the tongue but did when both stimuli were placed on the same side of the tongue.

At first glance, there might seem to be a contradiction between Kiesow's contrast and the suppression studies. However, Kiesow did not find simultaneous contrast between sucrose and quinine, the stimuli used by Lawless. In fact, Kiesow's experimental procedure was aimed at contrast and could have missed suppression. Kiesow selected near-threshold stimuli, then varied the concentration of the inducing stimulus to see whether the near-threshold stimulus would become clearly perceptible. Failure to observe contrast could reflect either no interaction or the presence of suppression.

NaCl and sucrose did produce contrast. Kiesow's most impressive example of contrast was the enhancement of sweetness induced by NaCl. Bujas (1934, 1937) concentrated on the reverse effect: sucrose as the inducing and NaCl as the test stimulus. He found that relatively weak sucrose lowered the threshold for NaCl but stronger sucrose elevated it to above the normal value. Perhaps the sweet taste of dilute NaCl (Renqvist, 1919) plays a role. Suppose some neurons that are predominantly sensitive to sucrose were to converge in such a way that neurons from homologous areas could pool onto CNS neurons. The sweetness of NaCl appears to be mediated by the same receptor sites as sucrose (Bartoshuk et al., 1978). When Kiesow put NaCl on one tongue edge and sucrose on the other, the response produced by NaCl in the fibers responsive to sucrose would summate with the response produced by the sucrose in the CNS. This would intensify the near-threshold sucrose in Kiesow's experiment (and also produce a water taste, as Kiesow observed). In Bujas's experiment (1937), the sucrose-inducing stimulus would summate with the sweet taste of dilute NaCl to lower the detection threshold for NaCl. Note, however, that in Gillan's experiment the inducing and test stimuli were quite strong (1.2 M sucrose, 0.7 M NaCl). Thus a small amount of sweet input from NaCl would not be expected to produce a noticeable increase in sweetness.

Incidentally, the task of the subject in Kiesow and Bujas's experiments is critical to the interpretation above. Kiesow noted anecdotally whether or not sweetness intensified. In Bujas's threshold studies (using the method of constant stimuli) subjects tasted a variety of concentrations and noted whether or not they produced a taste sensation. In each case, the response of the CNS neurons receiving the convergent sweet messages would be increased with NaCl on one tongue edge and sucrose on the other. However, there are other threshold procedures in which convergent input would actually impair threshold performance. Consider the modern two-alternative, forced-choice procedure (McBurney and Collings, 1977), in which the subject must choose between water and a weak taste solution. When the inducing stimulus imparts a taste to both water and test solution, the discrimination would become more difficult if the induced quality and the test stimulus quality were the same.

Obviously, the convergence described above is speculative. We do not yet know the nature of the CNS interactions between the two halves of the tongue. Our point here is to cast doubt especially on the simultaneous "contrast" phenomenon. We have suggested one way in which the apparent contrast between different qualities observed by Kiesow and Bujas might be explained as simple summation of common qualities. Öhrwall (1901) criticized Kiesow's data (in part because of across-subject variability), and McBurney (cited in Pfaffmann et al., 1971) was unable to replicate Bujas's demonstration of enhancement of NaCl induced by sucrose.

Suppression across stimuli with different taste qualities. We have already noted Lawless's important observation that suprathreshold sucrose and quinine suppress one another when the stimuli are on opposite sides of the tongue. In addition to this, Gillan (1982) showed mutual suppression between NaCl and sucrose when they were on the same side of the tongue but separated. This latter result could have a peripheral (Oakley's data cited above show branching over a distance as large as 18 mm in the cat and Gillan's stimuli were separated by 1–3 mm) and/or a central explanation. Interestingly enough, Collings et al. (1976) did not find interactions with stimuli that were similarly placed. However, the tasks of the subjects were very different: Gillan's (1982) subjects estimated the perceived intensities of suprathreshold concentrations placed on the two areas, whereas Collings et al. (1976) used a suprathreshold-inducing stimulus and measured the recognition thresholds for the test stimulus.

Suppression across stimuli with the same taste qualities. Bujas (1937) discovered that NaCl thresholds were elevated above normal if he adapted the other side of the tongue to NaCl before obtaining the thresholds. This raises interesting questions about what might be happening at CNS neurons during the adaptation. Bujas described the inducing area as *geschmacksfrei* (taste free) as a result of the adaptation. If adaptation is peripheral (as suggested by the work of Zotterman and his colleagues, e.g., Borg et al., 1967) and Bujas's adaptation were total, then one would expect that CNS neurons would be free of the input from the inducing stimulus and able to respond normally to the test stimulus. However, if any neural response induced by the NaCl adaptation were still present, we would expect that the threshold in the test area would go up because the threshold stimulus would have to represent a just noticeable difference above the background.

Collings et al. (1976) also tested for spatial interactions with the same taste stimulus on two areas (both on the same side of the tongue). They found that for NaCl and sucrose, thresholds were elevated, but for QHCl, thresholds decreased.

Obviously, there are still puzzles to be solved. However, there is ample evidence that spatial interactions occur within areas likely to be innervated by the same neurons and across areas known not to have an anatomical possibility of convergence until the CNS.

Adaptation Experiments

Release of mixture suppression. Adaptation experiments also provide insights into the locus of the mechanisms for mixture interactions. The logic is as follows. Taste adaptation appears to be largely peripheral (Borg et al., 1967). Thus after adaptation is complete, relatively little neural activity goes up the taste nerves to the CNS. If mixture suppression requires activity in the CNS from both components (i.e., the suppression mechanism is central to adaptation), then adaptation will prevent the suppression. On the other hand, if mixture suppression is peripheral to the locus of adaptation, then adaptation will have no effect on it. There are several studies that provide these data.

In the study discussed above (Bartoshuk, 1975), the mixtures were tasted after water adaptation and also after adaptation to each of the components used to construct the mixtures (with the exception of HCl which was too irritating to keep on the tongue throughout the adaptation period). In the two-component mixtures, adapting to one component tended to release the suppression of the other component; however, there was one notable exception. Adaptation to NaCl did not release the suppression of QHCl in the NaCl-QHCl mixture (Bartoshuk, 1979, 1980; Bartoshuk and Seibyl, 1982). Lawless (1978, 1980) demonstrated release of suppression in sucrose-QHCl mixtures both by adapting to sucrose and by reducing the sweetness of sucrose with *Gymnema sylvestre*, an Indian herb. He also got mixture suppression and release through adaptation with a mixture of quinine and an artificial sweetener (L-aspartyl pehylalanyl methyl ester). Gillan (1982) found release of suppression after adaptation for NaCl-sucrose mixtures.

Adaptation to mixtures. Adapting to mixtures and then testing the components also provides information about the locus of mixture suppression. If adaptation to the mixture fails (i.e., if after adaptation to the mixture either or both components taste stronger than after adaptation to themselves, then we have reason to believe that the mixture suppression is at least partly peripheral to the adaptation.

A variety of mixture adaptation studies have now been done, and we can examine these results along with the adaptation results above. The first results (Kroeze, 1978, 1979) showed that NaCl-sucrose mixtures cross-adapted the components as effectively as the components self-adapted. This implies a central locus for the mixture suppression just as the release of suppression results did. On the other hand, Lawless (1982b) found that adaptation to sucrose-quinine mixtures was less effective than self-adaptation. This supports a peripheral locus for the mixture suppression, whereas the release from suppression results discussed above supported a central locus. As it turns out (Lawless, 1982a), this kind of apparent contradiction relates in part to procedural variation across studies. Lawless showed that NaCl-sucrose mixtures cross-adapted sweetness to varying degrees depending on the method of stimulation. Thus again we find that in order to completely describe taste mixtures, we must know more than the component qualities; we must know the stimulation conditions as well.

Another point is worth noting when interpreting these adaptation experiments. Certain designs may be more sensitive than others. For example, in the mixture adaptation design, the subject adapts to the mixture and then judges one of the components. If cross-adaptation is successful, then the componenet is rendered nearly tasteless. On the other hand, if cross-adaptation fails even partially, the component will have a perceptible taste. Thus we compare a small or absent sensation with a definite sensation. In the release of mixture suppression design, the subject adapts to a component and then tastes the mixture. If the adaptation to one component is successful, it will essentially remove its own taste from the mixture and the subject will estimate the perceived intensity of the other component. We compare the perceived intensity of that component to the perceived intensity of the unmixed stimulus alone. This comparison is likely to be more variable than the former one. Thus we conclude that Lawless's (1982b) finding that adaptation to sucrose-quinine mixtures does not completely cross-adapt sucrose or quinine is powerful evidence in favor of a peripheral locus for part of the mixture suppression.

The psychophysical experiments discussed above have potential value beyond their role in the interpretation of mixture suppression. We are interested in their implications concerning multiple bitter-receptor mechanisms. In particular, adaptation to a QHCl-NaCl mixture fails to cross-adapt the bitterness of QHCl (Bartoshuk, 1979 and 1980; Bartoshuk and Seibyl, 1982), but adaptation to a phenylthiocarbonide (PTC)-NaCl mixture, tested only on "tasters" of PTC (Harris and Kalmus, 1949; Bartoshuk, 1979; Lawless, 1980), does cross-adapt the bitterness of PTC (Seibyl and Bartoshuk, unpublished data). This confirms the generally accepted view (Kalmus, 1958; Fischer and Griffin, 1963; McBurney et al., 1972; Hall et al., 1975) that the receptor mechanisms for PTC and QHCl are different. More importantly, it gives us another psychophysical tool for the study of the nature of bitterness. The QHCl-PTC comparison also provides another example of the pitfalls of searching for mixture laws stated only in terms of the qualities of the components.

Habituation Experiment: More Evidence for Central Mixture Suppression

Kroeze (1982) demonstrated a central locus for mixture interactions in a particularly novel way. He habituated subjects to either NaCl or sucrose by having them repeatedly taste that simulus. After a sequence of these habituation trials, he presented a NaCl-sucrose mixture. As the number of habituation trials went up, the amount of mixture suppression shown by the other component went down. Kroeze concluded "that NaCl/sucrose mixtures excite two independent neuronal taste centers which have inhibitory projections to each other."

Synthesis vs. Analysis

Can Fusion Be Demonstrated in Taste?

The old debate between Öhrwall and Kiesow is remarkably similar to the modern debate over the nature of taste quality except that with the advances in neurophysiology, the debate has also now moved into the realm of neural quality coding (Erickson, 1968, 1973, 1977, 1978, 1982). The nature of taste mixtures plays a role in the debate over taste quality. In synthetic systems, qualities fuse in such a way that a mixture tastes qualitatively different from its components. In analytic systems, qualities remain recognizable. With such a simple distinction how can there be controversy over the classification of taste? The stumbling block now, as in the days of Kiesow and Öhrwall, is the meaning of "qualitatively different." Suppose we mix quinine and sucrose. We describe the mixture as "bittersweet." The description sounds analytic since it implies recognition of both the bitter and sweet components. However, one can argue that the mixture really has a new taste quality that is similar to bitter and to sweet and that we simply lack a name for the new quality. We can draw an analogy with color vision. A mixture of red and yellow is orange. If we had no name for orange, we could describe it as red-yellow meaning that it is qualitatively similar to both.

Readers can come to their own conclusions about whether or not taste qualities fuse as colors do. We believe that they do not but we do not know of a convincing experimental way to discriminate between the analytic and synthetic positions. In particular, we believe that experiments concluding that taste mixtures show synthesis (Erickson, 1982; Erickson and Covey, 1980) are consistent with the analytic point of view as well. "It is shown that mixtures are often judged to be as singular as their separate components, indicating synthesis of new tastes different from the components, and thus the existence of more than the few primary tastes" (Erickson and Covey, 1980). Subjects in this study tasted single compounds (sucrose, fructose, quinine sulfate, $MgCl_2$, NH_4Cl, K_2SO_4, KCl, Na acetate, NaCl, HCl, and HNO_3) and two component mixtures constructed from these and responded "singular" or "more than one." A number of the single compounds (KCl, K_2SO_4, NH_4Cl, $MgCl_2$) were judged "more than one" on about half the trials and a number of mixtures (e.g., NaCl + NH_4Cl, NaCl + HCl, fructose + sucrose, HCl + quinine sulfate) were judged "singular" most of the time. To explain these results from the analytic point of view, we simply note that the mixtures producing the greatest percentages of "singular" judgments tended to have components that taste similar to one another (e.g., NaCl + NH_4Cl). Further, mixture suppression is a powerful phenomenon in taste. As discussed above, when two substances are mixed, even if their intensities are prematched (and those of the stimuli in the study of Erickson and Covey were not), one component is often more suppressed than the other. Thus in many mixtures only one taste quality is perceived. Finally, note that a subject could have perceived both components and still have responded "singular" if that subject believed in fusion as Kiesow did.

Do Qualities Ever Change in Mixtures?

Although most mixture research shows that subjects identify taste qualities in mixtures accurately, one can still ask if there are subtle changes in those qualities. Moskowitz (1973) noted that subjects described a qualitative shift in the sweetness of sugars when a second taste was present. Does this mean that a kind of subtle fusion goes on? Kuznicki and his colleagues have examined this possibility in a series of studies (Kuznicki and Ashbaugh, 1979, 1982; Kuznicki et al., 1983). In one (Kuznicki and Ashbaugh, 1982) they studied the sweetness of sucrose when mixed with NaCl or quinine sulfate. Subjects judged the sweetness of a mixture of NaCl and sucrose to be qualitatively different from the sweetness of a mixture of quinine and sucrose. When the mixture components were placed on opposite sides of the tongue, the qualitative difference decreased and when they were separated in time as well, the qualitative difference disappeared. Kuznicki and Ashbaugh suggested that "space-time coherence of taste sensations result in perceptual blending and a consequent failure of selective attention to any single component" and that "taste mixtures can be considered as single tastes even though component qualities can be reliably identified using other techniques."

The analysis by Kuznicki and his colleagues, using the tools of modern cognitive psychology (e.g., Kuznicki et al., 1983), are provocative and may help us understand the limits of our own introspections into the nature of taste mixtures.

References

Andersson, B., S. Landgren, L. Olsson, and Y. Zotterman. 1950. The sweet taste fibres of the dogs. *Acta Pysiol. Scand.* 21:105–119.

Bartoshuk, L. M. 1968. Water taste in man. *Percept. Psychophys.* 3:69–72.

———. 1975. Taste mixtures: is mixture suppression, related to compression? *Physiol. Behav.* 14:643–649.

———. 1978. The psychophysics of taste. *Am. J. Clin. Nutr.* 31:1068–1077.

———. 1979. Taste interactions in mixtures of NaCl with QHCl and sucrose with QHCl. Poster presented at the ninth Annual Meeting of the Society for Neuroscience, Atlanta, Georgia.

———. 1980. Sensory analysis of the taste of NaCl. In *Biological and Behavioral Aspects of Salt Intake.* M. R. Kare, M. J. Fregley, and R. A. Bernard, eds. New York: Academic Press, Inc., pp. 83–98.

Bartoshuk, L. M., and C. T. Cleveland. 1977. Mixtures of substances with similar tastes: a test of a new model of taste mixture interactions. *Sensory Processes.* 1:177–186.

Bartoshuk, L. M., and J. P. Seibyl. 1982. Suppression of bitterness of QHCl in mixtures: possible mechanisms. Poster presented at the fourth Annual Meeting of the Association for Chemoreception Sciences, Sarasota, Florida.

Bartoshuk, L. M., D. H. McBurney, and C. Pfaffmann. 1964. Taste of sodium chloride solutions after adaptation to sodium chloride: implications for the "water taste." *Science (Wash. DC).* 143:967–968.

Bartoshuk, L. M., C. Murphy, and C. T. Cleveland. 1978. Sweet taste of dilute NaCl: psychophysical evidence for a sweet stimulus. *Physiol. Behav.* 21:609–613.

Bartoshuk, L. M., K. Rennert, J. Rodin, and J. C. Stevens. 1982. Effects of temperature on the perceived sweetness of sucrose. *Physiol. Behav.* 28:905–910.

Beebe-Center, J. G., M. S. Rogers, W. H. Atkinson, and D. N. O'Connell. 1959. Sweetness and saltiness of compound solutions of sucrose and NaCl as a function of concentration of solutes. *J. Exp. Psychol.* 4:231–234.

Beidler, L. M. 1969. Innervation of rat fungiform papilla. In *Olfaction and Taste III* (Proceedings of the Third International Symposium). C. Pfaffmann, ed. New York: The Rockefeller University Press, pp. 352–369.

Borg, G., H. Diamant, B. Oakley, L. Ström, and Y. Zotterman. 1967. A comparative study of neural and psychophysical responses to gustatory stimuli. In *Olfaction and Taste II* (Proceedings of the Second International Symposium). T. Hayashi, ed. Oxford: Pergamon Press Ltd., pp. 253–264.

Bujas, Z. 1934. Quelques remarques sur le contraste et l'inhibition à la suite d'excitations gustatives simultanées. *C. R. Séances Soc. Biol.* 116:1304–1306.

———. 1937. Kontrast- und Hemmungserscheinungen bei disparaten simultanen Geschmacksreizen. *Acta Inst. Psycho. Univ. Zagrabiensis.* 2:3–12.

Cameron, A. T. 1947. The taste sense and the relative sweetness of sugars and other sweet substances. *Science Report*, series no. 9. New York: Sugar Research Foundation.

Collings, V. B., L. Lindberg, and D. H. McBurney. 1976. Spatial interactions of taste stimuli on the human tongue. *Percept. Psychophys.* 19:69–71.

Erickson, R. P. 1968. Stimulus coding in topographic and nontopographic afferent modalities: on the significance of the activity of individual sensory neurons. *Psychol. Rev.* 75:447–465.

———. 1973. Parallel "population" neural coding in feature extraction. In *The Neuroscienes: A Study Program*. G. Quarton, T. Melnechuk, and F. Schmitt, eds. New York: The Rockefeller University Press, pp. 155–169.
———. 1977. The role of "primaries" in taste research. In *Olfaction and Taste VI* (Proceedings of the Sixth International Syposium). J. Le Magnen and P. MacLeod, eds. London: IRL Press, pp. 368–376.
———. 1978. Common properties of sensory systems. In *Handbook of Behavioral Neurobiology*, vol. 1. R. B. Masterton, ed. New York: Plenum Publishing Corp., pp. 73–90.
———. 1982. The across-fiber pattern theory: an organizing principle for molar neural function. *Contrib. Sens. Physiol.* 6:79–110.
Erickson, R. P., and E. Covey. 1980. On the singularity of taste sensations: what is a taste primary? *Physiol. Behav.* 25:527–533.
Fabian, F. W., and H. B. Blum. 1943. Relative taste potency of some basic food constituents and their competitive and compensatory action. *Food Res.* 8:179–193.
Faurion, A., S. Saito, and P. MacLeod. 1980. Sweet taste involves several distinct receptor mechanisms. *Chem. Sens.* 5:107–126.
Filin, V. A., and A. I. Esakov, 1968. Interaction between taste receptors. *Bull. Eksp. Biol. Med. (Engl. Trans.) Byull. Eksp. Biol. Med.* 65:12–15.
Fischer, R., and F. Griffin. 1963. Quinine dimorphism: a cardinal determinant of taste sensitivity. *Nature (Lond.)*. 200:343–347.
Gent, J., and L. M. Bartoshuk. 1983. Sweetness of sucrose, Neohesperidin dihydrochalocone, and saccharin is related to genetic ability to taste the bitter substance 6-*n*-propylthiouracil. *Chem. Sens.* 7:265–272.
Gillan, D. J. 1982. Mixture suppression: the effect of spatial separation between sucrose and NaCl. *Percept. Psychophys.* 32:504–510.
Hall, M. J., L. M. Bartoshuk, W. S. Cain, and J. C. Stevens. 1975. PTC taste blindness and the taste of caffeine. *Nature (Lond.)*. 253:442–443.
Harris, H., and H. Kalmus. 1949. The measurement of taste sensitivity to phenylthiourea (P.T.C.). *Ann. Eugenics.* 15:24–31.
Indow, T. 1969. An application of the τ scale of taste: interaction among the four qualities of taste. *Percept. Psychophys.* 5:347–351.
Jakinovich, W. 1982. Stimulation of the gerbil's gustatory receptors by saccharin. *J. Neurosci.* 2:49–56.
Kalmus, H. 1958. Improvements in the classification of taste genotypes. *Ann. Hum. Genet.* 22:222–230.
Kamen, J. M., F. J. Pilgrim, N. J. Gutman, and B. J. Kroll. 1961. Interactions of suprathreshold taste stimuli. *J. Exp. Psychol.* 62:348–356.
Kiesow, F. 1894. Beiträge zur physiologische Psychologie des Geschmackssinnes. *Philos. Stud.* 10:329–368, 523–561.
———. 1896. Beiträge zur physiologische Psychologie des Geschmackssinnes. *Philos. Stud.* 12:25–278.
Kroeze, J. H. A. 1978. Taste of sodium chloride: masking and adaptation. *Chem. Sens. Flav.* 3:443–449.
———. 1979. Masking and adaptation of sugar sweetness intensity. *Physiol. Behav.* 22:347–351.
———. 1982. After repetitious sucrose stimulation saltiness suppression in NaCl-sucrose mix-

tures is diminished: implications for a central mixture suppression mechanism. *Chem. Sens.* 7:81–92.

Kuznicki, J. T., and N. Ashbaugh. 1979. Taste quality differences within the sweet and salty taste categories. *Sensory Processes.* 3:157–182.

———. 1982. Space and time separation of taste mixture components. *Chem. Sens.* 7:39–62.

Kuznicki, J. T., M. Hayward, and J. Schultz. 1983. Perceptual processing of taste quality. *Chem. Sens.* 7:273–292.

Kuznicki, J. T., and N. B. McCutcheon. 1979. Cross-enhancement of the sour taste on single human taste papillae. *J. Exp. Psychol.* 108:68–89.

Lawless, H. T. 1977. The pleasantness of mixtures in taste and olfaction. *Sensory Processes.* 1:227–237.

———. 1978. Evidence for neural inhibition in bittersweet mixtures. Ph.D. dissertation, Brown University.

———. 1979. Evidence for neural inhibition in bittersweet taste mixtures. *J. Comp. Physiol. Psychol.* 93(3):538–547.

———. 1980. A comparison of different methods used to assess sensitivity to the taste of phenylthiocarbamide (PTC). *Chem. Sens.* 5:247–256.

———. 1982a. Adapting efficiency of salt-sucrose mixtures. *Percept. Psychophys.* 32:419–422.

———. 1982b. Paradoxical adaptation to taste mixtures. *Physiol. Behav.* 25:149–152.

Lawless, H. T., and D. A. Stevens. 1983. Cross adaptation of sucrose and intensive sweetness. *Chem. Sens.* 7:309–315.

McBurney, D. H. 1969. Effects of adaptation on human taste function. In *Olfaction and Taste III* (Proceedings of the Third International Symposium). C. Pfaffmann, ed. New York: The Rockefeller University Press, pp. 407–427.

———. 1972. Gustatory cross-adaptation between sweet-tasting compounds. *Percept. Psychophys.* 11:225–227.

McBurney, D. H., and L. M. Bartoshuk. 1972. Water taste in mammals. In *Olfaction and Taste IV* (Proceedings of the Fourth International Symposium). D. Schneider, ed. Stuttgart: Wissenschaftliche Verlagsgesellschaft, pp. 329–335.

———. 1973. Interactions across stimuli with different taste qualities. *Physiol. Behav.* 10:1101–1106.

McBurney, D. H., and V. B. Collings. 1977. *Introduction to Sensation/Perception.* Englewood Cliffs, NJ: Prentice-Hall Inc.

McBurney, D. H., and J. F. Gent. 1979. On the nature of taste qualities. *Psychol. Bull.* 86:151–167.

McBurney, D. H., and C. Pfaffmann. 1963. Gustatory adaptation to saliva and sodium chloride. *J. Exp. Psychol.* 65:523–529.

McBurney, D. H., D. V. Smith, and T. R. Schick. 1972. Gustatory cross adaptation: sourness and bitterness. *Percept. Psychophys.* 11:228–232.

Meiselman, H. L. 1971. Effect of presentation procedure on taste intensity functions. *Percept. Psychophys.* 10:15–18.

Miller, I. J. 1976. Taste bud distribution and regional responsiveness of the anterior tongue of the rat. *Physiol. Behav.* 16:439–444.

Moskowitz, H. R. 1973. Effects of solution temperature on taste intensity in humans. *Physiol. Behav.* 10:289–292.

———. 1974. Models of additivity for sugar sweetness. In *Sensation and Measurement.* H. R.

Moskowitz, B. Sharf, and J. C. Stevens, eds. Boston: Reidel, pp. 379–388.
Moskowitz, H. R., and C. Dubose. 1977. Taste intensity, pleasantness and quality of aspartame, sugars, and their mixtures. *Canadian Institute of Food Science Technology Journal.* 10:126–131.
Norgren, R., and C. M. Leonard. 1973. Ascending central gustatory pathways. *J. Comp. Neurol.* 150:217–238.
Oakley, B. 1975. Receptive fields of cat taste fibers. *Chem. Sens. Flav.* 1:431–442.
Öhrwall, H. 1891. Untersuchungen über den Geschmackssinn. *Skand. Arch. Physiol.* 2:1–69.
———. 1901. Die Modalitäts- und Qualitätsbegriffe in der Sinnesphysiologie und deren Bedeutung. *Skand. Arch. Physiol.* 11:245–272.
Pangborn, R. M. 1960. Taste interrelationships. *Food Res.* 25:245–256.
———. 1961. Taste interrelationships. II. Supra-threshhold solutions of sucrose and citric acid. *J. Food Sci.* 26:648–655.
———. 1962. Taste interrelationships. III. Supra-threshold solutions of sucrose and NaCl. *J. Food Sci.* 27:495–500.
Pfaffmann, C. 1959. The sense of taste. *Handb. Physiol.* I:(sect. 1) 507–533.
———. 1969. Taste preference and reinforcement. In *Reinforcement and Behavior.* J. Tapp, ed. New York: Academic Press, Inc. pp. 215–240.
———. 1970. Physiological and behavioural processes of the sense of taste. In *CIBA Foundation Symposium on Taste and Smell in Vertebrates.* G. E. W. Wolstenholme and J. Knight, eds. London: J. & A. Churchill, Ltd., pp. 31–50.
Pfaffmann, C., L. M. Bartoshuk, and D. H. McBurney. 1971. Taste psychophysics. In *Handbook of Sensory Physiology, The Chemical Senses: Taste.* L. Beidler, ed. Vol. IV, pt. 2. Berlin: Springer-Verlag, pp. 75–101.
Pfaffmann, C., M. Frank, L. M. Bartoshuk, and T. C. Snell. 1976. Coding gustatory information in the squirrel monkey chorda tympani. *Prog. Psychobiol. Physiol. Psychol.* 6:1–27.
Rapuzzi, G., and C. Casella. 1965. Innervation of the fungiform papillae in the frog tongue. *J. Neurophysiol. (Bethesda).* 28:154–165.
Renqvist, Y. 1919. Über den Geschmack. *Skand. Arch. Physiol.* 38:97–201.
Rifkin, B., and L. M. Bartoshuk. 1980. The bitter taste of potassium chloride, sodium benzoate, and potassium benzoate is related to the genetic ability to taste 6-*n*-propylthiouracil. Presented at the Second Annual Meeting of the Association for Chemoreception Sciences, Sarasota, Florida.
Schiffman, S. S., H. Cahn, and M. G. Lindley. 1981. Multiple receptor sites mediate sweetness: evidence from cross adaptation. *Pharmacol. Biochem. Behav.* 15:337–388.
Schiffman, S. S., D. A. Reilly, and T. B. Clark. 1979. Qualitative differences among sweeteners. *Physiol. Behav.* 23:1–9.
Smith, D. V., and D. H. McBurney. 1969. Gustatory cross-adaptation: does a single mechanism code the salty taste? *J. Exp. Psychol.* 80:101–105.
Stevens, S. S. 1969. Sensory scales of taste intensity. *Percept. Psychophys.* 6:302–308.
Stone, H., and S. M. Oliver. 1969. Measurement of the relative sweetness of selected sweeteners and sweetener mixtures. *J. Food Sci.* 34:215–222.
Taglietti, V., C. Casella, and E. Ferrari. 1969. Interactions between taste receptors in the frog tongue. *Pflügers Arch. Eur. J. Physiol.* 312:139–148.
Van der Wel, H. 1972. Thaumatin, the sweet-tasting protein from *Thaymacoccus daniellii.* In *Olfaction and Taste IV* (Proceedings of the Fourth International Symposium). D. Schneider, ed.

Stuttgart: Wissenschaftliche Verlagsgesellschaft MBH, pp. 226–233.
Van der Wel, H., and K. Arvidson. 1978. Qualitative psychophysical studies on the gustatory effects of the sweet-tasting proteins thaumatin and monellin. *Chem. Sens. Flav.* 3:291–297.
von Skramlik, E. 1926. *Handbuch der Physiologie der niederen Sinne*. Leipzig, Germany: Georg Thieme Verlag.
Yamaguchi, S., T. Yoshikawa, S. Ikeda, and T. Ninomiya. 1970. Studies on the taste of some sweet substances. II. Interrelationships among them. *Agric. Biol. Chem.* 34:187–197.

10.
The Sense of Taste and the Study of Ingestion

Scientific study of the gustatory system derives from at least three perspectives, the developmental, sensory, and biological. Although some may object to the implication that the first two approaches are less biological than the third, a case can be made for the division based upon the rationale, implicit or explicit, underlying many of the experiments involved. One aspect of the developmental approach has little to do with taste as a sensory system per se, but uses gustatory receptor cells and taste afferent axons as a model for neurally supported cellular differentiation (Oakley, this volume).

As with vision, audition, and somesthesis, taste is a sensory experience amenable to introspective analysis and verbal report. Sensory analysis has this human capacity at its base, and psychophysics as a primary methodology. Electrophysiological recording of the response patterns from gustatory neurons usually aims at comprehending the neural code for taste quality or intensity, and thus has a goal similar to that of psychophysics, understanding how the nervous system elaborates sensory experience from stimulation of the taste buds. As with other sensory systems, the biological utility of this sensory experience usually is assumed, but often ignored as well.

Most scientists engaged in sensory analysis are not so cavalier as to disregard function entirely. Indeed, in areas such echolocation, the ecology of the species under study has provided the basis for impressive gains in understanding the organization of a sensory system (Suga, 1981). The contrast is between species, such as bats, that have evolved sensory systems with highly specialized functions, and humans whose sensory systems often are pictured as generalized, that is, not adapted to functioning in any one biological system at the expense of others. If a sensory system must serve many biological functions, then no one of them is likely to have influenced its evolution dramatically.

In humans and other mammals, the gustatory system serves only a limited range of functions related to regulating energy, water, and electrolyte balance. The receptor apparatus is sufficiently stable that analogous structures can be

Ralph Norgren, Pennsylvania State University School of Medicine
Hershey, Pennsylvania 17033

identified in most animal phyla. In vertebrates, these receptors are innervated by homologous cranial nerves and, other than in fishes, are confined to the oral cavity. In other words, the biological utility of this sensory system has changed little over an enormous span of evolution. This chemosensory apparatus adapted early on to fulfilling these limited, but important functions, and that adaptive value has maintained the system in a relatively rudimentary stage of differentiation. In common biological parlance, differentiation refers to changed morphology, but for the gustatory system, the relative lack of differentiation also appears to apply to its functional characteristics. Taste receptors are sensitive to a wide variety of chemicals, but apparently with only a limited number of response categories, perhaps as few as five, and then only if water sense is included (Frank, this volume). The major evolutionary variables seem to be not the number of response categories, but how sapid chemicals are parsed among them (Boudreau, 1980).

This evolutionary conservatism makes the gustatory system particularly amenable to the third scientific perspective identified above, for lack of a better term, as biological. From this viewpoint, taste is one facet of the sensory apparatus engaged in the regulation of energy, water, and electrolyte balance. In conjunction with other oral and nasal sensory systems, taste initiates and maintains ingestion or rejection behavior, triggers a variety of reflexes that function to prepare the gastrointestinal tract for a bout of anabolic activity, and elicits hedonic responses that will support the learned behavior necessary to further obtain or avoid the stimulus in question (Norgren, 1984). The olfactory and trigeminal somatosensory systems subserve many other functions, but taste is limited to influencing ingestion and rejection behaviors and their sequelae. This limited functional scope may account for the limited number of sensory categories apparent in gustation.

Regardless of the validity of this link, the relative paucity of sensory categories and response functions reduces the need for sensory analysis as a prerequisite to understanding the neural relationships between gustatory stimulus and ingestive response. Exemplars of gustatory stimulus categories—sucrose, NaCl, quinine—elicit reliable ingestion or rejection responses that provide the basic behavioral paradigm for use with the traditional central lesion and stimulation techniques of physiological psychology. As it happens, experimental neuroanatomy also has contributed significantly to this biological approach to gustatory function. Psychophysics and electrophysiology, methods critical to sensory analysis, however, have been largely adjunct in this enterprise. At present, the biological approach to the gustatory system uses different techniques to answer different questions. The goal is not understanding human perception of sapid stimuli, but the influence of these stimuli on a complex regulatory process. In this sense, this approach parallels the study of the baro- and chemoreceptors in the control of the cardiovascular system.

Taste and Feeding in the Brain

Twenty years ago, gustatory neurons and ingestive responses were far apart in the brain. Taste responses had been localized in the nucleus of the solitary tract in the medulla and adjacent to the ventrobasal complex, the major somatosensory relay in the thalamus (Pfaffmann et al., 1961). In the rodent, at least, the gustatory representation had an analogous relationship to the somatosensory homunuclus on the cortex (Benjamin and Pfaffmann, 1955). The neural mechanisms underlying feeding, drinking, and pleasure, the major response systems influenced by the gustatory system, were localized in the hypothalamus (Hoebel and Teitelbaum, 1962; Margules and Olds, 1962). Despite the obvious behavioral connections between taste and ingestion, the anatomy available did not provide a convincing neural basis for the connection.

Since then the problem has changed remarkably. Third-order gustatory neurons have been demonstrated that project to the hypothalamus, and to several other areas in the limbic forebrain, for that matter. The neural substrates for hunger, thirst, and pleasure, however, once concentrated in the hypothalamus, have been distributed, in Dr. Mountcastle's sense of the word, to the far corners of the brain and viscera (Norgren and Grill, 1982). At least at the theoretical level, we know more, but understand less about feeding and drinking behavior now than we did twenty years ago. The central gustatory and visceral afferent systems may provide part of the framework needed to recapture this understanding from the army of facts that have all but overwhelmed it.

Sapid stimuli influence neural activity in the hypothalamus. In one sample of almost 100 hypothalamic units recorded from awake, behaving rats, > 50% altered their ongoing activity when water, sucrose, NaCl, or quinine HCl was injected directly into the oral cavity (Table I of Norgren, 1970). These responses did not resemble those in traditional sensory relays, which usually are relatively brisk and synchronized with both onset and offset of the stimulus. Even in awake, behaving animals, hypothalamic neurons typically have low spontaneous rates and respond to sensory stimuli sluggishly, if at all. The response to sapid stimuli was seldom brisk, sometimes had a long latency, and could even be inhibitory. In addition, some responses appeared to be a function of the organisms' state of deprivation. In food-deprived rats, sucrose solution was more effective than water. In replete rats, just the opposite was true (Table I, *water* and *sucrose*). These response characteristics raised legitimate questions about hypothalamic gustatory responses. Perhaps these changes in neural activity reflected nonspecific activation or attentional factors rather than specific sensory influence. The absence of direct anatomical connections between the gustatory system and the hypothalamus made this alternative interpretation difficult to refute.

Gustatory Systems in the Forebrain

This possible alternative explanation for hypothalamic gustatory responses prompted Dr. Christiana Leonard and me to reexamine the axonal projections arising from the first central gustatory relay in the nucleus of the solitary tract. We localized this area electrophysiologically, then made small electrolytic lesions, and after a few days, stained the axons that were degenerating as a result of the lesion. We never observed any degenerating axons in the hypo-

Table 1. *Hypothalamic Unit Responses to Sapid Stimuli Recorded from Awake, Behaving Rats*

	Unit response		
Substance tested	*Ad libitum*	*Food deprived*	*Total*
Water	8	1	9
Sucrose	1	7	8
Quinine	5	6	11
Water-sucrose	3	5	8
Water-quinine	3	2	5
Sucrose-quinine	3	3	6
Water-sucrose-quinine	6	3	9
Light or tone	4	4	8
None	17	15	32
Total	50	46	96

Adapted from Norgren (1970) with permission.

thalamus, but neither did we find them in the thalamic gustatory relay where all previous evidence suggested second-order gustatory neurons should project. Instead, degenerating axons from the damaged medullary taste relay passed a short distance rostrally into the pons and appeared to terminate in the parabrachial nuclei. Although this area, which spans the gap between the mesencephalic and principal trigeminal nuclei, had never been associated with gustatory function, we quickly confirmed that neurons in the parabrachial nuclei responded promptly and vigorously to sapid stimuli applied on the anterior tongue. Small lesions at these locations in the parabrachial nuclei, in conjunction with the Fink-Heimer degeneration stain, this time revealed degenerating axons in the ipsilateral thalamic gustatory relay, but still nothing in the hypothalamus (Figure 1; Norgren and Leonard, 1971).

Further neuroanatomical experiments using both degeneration stains and tritiated amino-acid autoradiography, however, traced axons from parabrachial neurons not only into the hypothalamus, but into the amygdala and bed nucleus of the stria terminalis as well (Norgren and Leonard, 1973; Norgren, 1976). Antidromic invasion of parabrachial gustatory neurons via ventral forebrain stimulating electrodes demonstrated that some of these axons conveyed taste afferent information directly to the limbic systen (Norgren, 1974, 1976). The central gustatory system in the rat bifurcates at the third-order level into a more or less standard, though ipsilateral, thalamocortical sensory projection and a complex, multilevel limbic forebrain projection (Figure 2). The historic cognitive and emotional division of labor between the thalamocortical and limbic systems seduced us to speculate on a similar segregation of gustatory function with the sensory-perceptual elaboration set in cortex and the biological responses arising from the so-called ventral pathway (Pfaffmann et al., 1977).

Although the central gustatory system now seems much better understood than previously, it still is not that simple. The basic arrangement, medullary and pontine relays with a bifurcated projection to dorsal and ventral forebrain, has been observed in several vertebrate classes, besides mammals. Experimental evidence from hamster, rabbit, and cat, and confirmatory data for the rat establish a gustatory function for neurons in the parabrachial nuclei and a bifurcating projection to forebrain arising from these nuclei (Block and Schwartzbaum, 1983; see Norgren, 1984, for documentation). The parabrachial nuclei, however, serve as a major relay for all levels of the nucleus of the solitary tract, and thus potentially convey visceral afferent information related to respiratory, cardiovascular, and gastrointestinal function as well as taste (Norgren, 1978; Ricardo and Koh, 1978). Numerous recent electro-

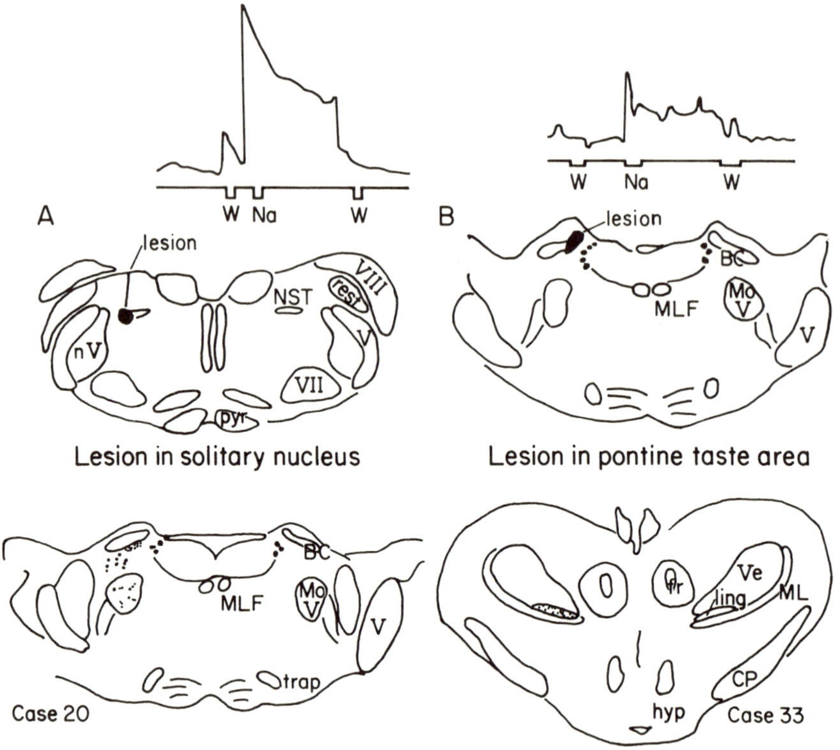

Figure 1. The central gustatory projection to the thalamus in the rat. (A) A lesion in the nucleus of the solitary tract (NST) at a point that responded to sapid stimuli produced terminal degeneration (fine dots, lower panel) in the parabrachial nuclei of the pons. The inset above the anatomical charts illustrates the integrated neural activity recorded from NST prior to the lesion. (B) A lesion in the parabrachial nuclei at a point that responded to sapid stimulation produced terminal degeneration in the ipsilateral thalamic gustatory and lingual somatsensory areas (ling). The inset above illustrates the integrated neural responses recorded from the parabrachial nuclei prior to the lesion. W, distilled water washed over the tongue; Na, 0.25 M NaCl solution washed over the tongue; V, trigeminal nerve and its spinal tract; nV, nucleus of the spinal trigeminal tract; VII, motor nucleus of the facial nerve; MLF, medial longitudinal fasciculus; MoV, motor nucleus of the trigeminal nerve; pyr, pyramidal tract; rest, restiform body; trap, trapezoid body; Ve, ventrobasal nucleus. Reprinted with permission from Norgren and Leonard (1971).

physiological, lesion, and stimulation studies attest to the importance of these nuclei in visceral function (e.g., Hamilton et al., 1981; Ciriello and Calarescu, 1980; Marovitch et al., 1982; Kannan et al., 1981). Nevertheless, the range of afferent information passing through the dorsal pons or its ultimate distribution in the forebrain remains poorly documented. In other words, the central gustatory system is both anatomically complicated and functionally complex. In fairness, it must be designated the central gustatory and visceral afferent system (Norgren, 1981).

In addition to the functional heterogeneity, this newfound limbic sensory system contains axons that skip synapses or even take altogether different routes. Ricardo and Koh (1978) demonstrated that some neurons in the caudal, visceral afferent solitary nucleus bypass the dorsal pontine synapse and project directly to the limbic system nuclei that serve as the primary synaptic targets for parabrachial neurons. Other presumed visceral afferent neurons from the vicinity of the caudal solitary nucleus project to noradrenergic cell groups in the ventral reticular formation. These noradrenergic neurons, in turn, project at least as far rostrally as the hypothalamus (see Swanson and Mogenson, 1981, and Sawchenko and Swanson, 1982, for details). Although a sensory function has not been demonstrated electrophysiologically for the pathway, some neurons in and near the parabrachial nuclei in rodents bypass the thalamic gustatory relay and terminate in the vicinity of gustatory cortex (Figure 2A; Saper and Loewy, 1980; Lasiter et al., 1982; Norgren 1983a). In addition, in primates, axons arising from the lateral, largely gustatory, division of the solitary nucleus bypass the parabrachial nuclei to project mono-

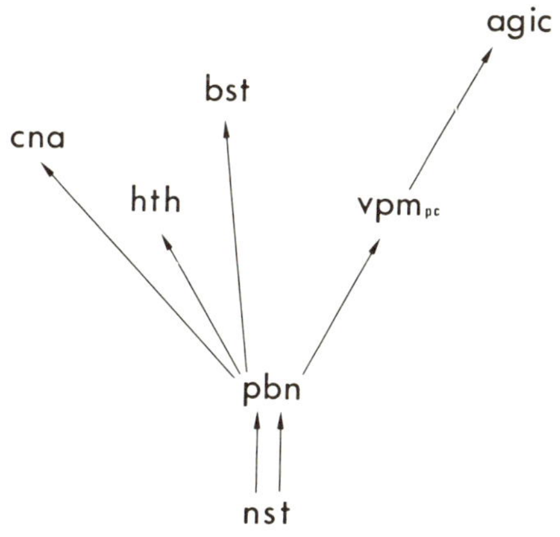

Figure 2. Schematic diagram of the central gustatory pathways in the rat. agic, agranular insular cortex; bst bed nucleus of the stria terminalis; cna, central nucleus of the amygdala; hth, hypothalamus; nst, nucleus of the solitary tract; pbn, parabrachial nuclei; vpm_{pc}, parvicellular division of the thalamic ventroposteromedial nucleus.

synaptically to the thalamic taste relay (Figure 3B; Beckstead et al., 1980; Norgren, 1981).

In all likelihood the central gustatory and visceral afferent systems have not yet settled down, but a few observations seem stable. First, although its functional organization is still being worked out, the thalamocortical axis of the gustatory system appears to be anatomically straightforward (Yamamoto et al., 1980; Norgren et al., 1982; Pritchard et al., 1983). Second, the ventral pathway may be both anatomically and functionally complex, but only three areas receive the bulk of the afferent input—the hypothalamo-preoptic continuum, particularly the paraventricular nucleus, the central nucleus of the amygdala, and the bed nucleus of the stria terminalis. These four areas, gustatory cortex and three ventral forebrain sites, share common characteristics relevant both to gustatory function and the control of ingestive behavior. In three of the four areas, lesions disrupt a variety of responses to sapid stimuli. Virtually no lesion-behavioral data exist for the fourth area, the bed nucleus of the stria terminalis. Table II summarizes the effects of localized lesions and monoamine depletion on the three most frequently examined responses to gustatory stimuli. In addition to these related functional characteristics, the

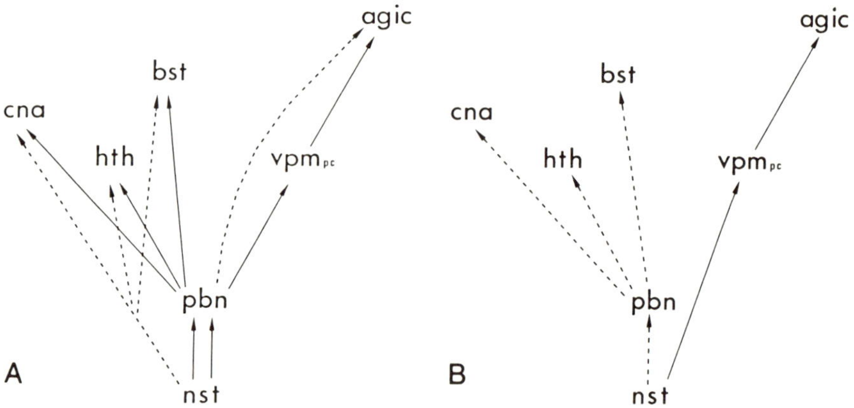

Figure 3. Complexities in central gustatory pathways. (A) Dashed arrows indicate anatomical pathways without proven gustatory function. Dashed projections arising from nst probably convey visceral afferent information; the function of those from pbn is unknown. (B) Gustatory pathways in monkey (solid arrows). As in A, the dashed arrows represent pathways without proven gustatory function that probably convey visceral afferent information. Abbreviations as in Figure 2.

major forebrain targets of the gustatory and visceral afferent system are anatomically interconnected, and they project caudally back to the vicinity of their afferent relays in the pons and medulla (Figure 4; Conrad and Pfaff, 1976; Hopkins and Holstege, 1978; Berk and Finkelstein, 1982; Norgren, 1983b; Sofroniew, 1983). The afferent modalities subserved by this limbic sensory system, its efferent return, and the physiology of the hypothalamus, all bespeak control of the autonomic nervous system. Indeed, some behavioral effects of hypothalamic lesions have been explained as secondary to disruption of autonomic regulation. Nevertheless, recent experiments examining the caudal brainstem control of ingestive behavior indicate these same ventral forebrain sensory systems may contribute to the behavioral as well as the autonomic and hormonal aspects of energy and fluid balance regulation.

Feeding Systems in the Hindbrain

We now have substantial anatomical and even some electrophysiological evidence that gustatory afferent activity reaches the hypothalamus, long thought to contain the neural substrates for hunger, thirst, and pleasure. While neuroanatomists have been busy getting gustatory and visceral afferent information into the hypothalamus and limbic system, however, other neuroscientists have been just as busy demonstrating that many functions previously ascribed to these structures are in fact distributed throughout the brain and even into the viscera. Although the functional relocation has reached virtually every corner of the CNS, the dispersion originally followed two major routes, the discovery and elucidation of the monoamines systems and the rediscovery of the functional importance of the hindbrain and visceral mechanisms in the control of energy and water balance. The major events in these discovery processes have been reviewed elsewhere (Norgren and Grill, 1982) and for present purposes we need only summarize the experiments related to gustatory function.

Presented with fluids injected through intraoral catheters, a rat responds with behaviors assembled from a limited lexicon of oral and whole-body movements that are highly stereotyped both within and between animals (Grill and Norgren, 1978a). Given single 50-μl injections of sucrose and quinine-HCl, the responses elicited derive from nonoverlapping components. Intraoral sucrose solution elicits mouth movements, tongue protrusions, and lateral tongue flicks, initially in a fixed order. The duration of the response is determined by the amount of fluid injected and the concentration of sucrose. At higher concentrations, intraoral quinine injections elicit a bout of mouth gaping followed by a series of whole-body movements that also occur in a

Table 11. *Effects of Monoamine Depletion and Localized Forebrain Lesions on Behavioral Responses to Gustatory Stimuli*

		Lesions		
Effects	*Monamine depletion*	*Hypothalamus*	*Amygdala*	*Taste cortex*
Preference/aversion functions	–	++	±	P ++/Av +
Sodium appetite	+	–[1]	–	+
Learned taste aversion				
Acquisition	+	–	+	+
Recall	+	+	–	–

(–) Absent, (+) present, (++) exaggerated, (±) equivocal. Derived from Braun et al. (1982), Pfaffmann et al. (1979), and Norgren (1984).
1. Deficit can be escaped with prior experience.

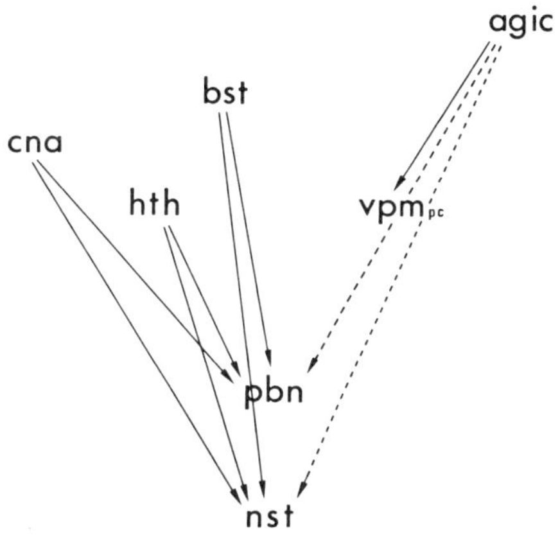

Figure 4. Reciprocal connections in the central gustatory system. Dashed arrows represent pathways not proven to arise from agranular insular cortex, i.e., gustatory cortex. Abbreviations as in Figure 2.

fixed sequence (Figure 5). The absence of fluid on the muzzle during sucrose injections and its presence on the muzzle, paws, and cage floor after quinine injections provided circumstantial evidence that these response patterns were the behavioral concomitants of ingestion and rejection, respectively. This supposition has been confirmed recently in the course of an electromyographic analysis of muscle activity during these responses to sapid stimuli (Travers and Norgren, 1983b). The response to sucrose was accompanied by swallowing at regular intervals, but at least during the oral phase of the quinine response, swallowing was inhibited. Using other sapid stimuli and a continuous rather than episodic intraoral infusion, more variable responses can be obtained, but it is only the admixture that changes, not the form or number of components in the lexicon (Berridge et al., 1981).

This taste reactivity test is useful on at least two dimensions for the analysis of energy and water balance. First, in the absence of other sensory cues, rodents respond to sucrose as a food, and water and sodium ions are adquate stimuli for restoring hydromineral balance. Thus, these sapid stimuli and the orofacial responses they elicit provide a restricted behavioral paradigm for testing the known controls on, and challenges to, these regulatory systems. Second, because this test uses fluids injected directly into the oral cavity, it circumvents the normal appetitive phase of feeding and drinking behavior. We developed the test to examine the ingestive behavior of chronically decerebrate rats, because these animals never exhibit spontaneous ingestion of food or fluid. When sapid stimuli are injected directly into their mouths, however, chronically decerebrate rats respond with ingestion and rejection behavior that is virtually normal (Grill and Norgren, 1978b).

Figure 5. The lexicon of nine response components that comprise the pure ingestion and rejection sequences. Ingestion consists of mouth movements, tongue protrusions, and lateral tongue flicks. Rejection begins with gapes followed by a series of whole-body movements—chin rubs, head shakes, face washes, forelimb shakes, and paw wipes. These nonoverlapping sequences can be elicited with sapid stimuli that are either strongly preferred or rejected in a standard consumption test. Adapted with permission from Grill (1980).

Experiments with the chronically decerebrate rat underscore an important functional dichotomy in the behavior associated with energy and fluid balance regulation. Given appropriate stimuli, the caudal brainstem contains neural apparatus sufficient to generate virtually normal ingestion and rejection responses. The behavior normally necessary to locate and contact these stimuli, however, clearly requires the forebrain for normal expression. Deprived of food and water, chronic decerebrate rats exhibit increased spontaneous activity, but they never orient toward food or water. In other words, with only the caudal brainstem available, the internal humoral, hormonal, and neural monitors of energy and water balance can activate, but not direct, behavior.

Under normal circumstances, food and water deprivation not only initiates appetitive behavior that leads to sustenance, but also increases the amount of food or fluid that will be ingested. Because virtually normal discriminative ingestion and rejection responses are organized by the caudal brainstem, perhaps deprivation signals influence these consummatory responses via the same mechanisms. This hypothesis has been tested with standard laboratory challenges of food, water, and electrolyte balance. Sodium depletion has no effect on the quantity of NaCl solution ingested by a chronically decerebrate rat Grill, personal communication). Similarly, none of the hydrational challenges that induce drinking in normal rats has any influence on the amount of water ingested by decerebrate animals (Grill and Miselis, 1981). Food deprivation, on the other hand, increases intake of sucrose solution equivalently in both normal and chronically decerebrate rats (Grill and Norgren, 1978c; Grill, 1980).

Although the behavioral index is identical in these three systems, i.e., some internal stimulus alters the response to an oral stimulus from one of ingestion to one of rejection or vice versa, it appears that the mechanisms for inducing food satiety exist in the caudal brainstem, whereas those triggering water and sodium intake require forebrain involvement. An alternative hypothesis is equally valid at present and requires further examination. There may be only one mechanism for switching from ingestion to rejection that exists in the caudal brainstem. What differs between water and electrolyte regulation and energy balance may be the location of the most important deficit or surfeit detectors. The most important neural detectors of fluid deficits are in the preoptic area and the circumventricular organs of the anteior ventricles (Epstein, 1982). If the major signals for sodium deficiency are related hormonally to the fluid deficit system, as has recently been hypothesized (Epstein and Jameson, 1982), then the anterior circumventricular organs might be critical in this system as well. In rats, at least, some aspects of food satiety also are mediated hormonally, but the effect requires sensory components of the subdiaphragmatic vagus nerve that terminates in the medulla (Norgren and Smith, 1983; Smith et al., 1983). Exogeneous cholecystokinin is as effective in inducing satiety in chronic decerebrate rats as it is in normal ones (Grill et al., 1983).

The caudal brainstem harbors neural mechanisms adequate for inducing discriminative ingestion and rejection responses. The second hypothesis extends this evidence by asserting that these same systems are capable of conditional responses—in a deprived animal, an oral stimulus elicits ingestion; in a sated animal, rejection. In each of the regulatory systems that require ingestion (water, electrolyte, and energy balance), both its onset and offset can be triggered by internal deficit and surfeit signals (even if their exact identity is not known). Normally, the onset of ingestive behavior is preceded by some noningestige, appetitive behavior. Because these behaviors apparently require the integrity of the forebrain, the onset of feeding may be a forebrain function. Satiety, however, also is an active process, because it usually occurs before physiological repletion. Satiety does not require appetitive behavior; thus the rudimentary mechanisms present in the caudal brainstem may be more easily demonstrable.

Although these hypotheses cannot be properly addressed (or even formulated) until we know more about the neural and humoral signals that initiate ingestion and those that eventually inhibit it, the evidence from which they arise indicates that the caudal brainstem is far from a passive recipient of detailed instructions from the forebrain. The medulla and pons contain the motor neuron pools that provide the final common path to ingestive behavior and receive the afferent sensory information that controls it. The interneurons that project to these oral motor neurons extend through the midlateral medullary and pontine reticular formation with concentrations at either end of the distribution, one associated with the nucleus of the solitary tract, the other with the pontine parabrachial nuclei (Travers and Norgren, 1983a). The solitary, parabrachial, and trigeminal nuclei all contain neurons that project into these reticular zones providing an anatomical basis for the gustatory, oral somatosensory, and visceral afferent influence on ingestion and rejection behavior (Norgren, 1976, 1978; Saper and Loewy, 1980; Travers and Norgren, 1983a). In addition to this local sensory influence, as noted earlier, the forebrain systems that influence ingestion all project into the caudal brainstem. Most investigators emphasize the forebrain projections to the sensory relays, particularly the solitary and parabrachial nuclei, but the anatomical evidence usually indicates substantial ramifications of these axons within the adjacent reticular formation as well.

Gustatory stimuli are not required for eliciting adequate ingestive behavior. (Sodium and salt appetite are important exceptions.) It is oral somatosensory stimulation that appears to be the most important sensory component for initiating food-seeking behavior (Jacquin and Zeigler, 1982, 1983; Jacquin, 1983; Miller, 1981). Nevertheless, the gustatory system serves as an excellent tool for examining the neurology of feeding behavior. The central gustatory and visceral afferent system provides an anatomical framework that links most of the forebrain areas implicated in the control of intake with the caudal brainstem systems that actually generate the ingestion and rejection behavior. Responses to gustatory stimuli provide a behavioral paradigm for testing the capacity of relatively isolated neural systems to regulate the onset and offset of intake. At a conceptual level, this framework cannot replace the all but defunct hypothalamic theory of feeding and drinking control. The evidence is fragmentary at best, and often of the wrong order, that is, anatomical when it should be functional, or functional (behavioral) when electrophysiology could be more convincing. As the word implies, a framework provides structure on which to display evidence and perhaps stretch a hypothesis or two. Explicit neural mechanisms are lacking, but standing on this scaffold, we can venture a prediction. Understanding the forebrain mechanisms controlling ingestive behavior will require analyzing the capacities of the hindbrain, but the reverse may not be true.

References

Beckstead, R., J. Morse, and R. Norgren. 1980. The nucleus of the solitary tract in the monkey: projections to thalamus and other brainstem nuclei. *J. Comp. Neurol.* 90:259–282.

Benjamin, R. M., and C. Pfaffmann. 1955. Cortical localization of taste in the albino rat. *J. Neurophysiol. (Bethesda).* 18:56–64.

Berk, M. L., and J. A. Finkelstein. 1982. Efferent connections of the lateral hypothalamic area of the rat: an autoradiographic investigation. *Brain Res.* 8:511–526.

Berridge, K., H. J. Grill, and R. Norgren. 1981. The relation of consummatory responses and preabsorptive insulin release to palatability and learned taste aversions. *J. Comp. Physiol. Psychol.* 95:363–382.

Block, C. H., and J. S. Schwartzbaum. 1983. Ascending efferent projections of the gustatory parabrachial nuclei in the rabbit. *Brain Res.* 259:1–9.

Boudreau, J. C. 1980. Taste and the taste of foods. *Naturwissenschaften.* 67:14–20.

Braun, J. J., P. S. Lasiter, and S. W. Kiefer. 1982. The gustatory neocortex of the rat. *Physiol. Psychol.* 10:13–45.

Ciriello, J., and F. R. Calarescu. 1980. Monosynaptic pathway from cardiovascular neurons in the nucleus tractus solitarii to the paraventricular nucleus in the cat. *Brain Res.* 193:529–533.

Conrad, L., and D. W. Pfaff. 1976. Efferents from medial basal forebrain and hypothalamus in the rat. II. An autoradiographic study of the anterior hypothalamus. *J. Comp. Neurol.* 169:221–261.

Epstein, A. 1982. The physiology of thirst. In *The Physiological Mechanisms of Motivation.* D. W. Pfaff, ed. New York: Springer-Verlag, pp. 164–214.

Epstein, A. N., and E. C. Jameson. 1982. Angiotensin of cerebral origin is the optimal synergist with DOCA for arousal of a sodium appetite. *Soc. Neurosci. Abstr.* 8:903.

Frank, M. E. 1984. On the neural code for sweet and salty tastes. In *Taste, Olfaction, and the Central Nervous System.* D. Pfaff, ed. New York: The Rockefeller University Press, pp. 107–128.

Grill, H. J. 1980. Production and regulation of ingestive consummatory behavior in the chronic decerebrate rat. *Brain Res. Bull.* 5:79–87.

Grill, H. J., D. Ganster, and G. P. Smith. 1983. CCK-8 decreases sucrose intake in chronic decerebrate rats. *Soc. Neurosci. Abstr.* 9:903.

Grill, H. J., and R. R. Miselis. 1981. Lack of ingestive compensation to osmotic stimuli in chronic decerebrate rats. *Am. J. Physiol.* 240:R81–R86.

Grill, H. J., and R. Norgren. 1978a. The taste reactivity test. I. Mimetic responses to gustatory stimuli in neurologically normal rats. *Brain Res.* 143:263–279.

———. 1978b. The taste reactivity test. II. Mimetic responses to gustatory stimuli in chronic thalamic and chronic decerebrate rats. *Brain Res.* 143:281–297.

———. 1978c. Chronic decerebrate rats demonstrate satiation but not baitshyness. *Science (Wash. DC).* 201:267–269.

Hamilton, R. B., H. Ellenberger, D. Liskowsky, and N. Schneiderman. 1981. Parabrachial area as mediator of bradycardia in rabbits. *J. Auton. Nerv. Syst.* 4:261–281.

Hoebel, B. G., and P. Teitelbaum. 1962. Hypothalamic control of feeding and self-stimulation. *Science (Wash. DC).* 135:375–377.

Hopkins, D. A., and G. Holstege. 1978. Amygdaloid projections to the mesencephalon, pons and medulla oblongata in the cat. *Exp. Brain Res.* 32:529–547.

Jacquin, M. F. 1983. Gustation and ingestive behavior in the rat. *Behav. Neurosci.* 97:98–109.
Jacquin, M. F., and H. P. Zeigler. 1982. Trigeminal orosensory deafferentation disrupts feeding and drinking mechanisms in the rat. *Brain Res.* 238:198–204.
———. 1983. Trigeminal orosensation and ingestive behavior in the rat. *Behav. Neurosci.* 97:62–97.
Kannan, H., K. Yagi, and Y. Sawaki. 1981. Pontine neurones: electrophysiological evidence of mediating carotid baroreceptor inputs to supraoptic neurones in rats. *Exp. Brain Res.* 42: 362–370.
Lasiter, P. S., D. L. Glanzman, and P. Mensah. 1982. Direct connectivity between pontine taste areas and gustatory neocortex in rat. *Brain Res.* 234:111–121.
Margules, D., and J. Olds. 1962. Identical "feeding" and "reward" systems in the lateral hypothalamus of rats. *Science (Wash. DC).* 135:374–375.
Miller, M. G. 1981. Trigeminal deafferentation and ingestive behavior in the rat. *J. Comp. Physiol. Psychol.* 95:252–269.
Mraovitch, S., M. Kumada, and D. J. Reis. 1982. Role of the nucleus parabrachialis in cardiovascular regulation in cat. *Brain Res.* 232:57–75.
Norgren, R. 1970. Gustatory responses in the hypothalamus. *Brain Res.* 21:63–71.
———. 1974. Gustatory afferents to ventral forebrain. *Brain Res.* 81:285–295.
———. 1976. Taste pathways to hypothalamus and amygdala. *J. Comp. Neurol.* 166:12–30.
———. 1978. Projections from the nucleus of the solitary tract in the rat. *Neuroscience.* 3:207–218.
———. 1981. The central organization of the gustatory and visceral afferent systems in the nucleus of the solitary tract. In *Brain Mechanisms of Sensation.* Y. Katsuki et al., eds. New York: John Wiley & Sons, Inc., pp. 143–160.
———. 1983a. The gustatory system in mammals. *Am. J. Otolaryngol.* 4:234–237.
———. 1983b. Afferent connections of cranial nerves involved in ingestion. *J. Auton. Nerv. Syst.* 9:67–77.
———. 1984. Section 1. The Nervous System. Vol. 3. Sensory Processes, pt. 2. I. Darian-Smith, vol. ed.; J. M. Brookhart and V. B. Mountcastle, sect. eds. Bethesda, MD: American Physiological Society. pp. 1087–1128.
Norgren, R., and H. J. Grill. 1982. Brain stem control of ingestive behavior. In *Physiological Mechanisms of Motivation.* D. W. Pfaff, ed. New York: Springer-Verlag, pp. 99–131.
Norgren, R., E. Kosar, and H. J. Grill. 1982. Gustatory cortex in the rat delimited by the thalamocortical projections, physiological properties, and cytoarchitecture. *Soc. Neurosci. Abstr.* 8:201.
Norgren, R., and C. M. Leonard. 1971. Taste pathways in rat brainstem. *Science (Wash. DC).* 173:1136–1139.
———. 1973. Ascending central gustatory pathways. *J. Comp. Neurol.* 150: 217–237.
Norgren, R., and G. P. Smith. 1983. The central distribution of vagal subdiaphragmatic branches in the rat. *Soc. Neurosci. Abstr.* 9:611.
Oakley, B. 1984. Trophic competence in mammalian gustation. In *Taste, Olfaction, and the Central Nervous System.* D. Pfaff, ed. New York: The Rockefeller University Press, pp. 92–103.
Pfaffmann, C., R. Erickson, G. Frommer, and B. Halpern. 1961. Gustatory discharges in the rat medulla and thalamus. In *Sensory Communication.* W. Rosenblith, ed. New York: John Wiley & Sons, Inc., pp. 455–473.

Pfaffmann, C., M. Frank, and R. Norgren. 1979. Neural mechanisms and behavioral aspects of taste. *Annu. Rev. Psychol.* 30:283–325.

Pfaffmann, C., R. Norgren, and H. J. Grill. 1977. Sensory affect and motivation. In *Tonic Functions of Sensory Systems.* B. M. Wenzel and H. P. Zeigler, eds. *Ann. NY Acad. Sci.* 290:18–34.

Price, J. L., and D. G. Amaral. 1981. An autoradiographic study of the projections of the central nucleus of the monkey amygdala. *J. Neurosci.* 1:1242–1259.

Pritchard, T., R. Hamilton, J. Morse, and R. Norgren. 1984. Gustatory and lingual areas in primate cortex. *Proc. Intl. Union Physiol. Sci.* 15:198.

Ricardo, J., and E. Koh. 1978. Anatomical evidence of direct projections from the nucleus of the solitary tract to the hypothalamus, amygdala and other forebrain structures in the rat. *Brain Res.* 153:1–26.

Saper, C. B., and A. D. Loewy. 1980. Efferent connections of the parabrachial nucleus in the rat. *Brain Res.* 197: 291–317.

Sawchenko, P. E., and P. E. Swanson. 1982. The organization of noradrenergic pathways from the brainstem to the paraventricular and supraoptic nuclei in the rat. *Brain Res. Rev.* 4:275–325.

Smith, G. P., C. Jerome, and R. Norgren. 1983. Vagal afferent axons mediate the satiety effect of CCK-8. *Soc. Neurosci. Abstr.* 9:902.

Sofroniew, M. V. 1983. Direct reciprocal connections between the bed nucleus of the stria terminalis and dorsomedial medulla oblongata: evidence from immunohistochemical detection of tracer proteins. *J. Comp. Neurol.* 213:399–405.

Suga, N. 1981. Neuroethology of the auditory system of echolocating bats. In *Brain Mechansims of Sensation.* Y. Katsuki, R. Norgren, and M. Sato, eds. New York: John Wiley & Sons, Inc., pp. 45–60.

Swanson, L., and G. Mogenson. 1981. Neural mechanisms for functional coupling of autonomic, endocrine and skeletomotor responses in adaptive behavior. *Brain Res. Rev.* 3:1–34.

Travers, J. B., and R. Norgren. 1983a. Afferent projections to the oral motor nuclei of the rat. *J. Comp. Neurol.* 220:280–298.

———. 1983b. Electromyographic analysis of ingestion and rejection responses elicited by sapid stimuli in the rat. *Soc. Neurosci. Absts.* 9:189.

Yamamoto, T., R. Matsuo, and Y. Kawamura. 1980. Localization of cortical gustatory area in rats and its role in taste discrimination. *J. Neurophysiol.* (*Bethesda*). 4:440–455.

Neurophysiological Coding of Olfactory Information

II.
Peripheral Mechanisms in the Olfactory Process

Odorant Differentation at the Olfactory Mucosa: Adrian's Legacy

Lord E. D. Adrian has to be considered the father of the modern study of olfaction, especially in regard to the mechanisms which, at the level of the olfactory mucosa, underlie odorant discrimination. Based upon his recordings from the olfactory bulb, Adrian (1950, 1953, 1954) proposed three possible discrimination mechanisms at the mucosal level: (1) different receptor cells might be selectively tuned to different odorants; (2) the molecules of different odorants, in accordance with their physicochemical properties, might spread in different space-time distribution patterns across the mucosal surface; and (3) different regions of the mucosa might be selectively sensitive to different odorants due to a mucosal clustering of similarly tuned receptor cells.

In the first mechanism, each receptor cell would signal how well the molecules of different incoming odorants match its particular sensitivity. Thus, each odorant would establish a different pattern of relative discharge levels across the entire ensemble of receptor cells. Note that this would not give rise to any spatially organized mucosal activity patterns if the receptor cells of like sensitivity were simply scattered across the entire mucosal sheet.

In the second mechanism, each receptor cell would not need to respond selectively to particular odorants but could instead signal how rapidly and in what amounts its position on the mucosa had been reached by the incoming molecules of different odorants. Thus, depending upon those physicochemical properties that could influence the mucosal spread of incoming molecules, different odorants could establish activity patterns across the mucosa differing from each other both spatially and temporally.

The third mechanism combines attributes of the first two. Although, as in the first mechanism, a receptor cell would still signal how well odorant molecules match its particular sensitivity, a regional clustering of similarly selective cells would allow different incoming odorants to establish, as in the second mechanism, different space-time activity patterns across the mucosa.

Maxwell M. Mozell and David E. Hornung, Physiology Department
Upstate Medical Center, SUNY Syracuse, New York 13210

Of these three mechanisms originally proposed by Adrian, the possibility of selectively sensitive receptor cells has received the most attention. A number of studies have recorded extracellularly from single olfactory receptor cells (Gesteland et al., 1963; Takagi and Omura, 1963; Gesteland et al., 1965; Shibuya and Tucker, 1967; Mathews, 1972; Getchell, 1974; Duchamp et al., 1974; Revial et al., 1978; Chastrette, 1981; Revial et al., 1982). With some differences in specific details, these studies have reported the same general findings. Each receptor cell is indeed selectively sensitive to a particular group of odorants, but few, if any, cells appear sensitive to the same total group of odorants. Thus, the receptor cells have so far resisted a sharp classification into basic receptor cell types. Apparently each receptor cell responds to the world of odorants more or less in its own characteristic way. Although our own laboratory has contributed to this line of investigation (O'Connell and Mozell, 1969), it has not been our primary focus. We have concentrated instead upon the possibility of spatiotemporal activity patterns across the mucosa. It is this concept of Adrian's that we will now examine.

The Differential Distribution of Odorant Molecules across the Mucosal Surface: Electrophysiological Approach

To test for mucosal spatiotemporal activity patterns, we (Mozell, 1964a, 1966) recorded the summated multiunit discharges (Beidler, 1953) from two branches of the olfactory nerve, which supplied two widely separated regions of the olfactory mucosa on the roof of the bullfrog's olfactory sac (Figure 1). One branch, the most medial (MB), supplied a region of the mucosa near the external naris. The other branch, the most lateral (LB), supplied a region overhanging the internal naris. Thus, the summated MB discharge reflected the activity of a mucosal region first contacted by the odorized air entering the olfactory sac, and the LB summated discharge reflected the activity of a mucosal region farther along the flow path. To compare the activity of these two mucosal regions in response to a sniff of odorized air, we took the LB/MB ratio of the summated discharges from the two nerve branches. The smaller this ratio, the steeper would be the gradient in the falloff of activity from the mucosal region at the external naris to the mucosal region overhanging the internal naris.

The array of responses shown in Figure 2 is one of 30 such arrays recorded from 10 different animals (Mozell, 1966). Each odorant produced a characteristic median LB/MB ratio that remained rather constant from animal to animal. Thus, *d*-limonene showed almost no gradient in activity across the mucosa (LB/MB $\cong$ 1), whereas the activity gradients for geraniol and citral

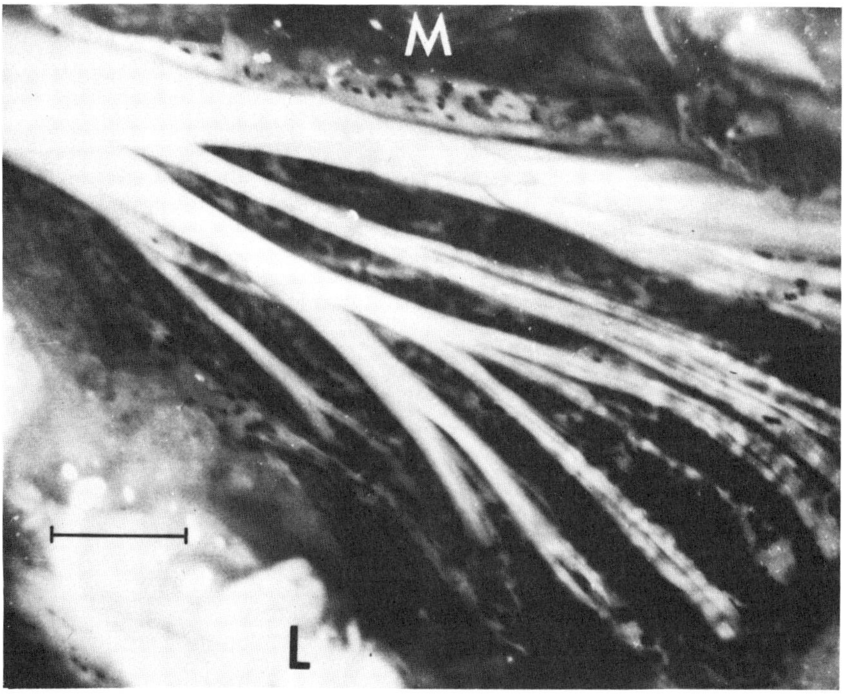

Figure 1. A bullfrog's (Rana catesbeiana) right olfactory nerve as it branches over the roof of the olfactory sac. L. lateral aspect; M, medial aspect. The calibration line represents 1 mm. (Mozell, 1964a; by permission from the American Association for the Advancement of Science.)

were appreciably steeper (LB/MB $\simeq$ 0.5). (Note that although according to the one array of Figure 2 the LB response for octane appears larger than the MB response, octane's median LB/MB ratio for all 30 arrays approached 1.00.)

In addition to the LB/MB ratio differences for different odorants, we also observed that each odorant produced a characteristic time interval between the onset of the summated multiunit discharges recorded from the two nerve branches. This was true in spite of the fact that all odorants were drawn into the olfactory sac at the same flow rates. The time intervals added a temporal component to the spatial activity gradients, thus giving, as Adrian had predicted, a spatiotemporal differentiation of odorants across the mucosa. Furthermore, the greater the time interval between the onset of the nerve branch

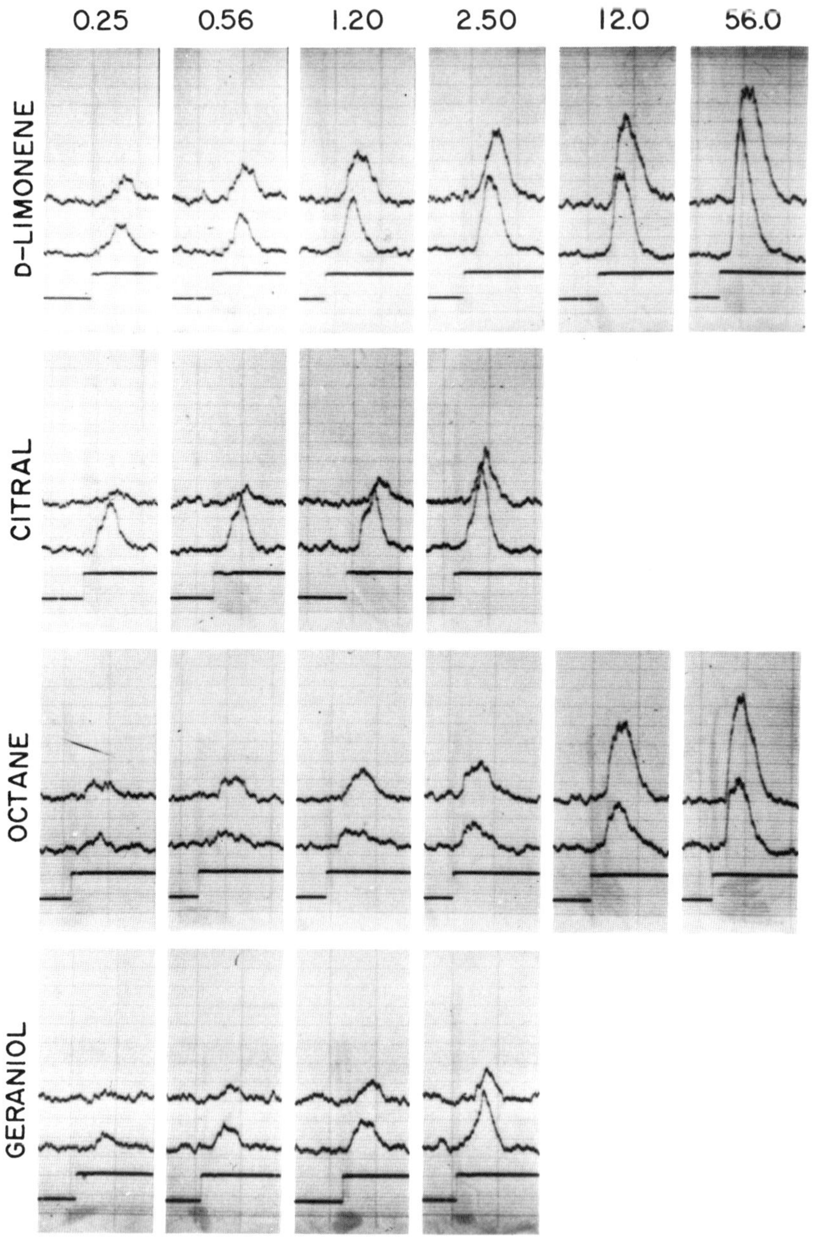
0.25
0.56
1.20
2.50
12.0
56.0
D-LIMONENE
CITRAL
OCTANE
GERANIOL

discharges for a given odorant, the smaller was that odorant's LB/MB ratio. This close relationship suggested that the time intervals and the LB/MB ratios were two measures of the same underlying mechanism.

As discussed above, Adrian had already suggested two possible candidates for this underlying mechanism. First, the sensitivity of the receptor cells might be greater for some odorants around the external naris than around the internal naris. This might be true, for instance, for citral and geraniol. For other odorants like octane and *d*-limonene, the receptor cells around the external and internal nares might be more equally sensitive. This type of regional sensitivity difference could explain the small citral and geraniol LB/MB ratios and the near-unity ratios for *d*-limonene and octane. On the other hand, the same observations could be explained by Adrian's other suggestion that the mucosa attracts and sorbs the molecules of some odorants (e.g., citral and geraniol) more strongly than it attracts and sorbs the molecules of other odorants (e.g., octane and *d*-limonene). In the time-frame of a given sniff, the molecules of the more strongly attracted odorants would be sorbed around their initial mucosal contact at the external naris, leaving few to reach the region of the internal naris. The molecules of the less-attracted odorants would remain more readily in the air phase so that when sorbed they would cover the mucosa more evenly.

To choose between these two possibilities, we reversed the flow direction of the odorized air through the olfactory sac (Mozell, 1964b). Instead of introducing the odorized air through the external naris, we introduced it through the internal naris. We reasoned that if the activity gradients (LB/MB ratios) were due to mucosal regions of differing selective sensitivity, they would not vary much with flow direction across the mucosa. That is, without some rather contrived mechanism, the receptor cells in various regions would be expected to maintain their particular selective sensitivities regardless of the direction from which they were approached. On the other hand, if the mucosal activity gradients were due to differences in the attraction and sorption of different odorants, they would be expected to vary considerably with the

---

Figure 2. Traces of summated multiunit discharges showing the responses of a bullfrog to one complete stimulus array consisting of four odorants, each at several concentrations expressed in partial pressure $\times 10^{-2}$ mm Hg. The upper response in each pair is recorded from the lateral nerve branch, and the lower response is recorded from the medial nerve branch. The stimulus marker shows only the onset of the stimulus. Vertical time lines occur once every 10 s. Stimulus duration was 3 s and stimulus volume was 0.4 cm^3. (Mozell, 1966; by permission of The Rockefeller University Press.)

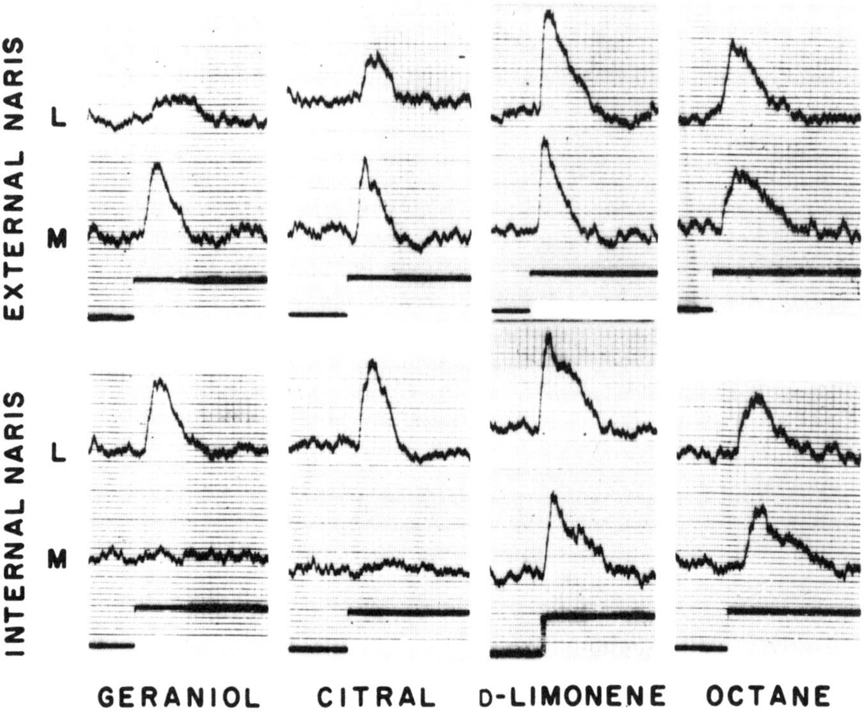

Figure 3. Traces of summated multiunit discharges showing the difference in the responses of the medial (M) and lateral (L) nerve branches to different odorants presented through the internal naris and through the external naris. Vertical time lines occur once every second. (Mozell, 1964b; reprinted by Macmillan Journals Ltd.)

direction of flow. In this case, the molecules of the strongly attracted odorants would pile up near the naris they entered, with only a few migrating farther down the flow path. Consequently, a reversal of flow direction for these odorants would be expected to reverse the direction of their mucosal activity gradients. As can be seen in Figure 3, this is exactly what occurred. When introduced through the external naris, both geraniol and citral gave the larger discharge on the medial nerve branch, whereas the lateral nerve branch gave the larger response when they were introduced through the internal naris. Therefore, the differential activity gradients seemed more explicable in terms of differential sorption and distribution of molecules than in terms of mucosal regions of selective sensitivity.

The differing time intervals between the onset of the discharges on the two nerve branches also seemed more explicable in terms of differential sorption. It can be argued that the more an odorant's molecules are sorbed by the mucosa, the longer it will take them to reach the more distant regions along the flow path. Thus, differential sorption appeared to be an explanation for both the LB/MB ratios and the time intervals between the nerve branch discharges; this seems to explain the correlation between the two measures.

If the differential sorption of odorants by the mucosa is a basis for differential odorant migration patterns across the mucosa, an analogy can be drawn between olfaction and gas chromatography (Figure 4). Gas chromatography also separates chemicals by passing them through a system in which their molecules can be partitioned between a moving phase and a stationary phase. It is this partitioning that is the basic principle of gas chromatography. The more the partitioning of a given chemical favors the stationary phase because of its attraction to that phase, the more its molecules will be held up as they pass through the system. The measurement generally taken of this process is the time (called the retention time) required by the molecules to reach a detector positioned at the end of the flow path (i.e., the end of the chromatographic column that contains the stationary phase and is the conduit for the moving phase). In the olfactory system, however, the carrier gas cannot flow for long periods of time in one direction transporting the molecules to the end of the flow path. Nonetheless, the olfactory system could still use the parti-

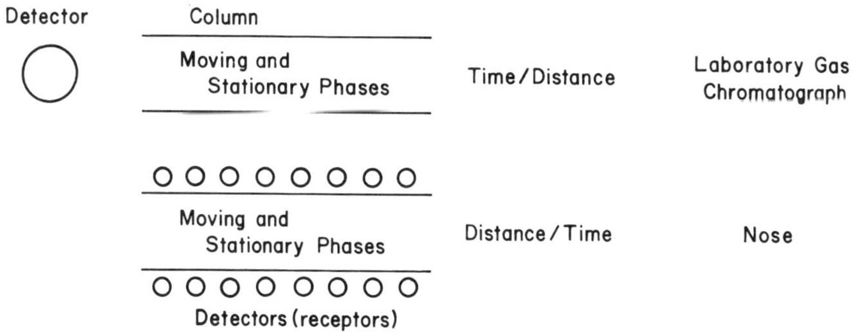

Figure 4. A schematic comparison of laboratory gas chromatographs (above) and the olfactory region of the nose (below). Both have moving and stationary phases, but the gas chromatograph has only one detector at the end of the flow path (column), whereas the nasal olfactory region has millions of detectors (receptors) along the entire flow path. See text for discussion.

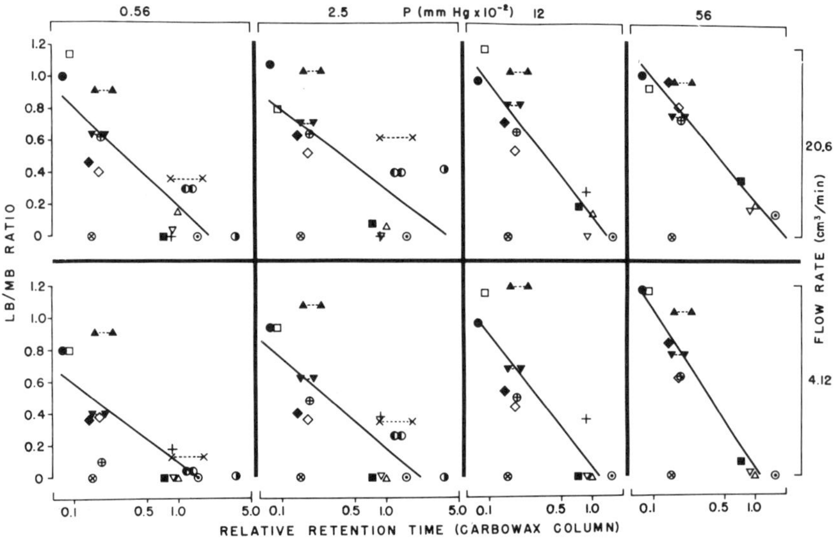

Figure 5. LB/MB ratio as a function of retention time on a Carbowax 20M column for 16 different odorants at four different partial pressures and two different flow rates. Each of the eight graphs plots the ratios (ordinate) for the odorants (● , octane; □ , nonane; ◆ , butyl acetate; ⊗ , butanol; ▼, amyl acetate; ◇ , 4-heptanone; ⊕ , heptaldehyde; ▲ , D-limonene; ■ , benzaldehyde; +, isovaleric acid; ▽, furfurol; △ , methyl benzoate; ◐, citral; ×, geraniol; ⊙, carvone; ◑ , diphenyl oxide) at one partial pressure (P; in parentheses at top) and one flow rate (in parentheses at right). Due to their comparatively low vapor pressures at 23°C, geraniol, citral, and diphenyl oxide could not be presented at 12.0 and 56.0 $\times$ 10^{-2} mm Hg. The longest and shortest retention times (connected by dashed lines) are shown for those odorants having multipeaked chromatograms. The straight lines fitted to these data do not include the butanol point (see original paper for justification). (Mozell, 1970; by permission of The Rockefeller University Press.)

tioning principle to analyze odorants because it does not have only one detector at the end of the flow path but has, instead, millions of detectors along its whole extent. Therefore, any effect of the partitioning between the moving and stationary phases (the basic principle in chromatography) can be monitored immediately along the entire pathway. Thus, the measurement can be the distance the molecules travel in a given time rather than, as in standard gas chromatography, the time required to travel a given distance. The principle is the same; only the measurement differs. Thus, the analogy proposes that, in the time-frame of a given sniff, the more readily an odorant is sorbed by the mucosa, the less distance its molecules will cover along the mucosal surface and the more likely it is that most of its molecules will be concentrated at the point of initial mucosal contact.

We pursued this analogy, using 16 different odorants chosen in accordance with their different retention times on a gas chromatographic Carbowax 20M column (Mozell, 1970). We recognized that if the activity gradients were to contribute to the encoding of different odorants, they would have to be relatively independent of some of the variations that could occur in olfactory stimulation, such as changes in odorant concentration and alterations in the flow rate of the sniff. Thus we presented the odorants at four different concentrations spaning a 100-fold increase and at two different flow rates in a 5:1 ratio. The results (Figure 5) showed that the 16 different odorants could produce a range of LB/MB ratios and that an odorant's LB/MB ratio was characteristic of that odorant in any one animal and from animal to animal. Although the LB/MB ratios of some odorants increased somewhat as concentration and/or flow rate increased, we were, at that time, more impressed by the overall stability of the LB/MB ratios. We suggested that the changes that did occur might possibly explain why the odor qualities of some odorants have been observed to vary with, for instance, concentration. Most notable in these results, however, was the strong relationship between an odorant's retention time on a Carbowax column of a gas chromatograph and its LB/MB ratio: as the retention times increased, the LB/MB ratios decreased. In other words, as the facility with which the molecules passed through the chromatographic column decreased, so did their facility to pass along the olfactory mucosa. Perhaps, we suggested, similar processes are occurring in both systems.

The Differential Distribution of Odorant Molecules across the Mucosal Surface: Gas Chromatographic and Radioisotopic Approaches

We gained further confidence in this model by using non-electrophysiological techniques. We recognized that our electrophysiological recordings monitored neural events several steps beyond the molecular migration events they were supposed to reflect. Thus, only by inference did our electrophysiological technique give evidence of differential molecular migrations across the olfactory mucosa. A more direct measure was needed.

In keeping with our gas chromatographic analogy, we decided to measure

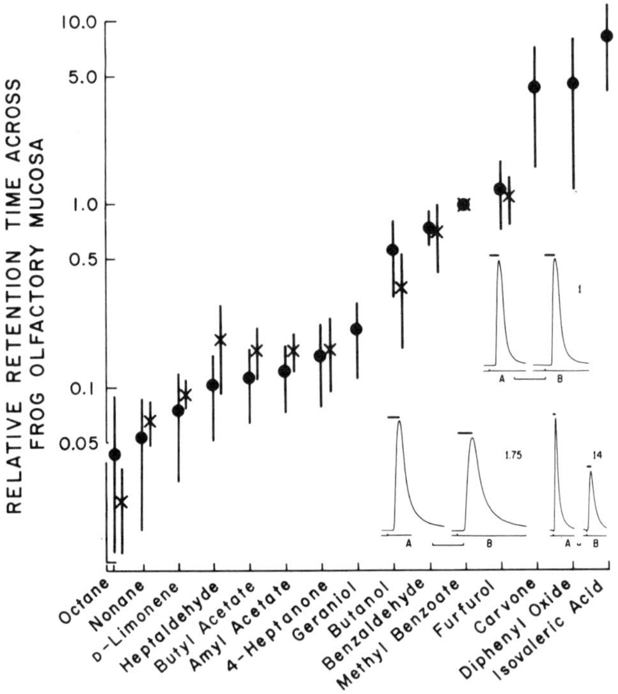

Figure 6. Mean relative retention times and standard deviations for different odorants across the olfactory mucosa. This was done under two conditions: (1) when each odorant was presented at the same partial pressure (×, ≅ 0.56 mm Hg), and (2) when each odorant was presented at its highest partial pressure at room temperature (●). For this figure, the retention times of the odorants were all made relative to the retention time of methyl benzoate, but the wide range of retention times is still obvious. With the carrier gas flowing at 25 cm^3/min, the shortest actual retention time (octane) was 1.2 s and the longest (isovaleric acid) was 274 s. (inset) Retraced chromatograms of three representative odorants: octane (top), 4-heptanone (left), benzaldehyde (right). See the original paper for a more detailed discussion. (Mozell and Jagodowicz, 1973; by permission from the American Association for the Advancement of Science.)

the retention times of different odorants as their molecules moved across the frog's olfactory mucosa just as we would measure them in a gas chromatograph. Thus, we replaced the column of a standard gas chromatograph with the intact olfactory sac of anesthetized frogs (Mozell and Jagodowicz, 1973, 1974). As shown in Figure 6, we observed that in spite of the mucosa's short flow path (compared with standard chromatographic columns), the different odorants migrated across the mucosa with very different individual retention times. The range of these retention times, which showed a 220-fold increase from the shortest to the longest, was quite impressive. Furthermore, there was a very strong correlation between these retention times and the previously measured LB/MB ratios; the longer the retention time, the smaller the LB/MB ratio. This was exactly what would be expected if the chromatographic analogy held. That is, the longer it takes an odorant's molecules to migrate across the mucosa, the more, in the time-frame of a sniff, its molecules would be retained near the external naris and the smaller its LB/MB ratio would be.

Although we could predict from their mucosal retention times how (in general) the molecules of different odorants would probably be distributed, a more detailed and certain appreciation of this distribution required a technique giving a more direct display of the sorbed molecules along the mucosal sheet. This was achieved with odorants radioactively labeled either with tritium, as in our initial work, or with carbon-14, as in our more recent work (Hornung et al., 1975; Hornung and Mozell, 1977a and 1981). In this procedure, anesthetized bullfrogs were given a sniff of radioactive odorant and were then immediately quick-frozen in liquid nitrogen to prevent any further diffusion of the odorant molecules. While still frozen, the roof of the olfactory sac was removed. This piece, together with its mucosal lining, was then cut into five 2-mm-wide sections running in sequential order from front to back. By liquid scintillation counting, we determined the radioactivity in each of the sections, and from these determinations we estimated the number of molecules per square millimeter in each section.

We have so far mapped the mucosal distribution of three odorants: butanol, octane, and butyl acetate. We chose butanol because, like geraniol, it has a small LB/MB ratio and a moderately long mucosal retention time, which led us to predict that the concentration of its molecules would show a sharply decreasing gradient from the external naris to the internal naris. From the near-unity LB/MB ratio and very short mucosal retention time of octane, one would predict a more even distribution of its molecules across the mucosa. Finally, butyl acetate's LB/MB ratio and retention time falls between those of octane and butanol, which suggests that its concentration gradient across the mucosa could also fall between octane and butanol. As can be seen in Figure 7, these predictions were indeed fulfilled as butanol showed the steepest gradient, octane showed no gradient, and butyl acetate showed a gradient between the other two. It should also be noted in this figure that, as the above line of reasoning predicted, the more readily an odorant is sorbed by the mucosa (as measured by the area under its curve), the greater is its concentration gradient from external naris to internal naris.

In analogy to the earlier electrophysiological studies, we examined, using radioactive butanol, the effects of varying odorant concentration, sniff flow rate, and sniff volume upon the mucosal concentration gradients. Likewise, the effect of reversing the flow direction was also examined (Hornung and Mozell, 1977a). Although we were impressed by the stability of the gradients as the sniffing variables were altered, we did report some quantitative changes in the gradient at the highest presented, physiologically relevant flow rate and the lowest presented, physiologically relevant volume. A 100-fold increase in the concentration, on the other hand, produced no change in the gradient. As would be predicted from the electrophysiological results, reversing the flow direction of the radioactive butanol through the olfactory sac reversed the regions of greater and lesser sorption. Therefore, the possibility of an odorant-dependent differential distribution of molecules across the olfactory mucosa,

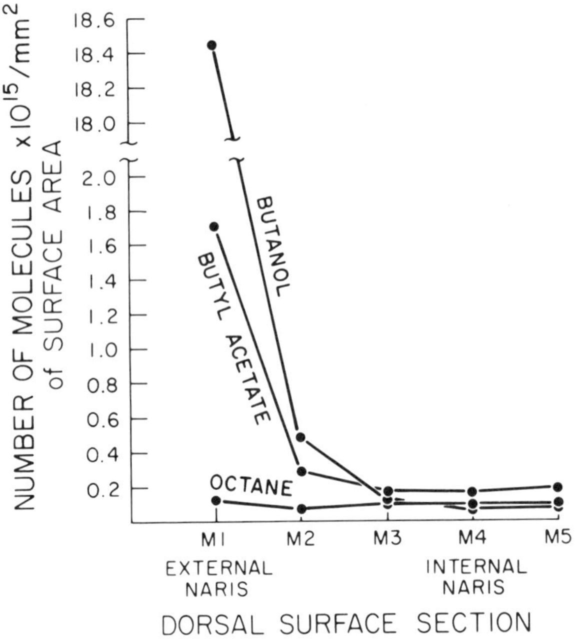

Figure 7. The surface area concentrations for each mucosal section following a sniff of radioactively labeled butanol, butyl acetate, or octane. M1–M5 denote 2-mm-wide sections of the mucosa on the roof of the olfactory sac running from the external naris to behind the internal naris. (Hornung and Mozell, 1981; by permission from Academic Press, Inc.)

earlier deduced from electrophysiological data, seemed confirmed by these more direct radioisotopic observations.

As noted above, the principles of gas chromatography would suggest that these odorant-dependent concentration gradients across the mucosa result from the partitioning of the molecules between the moving air phase and the stationary mucosal phase. We were able to pursue this suggestion by using radioactive odorants to determine mucosa/air partition coefficients (Hornung et al., 1980). We placed samples of olfactory mucosa into sealed vials containing air continually saturated with a given radioactive odorant. After equilibration, the mucosa/air partition coefficient for that odorant was given by the ratio of the radioactivity per gram of mucosa to the radioactivity per cubic centimeter of air. The mucosa/air partition coefficients for butanol, butyl acetate, and octane were 2.98×10^3, 0.62×10^3, and 0.00021×10^3, respectively. Therefore, the more the molecules of an odorant favored the mucosa phase (butanol $>$ butyl acetate $>$ octane), the steeper was that odorant's concentration gradient across the mucosa (butanol $>$ butyl acetate $>$ octane), the longer was its mucosal retention time (butanol $>$ butyl acetate $>$ octane), and the steeper was its mucosal activity gradient as shown by decreasing LB/MB ratios (butanol LB/MB ratio $<$ butyl acetate LB/MB ratio $<$ octane LB/MB ratio). All of this consistency seemed to give credence to an analogy between mucosal olfactory events and a chromatograph-like processing.

The importance of this partitioning of odorants between the air phase and the mucosal phase became apparent from still another perspective, viz., the rapidity with which odorant molecules are removed from the mucosa after presentation. Using radioactive odorants, we followed their removal by three pathways: desorption back into the air, mucus flow, and uptake into the bloodstream (Hornung and Mozell, 1977b, 1980). The removal of butanol by desorption alone is shown in Figure 8. These experiments were similar to the mapping experiments except that between the sniff of radioactive odorant and the freezing of the animal in liquid nitrogen, nonodorized air was drawn through the olfactory sac for specified times. Figure 8 shows that as the nonodorous air flow increased in time, the butanol molecules were moved along the mucosa from the region around the external naris to regions farther along the flow path and eventually into a trap at the exit from the olfactory sac just beyond the internal naris. This progressive migration of the molecules along the mucosa suggested that there is repeated sorption and desorption of the molecules between a stationary phase and a moving phase much like that which occurs in a gas chromatographic column. Note further that even after 30 min of such treatment, 75% of the butanol still remained in the mucosa.

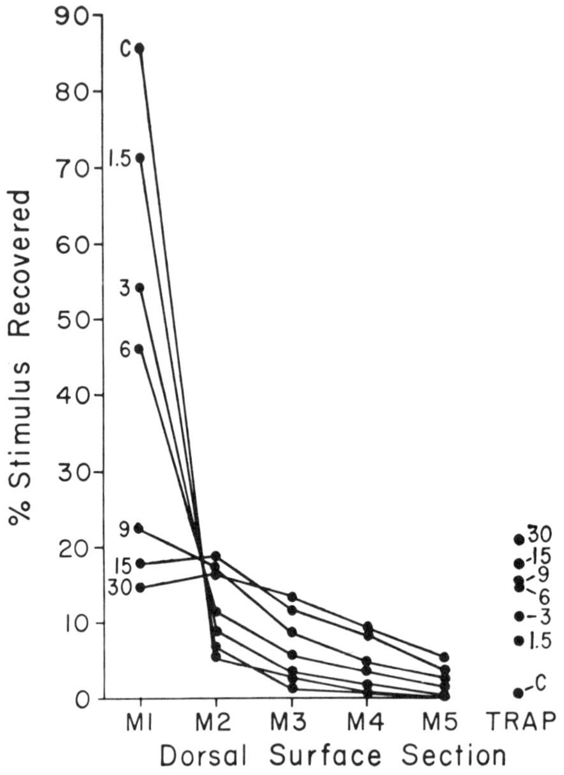

Figure 8. A family of curves showing butanol distribution gradients across the roof of the olfactory sac with time of nonodorized air flow as a parameter. The small numbers under each curve represent the time interval between the sniff of radioactive odorant and the freezing of the frog, i.e., the duration of the nonodorized air flow. C represents the control run, which is the same situation as in Figure 7, viz., the animal was frozen immediately after the sniff. M1–M5 denote the 2-mm sections of the roof of the olfactory sac with M1 including the external naris, M4 overhanging the internal naris, and M5 just rostral to the internal naris. For each time period the mean percent of the stimulus recovered from each section is given on the ordinate. At the far right of the graph the percent of the stimulus recovered from a trap at the internal naris is plotted for each time period. (Hornung and Mozell, 1977b; by permission from Elsevier Scientific Publishing Co.)

On the other hand, octane, an odorant having a mucosa/air partition coefficient strongly favoring the air phase, was rapidly removed from the mucosa by desorption back into the air. Less than 1% of the presented molecules were still found in the mucosa 3 min after a single octane sniff. Whether the difference in removal plays a role in adaptation remains to be investigated, but the fact that odorant removal, like odorant distribution, depends upon the mucosa/air partition coefficient further highlights the possibly pervasive role played by this partitioning in the peripheral olfactory process.

Mucosal Regions of Differing Selective Sensitivity

To explain the different mucosal activity patterns produced by different odorants, Adrian's proposed mechanism of a differential distribution of molecules across the mucosa seems well supported by evidence coming from several very different experimental approaches. However, from a series of studies by Moulton and his students (Kauer and Moulton, 1974; Kubie et al., 1980; Mackay-Sim and Kubie, 1981; Mackay-Sim et al., 1982), we now have evidence supporting the other mechanism proposed by Adrian as possibly underlying differential activity patterns across the mucosa. These studies have shown regional differences in the mucosa's sensitivity to different odorants. For the most part salamanders rather than frogs were used, and although one study (Kauer and Moulton, 1974) recorded the discharge from single bulbar units in response to the punctate stimulation of the mucosa, the more common approach was to record EOGs (electro-olfactograms) from the mucosal surface. One wall of the olfactory sac was removed, allowing an electrode to be placed directly upon the mucosa in different specified regions. The odorant was puffed directly onto the recording site. Note that this maneuver could produce a response at the recording site without having the molecules migrate toward that site across the mucosal surface. This would circumvent the possibility that the discharge recorded at any site would reflect a loss of molecules due to their sorption by the mucosa along the flow path from the point of introduction to the point of recording. In other words, this procedure would circumvent the chromatograph-like effect so that differences in the EOGs recorded at different sites for the same odorant could be considered the effect of differences in the sensitivity and/or the number of selectively sensitive receptors at those sites. (There are a number of other possibilities such as differences in the composition of the mucus, differences in the proximity of the receptors to the mucus-air interface, etc., but there is as yet no definitive data supporting such possibilities.)

An example of the regional sensitivity differences observed in these experiments is given in Figure 9. It is clear that the sensitivity to butanol was greater in the anterior mucosa than in the posterior, whereas the reverse was true for limonene. Several other odorants were also found to have an anterior-posterior sensitivity difference, whereas still others showed a more uniform sensitivity across the mucosal surface. Furthermore, within any region of increased sensitivity for a given odorant, there was often a smaller focus of even greater sensitivity. In general, these regional sensitivity maps remained fairly stable as the concentration of the odorants was varied, but for at least one odorant, amyl acetate, the region of greatest sensitivity shifted somewhat as the concentration went from low to high (Mackay-Sim et al., 1982). In summary, there nevertheless seems to be ample evidence to include both regional differences in

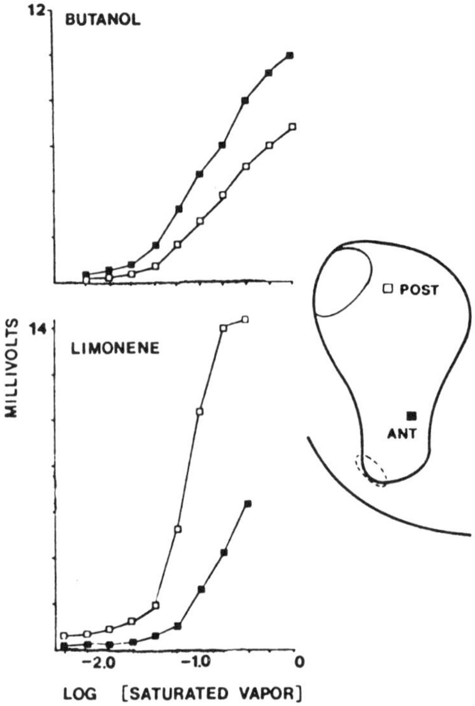

Figure 9. Magnitude of the electro-olfactogram recorded from an anterior and a posterior site on the salamander's olfactory mucosa in response to butanol and limonene, each at several concentrations. The anterior site (■) is more responsive to butanol at all concentrations, whereas the posterior site (□) is more responsive to limonene at all concentrations. (Mackay-Sim and Kubie, 1981; by permission from IRL Press.)

mucosal sensitivity and differential molecular sorption as possible contributors to the differential activity patterns across the mucosa. However, it should be noted that although differential sorption can explain, as discussed above, both the spatial and temporal aspects of the space-time activity patterns, regional sensitivity differences would appear considerably more adept at explaining the spatial aspects than the temporal aspects.

This evidence for regional differences in mucosal sensitivity seems contradictory to the results of our previously described reversed-flow experiments. Note, however, that we did not interpret the reversed-flow experiments as disproving the *existence* of regional differences in odorant sensitivity. Instead, we used these experiments only to choose between regional sensitivity differences and differential molecular distribution as the basis for certain observed mucosal activity patterns. Nonetheless, these data supporting regional differences in selective sensitivity must make us consider why these regional differences did not surface in our reversed-flow experiments.

One possibility could have simply been that the LB and MB electrodes did not sample regions of the bullfrog's mucosa that differed in sensitivity. In the salamander, the differences in sensitivity to butanol and octane, for instance, were described as occurring between the anteior and posterior mucosal regions (Figure 9). Assuming that the same anterior-posterior differences for butanol and limonene also occur in bullfrogs, we may not have observed them because the region sampled by LB is as far lateral to the region sampled by MB as it is posterior. Therefore, not only is there a species difference between the two studies, this difference may have been further compounded by not recording from analogous regions.

Another, more subtle reason for our not having observed regional sensitivity differences in the reversed-flow study may have been the strength of the sorption effect. Perhaps it was so strong for citral and geraniol that it masked the regional differences. Note first that these regional differences in sensitivity are not all-or-none, since, as shown in Figure 9, the butanol-sensitive region can give a respectable response to limonene and vice versa. Secondly, recall that the molecules of odorants highly sorbed by the mucosa almost all pile up at the region of initial mucosal contact, leaving very few, if any, to contact regions farther along the flow path (Figure 7). These odorants would give their largest responses where they first make contact, since even if there is a more sensitive region farther down the flow path, this region's response to so few molecules would be quite small. Thus, without some particular strategy to highlight the effect of regional sensitivity differences, it could have gone unnoticed in our reversed-flow study.

Parceling out the Effects of Differential Sorption and Regional Differences in Selective Sensitivity

Figure 10 schematizes a strategy (Mozell and Hornung, 1981) that could parcel out the individual contributions of regional sensitivity and molecular sorption in determining the activity patterns produced by different odorants across the mucosa. The arrowheads show the flow direction of the odorized air, and the hooks, representing electrodes, identify the nerve branch discharge being sampled. The subscripts designate the direction of flow giving rise to the recorded discharge (LB or MB). S designates the usual sniff direction from external naris to internal naris, and R represents the reverse direction from internal naris to external naris. Thus, for instance, LB_S denotes the discharge of the lateral nerve branch when the odorized air enters the olfactory sac through the external naris and flows toward the internal naris.

The diagram on the extreme left of Figure 10 is the paradigm for determining the LB/MB ratios which, according to our new designation, becomes the LB_S/MB_S ratio. In this paradigm, the discharges from both the lateral and medial nerve branches are recorded simultaneously as the air flows across the mucosa from the external naris to the internal naris. The LB_S/MB_S ratio would reflect the chromatographic effect because the molecules can be sorbed along the flow path to the lateral nerve branch. In addition, the LB_S/MB_S ratio would reflect the effect of regional sensitivity because different regions of the mucosa are compared. The effects of both these mechanisms are, therefore, combined in the LB_S/MB_S ratio.

In the middle diagram, however, only one mechanism can have an effect. In

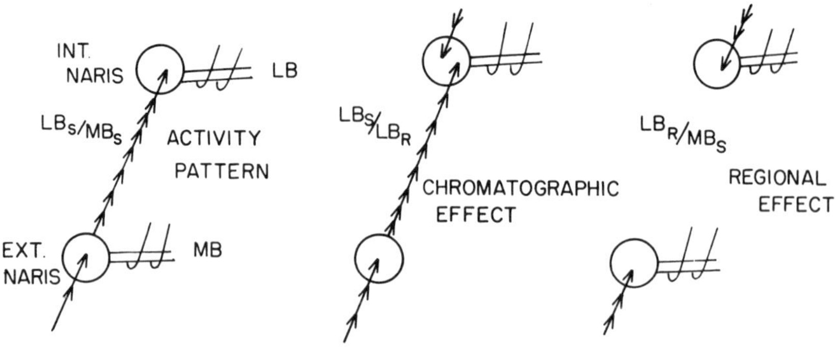

Figure 10. The strategy to parcel out the contributions of regional sensitivity and molecular sorption in determining mucosal activity patterns. See text for description.

this case LB is recorded alone under two conditons: (1) when the odorized air is brought directly to the LB region of the mucosa through the internal naris (LB_R) with no flow over the mucosa and thus with no chance for a mucosal chromatographic effect, and (2) when the air flows to the LB region over the surface of the mucosa (LB_S), allowing an opportunity for the development of a chromatographic effect. Since both recordings come from the same nerve branch, their comparison (the LB_S/LB_R ratio) cannot reflect a difference of regional sensitivities. Since, however, this ratio compares LB responses with and without molecular migration across the mucosa, it can be taken as an index of the chromatographic effect uncontaminated with the effect of regional differences in sensitivity. The smaller an odorant's LB_S/LB_R ratio, the greater is its chromatographic effect.

In the diagram on the extreme right, the lateral nerve discharge is recorded in response to direct flow through the internal naris (LB_R), and the discharge of the medial nerve branch is recorded in response to direct flow through the external naris (MB_S). In neither case does the odorized air flow over the mucosa, thus removing the opportunity for a mucosal chromatographic effect. Therefore, the LB_R/MB_S ratio can be considered an index of the regional sensitivity effect uncontaminated by the chromatographic effect. The smaller an odorant's LB_R/MB_S ratio, the greater is the mucosa's sensitivity to that odorant around the internal naris, as compared with its sensitivity around the external naris.

Note that the product of the chromatographic effect (LB_S/LB_R) and the regional sensitivity effect (LB_R/MB_S) gives the overall mucosal activity pattern (LB_S/MB_S). Note also that the arrows on the extreme left diagram and on the middle diagram can be drawn from the internal naris to the external naris rather than from external to internal. Moreover, in the middle diagram the medial nerve branch could be sampled rather than the lateral nerve branch. Under these conditions, the chromatographic effect will be indexed by the MB_R/MB_S ratio, the regional sensitivity effect will be indexed by the MB_S/LB_R ratio, and the combined overall activity pattern will be the MB_R/LB_R ratio. As compared with the earlier group of ratios, this latter group of ratios covers the chromatographic effect from the internal naris to the external naris rather than from the external naris to the internal naris. *A priori*, one would expect the chromatographic effects across the mucosa to be the same in both directions.

In our initial attempt to use the above strategy to parcel out the effects of regional sensitivity and differential sorption, we applied that strategy to the data of the 1964 reversed-flow study (Mozell, 1964b). There were six animals involved, and although our techniques and equipment did not have the finesse and precision which we can bring to bear on the same problem today, the necessary determinations (MB_S, LB_R, LB_S, and, if required, MB_R), as crude as they might have been, were still available. The results of this reanalysis of the 1964 data for three of the our odorants used are given in Figure 11. *It is to be emphasized, however, that these data (averaged over the six animals) must be considered, at best, preliminary and are presented here only as illustrative of a strategy yet to be fully developed.* The LB_S/LB_R ratios for geraniol and citral are relatively small, which, from the reasoning of the above strategy, demonstrates a rather strong chromatographic effect. This conclusion was also reached in 1964, but using the LB_R/MB_S ratios, we can now draw inferences beyond those suggested in the report of the original 1964 study. For both citral and geraniol these ratios are appreciably less than unity, which by the above strategy demonstrates a difference in regional sensitivity, with the mucosal region around the external naris being more sensitive than the region around

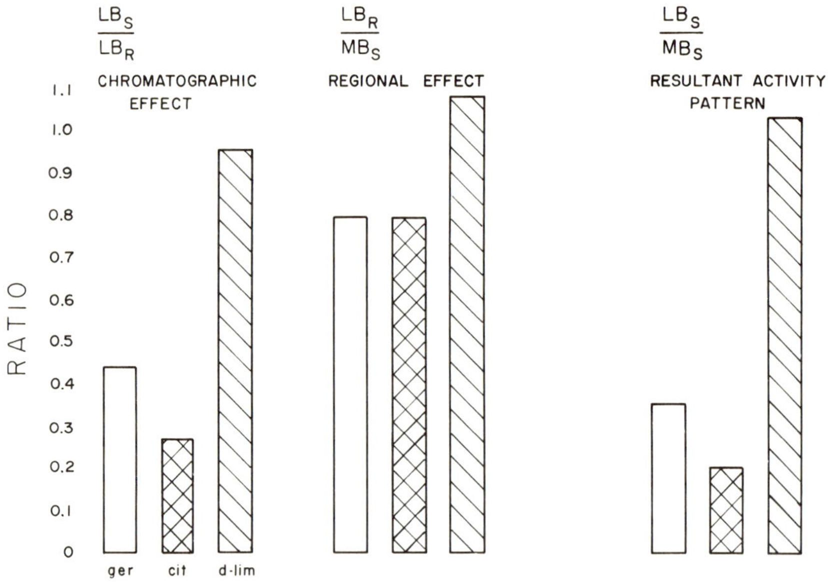

Figure 11. The ratios indexing the contributions of the chromatographic effect and regional sensitivity upon the resultant mucosal activity patterns for three odorants: geraniol, citral, and *d*-limonene. These results came from a reanalysis of earlier data (Mozell, 1964b) by the strategy depicted in Figure 10.

the internal naris. With the caution noted earlier, this can be compared with the salamander, in which the greatest sensitivity to geraniol, an odorant common to both studies, was in the anterior mucosa (Kubie et al., 1980). From this analysis it appears that the geraniol and citral mucosal activity patterns as measured by summated multiunit discharges depend upon regional differences in sensitivity as well as differential sorption patterns across the mucosa.

The *d*-limonene result is less clear. The LB_S/LB_R ratio is 0.95, which, considering the preliminary status and variability of the data, is not sufficiently different from 1.0 to be interpreted as showing a chromatographic effect. Likewise, the LB_R/MB_S ratio (1.08) cannot be interpreted as showing a difference in regional sensitivity. However, in order to make an illustrative point later on, we will exaggerate by saying that *d*-limonene shows a very slight chromatographic effect and that the mucosal region around the internal naris is slightly more sensitive to *d*-limonene than is the mucosal region around the external naris.

This reanalysis of the 1964 reversed-flow data is useful in illustrating how the chromatographic effect and the regional sensitivity differences can combine to give the activity gradients across the mucosa. The data indicates that, in a given sniff of citral or geraniol, the chromatographic effect will establish a steeply decreasing mucosal concentration gradient from the external naris to the internal naris. These decreasing concentration gradients should by themselves establish decreasing activity gradients across the mucosa. In addition, however, these decreasing concentration gradients fall upon a mucosa in which the sensitivity for geraniol and citral also decreases from the external naris to the internal naris. Like the decreasing concentrations, these decreasing sensitivities could by themselves establish decreasing activity gradients. Thus, the activity gradients due to the chromatographic effect (LB_S/LB_R) and those due to the regional sensitivity effect (LB_R/MB_S) could reinforce each other, giving, as illustrated in Figure 11, resultant activity gradients steeper than either effect could give alone. For citral, the activity gradients were 0.27 and 0.79 for the chromatographic and regional sensitivity effects, respectively, whereas the resultant activity gradient for both effects was 0.21. Similarly, for geraniol, $LB_S/LB_R = 0.44$, $LB_R/MB_S = 0.79$, and $LB_S/MB_S = 0.35$.

However, it is not necessary that the two effects always reinforce each other. As shown by the "exaggerated" *d*-limonene data, a decreasing concentration gradient from the external naris to the internal naris can fall upon an increasing mucosal sensitivity in the same direction. Thus, the chromatographic effect and the effect of regional sensitivity can counterbalance each other so that the resultant activity gradient falls between those established by each effect alone. Whether counterbalancing or reinforcing, the chromatographic and regional sensitivity effects together determine the resultant mucosal activity patterns. These patterns could vary from odorant to odorant as the weights of the two effects vary.

Another Mechanism Possibly Affecting Molecular Distributions across the Mucosa

This recent reanalysis of the 1964 reversed-flow data, along with the early results of our current reversed-flow experiments based upon the same strategy, has raised the possibility of other processes that, at the level of the olfactory mucosa, might act in concert with the chromatographic effect. These pilot data seem to indicate that the LB_S/LB_R ratio (which indexes the chromatographic effect from the external naris to the internal naris) tends to be somewhat larger than the MB_S/MB_R ratio (which indexes the chromatographic effect from internal naris to the external naris). We are considering a number of different possibilities to explain this apparent inconsistency, but one in particular is that the flow paths taken by the odorized air in the two directions might differ. For instance, the eminentia, a moundlike structure with asymmetric slopes rising from the floor of the olfactory sac, might force the air entering through the internal naris to take a longer path across the mucosa than the air entering through the external naris. In addition, the eminentia, which itself bears olfactory mucosa, could by shape and position offer a better sorbing surface to the air stream entering through the internal naris than to that entering through the external naris. In both these cases, one might expect a greater sorption of molecules going from the internal to the external naris than going in the opposite direction. Therefore, the flow from the internal naris to the external naris would be expected to give the bigger difference between the LB and MB discharges. There are, of course, more complicated possibilities, such as turbulent flow and eddy currents, which, due to the morphology of the olfactory sac, might occur more readily in one flow direction than the other. This could differentially influence the amount of odorant sorbed by different regions. Therefore, in emphasizing, as we do, that the distributions of the molecules across the mucosa depend upon the mucosa/air partition coefficient, we must not lose sight of the fact that these distributions can also depend upon how the air carrying the molecules flows over the mucosa.

We might also speculate how differing flow paths might explain another finding in our reanalysis of the 1964 reversed-flow experiment. The fourth odorant used, but not plotted in Figure 11, was octane. Octane's LB_S/LB_R ratio was 1.18, indicating that the response it produced after its molecules migrated across the mucosal surface was greater than the response it produced when such migration did not occur. This, in turn, suggests that more octane molecules reach the LB mucosal region, which overhangs the internal naris, when the air is drawn through the external naris than when it is drawn through the internal naris. A case can be made for this apparently paradoxical possibility by recalling that octane's mucosa/air partition coefficient greatly favors the air phase and by comparing the possible flow paths of the air drawn through the olfactory sac via the internal and external nares. Since octane molecules are so poorly sorbed by the mucosa, the number of these molecules arriving at the LB region would not be influenced much by the naris entered. With so little effect due to sorption, the effect of flow path could become more apparent.

Recall first that the LB region is situated on the roof of the olfactory sac directly above the internal naris and that the internal naris can be likened to a hole punched through the floor of the olfactory sac at its most lateral boundary. Thus, the flow of the airstream drawn through the internal naris toward the external naris might bend away from the direct vertical path needed to reach the LB region. Furthermore, the position of the LB region is probably too early in the flow path from internal naris to external naris for baffles such as the eminentia either to deflect the airstream toward the LB region or to mix the incoming odorized air with that already present in the LB vicinity. On the other hand, air drawn through the external naris toward the internal naris must travel the entire length of the sac to reach the LB region. In this case, the eminentia and other possible baffles might deflect the airstream enough and provide sufficient mixing to get a somewhat larger number of octane molecules to the LB region than does the flow through the internal naris. Considering the preliminary nature of the data, this detailed discussion of airflow through the olfactory sac, though it may offer a possible explanation for octane's LB_S/LB_R ratio, is, of course, conjecture. However, it is not our intention in this discussion to draw conclusions or to give convincing arguments concerning specific airflow patterns. Instead, our intention is to motivate, by giving sufficient reason, a deeper consideration of the odorant flow path, which, when combined with molecular sorption, might be central to our understanding of the olfactory process.

A Word of Caution

In reviews of this type we traditionally add a word of caution. It must be stated that the differential sorption gradients across the mucosa which we have demonstrated may simply be the result of physicochemical processes that must occur whenever molecules pass over sorbing surfaces and might not themselves represent one of the bases for odorant discrimination. (This latter possibility seems refuted by at least one behavioral experiment [Bennett, 1972].) However, even if this sorption is not itself a basis for the analysis of odorants, it could very likely affect the analysis made by the other mechanisms, since it influences the percent of the available odorant molecules accessing each of the receptor cells and each of the selectively sensitive regions. It also influences the removal of the molecules. Thus, at one extreme, the olfactory system may have to neutralize the chromatographic effect so that other mechanisms can work efficiently. At the other extreme, the olfactory system may adapt to the chromatographic effect by, for instance, having the receptor cells of particular selective sensitivity positioned where the molecules matching that sensitivity are likely to sorb. (It cannot go unnoticed in this regard that all five of the odorants to which Moulton [1981] found the posterior of the mucosa most sensitive were also "insoluble" in water. If instead they were water soluble, they could be taken up by the mucus in the anterior regions before even reaching the posterior regions.) Therefore, whether differential sorption is itself a basis for odorant differentiation or whether it merely has an impact upon those other mechanisms that really do differentiate odorants, it would seem that the olfactory process would require some stability, or at least some control, in the molecular distribution patterns established by different odorants across the mucosa. Perhaps the sniff plays a role in this regard. At any rate, we have given particular attention to those variables that might possibly affect these distributions. These variables include the flow rates and volumes of the sniffs and the concentrations of the odorants. As discussed above, we have observed some changes with the higher flow rates and the lowest volume. Although the highest flow rate so far tested was well within the frog's range of normal flow rates, we have not yet tested the highest instantaneous flow rates frogs can achieve. It is an interesting thought, however, that if animals can alter the distribution of odorants across their mucosas by altering sniff variables, they might use this technique to provide their mucosas with different scanning perspectives of the odorous environment.

Regional Sensitivity Differences and Differential Sorption: Their Possible Interplay in Discriminating Odorants

By way of summarizing the possible interplay of the chromatographic effect and the effect of regional sensitivity in establishing different activity patterns across the mucosa to differentiate odorants, we cite Kauer's (1980) diagram (Figure 12) representing different regional sensitivities across the mucosa. As described in Figure 12, each region is sensitive to the same odorants but to different degrees. Let the external naris be at the top of the figure. Consider two odorants from the response spectrum of the D-receptor cells, which have the same small mucosa/air partition coefficient. When presented, they would both establish the same increasing activity gradient beginning at the external naris. Thus, they would not be differentiated. However, if one of the odorants had a large mucosa/air partition coefficient (and thus a strong chromatographic effect), its decreasing concentration gradient across the mucosa would counterbalance the increasing sensitivity. This would produce a less steep or even reversed activity gradient by which the odorants could be separated. On

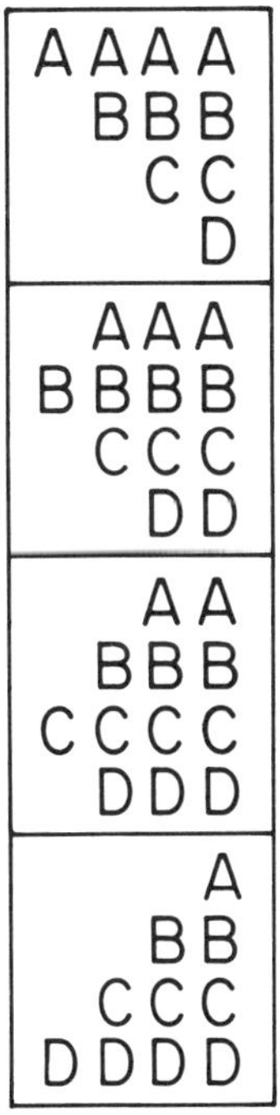

Figure 12. A schematic diagram for the regional differences in mucosal sensitivity. A given letter (A, B, C, or D) represents receptor cells responding to a similar, though not necessarily an identical, spectrum of odorants. Each square represents a different region of the mucosa. All four receptor cell spectra are represented in each region but to varying degrees, as indicated by the repetition of the letters. Thus, mucosal regions differ in their relative sensitivities to the same odorants. (Kauer, 1980; by permission from IRL Press.)

the other hand, two odorants having similar mucosa/air partition coefficients, but having different regions to which they are relatively most sensitive, will also be differentiated. Thus, the interplay of these two mechanisms (and perhaps others not yet sufficiently documented) can give a finer base for the differentiation of odorants than can each one alone.

Acknowledgment

The preparation of this manuscript and many of the studies reported herein were supported by NIH grant NS 03904.

References

Adrian, E. D. 1950. Sensory discrimination with some recent evidence from the olfactory organ. *Br. Med. Bull.* 6:330–333.

———. 1953. The mechanism of olfactory stimulation in the mammal. *Adv. Sci.* 9:417–420.

———. 1954. The basis of sensation—some recent studies of olfaction. *Br. Med. J.* 1:287–290.

Beidler, L. M. 1953. Properties of chemoreceptors of tongue of rat. *J. Neurophysiol. (Bethesda).* 16:595–607.

Bennett, M. H. 1972. Effects of asymmetrical mucosal stimulation upon two-odor discrimination in the rat. *Physiol. Behav.* 9:301–306.

Chastrette, M. 1981. An approach to a classification of odours using physiochemical parameters. *Chem. Sens.* 6:157–163.

Duchamp, A., M. F. Revial, A. Holley, and P. MacLeod. 1974. Odor discrimination by frog olfactory receptors. *Chem. Sen. Flav.* 1:213–233.

Gesteland, R. C., J. Y. Lettvin, W. H. Pitts, and A. Rojas. 1963. Odor specificities of the frog's olfactory receptors. In *Olfaction and Taste* (Proceedings of the First International Symposium). Y. Zotterman, ed. Oxford: Pergamon Press Ltd., pp. 19–44.

———. 1965. Chemical transmission in the nose of the frog. *J. Physiol. (Lond.).* 181:525–559.

Getchell, T. V. 1974. Unitary responses in frog olfactory epithelium to sterically related molecules at low concentrations. *J. Gen. Physiol.* 64:241–261.

Hornung, D. E., R. D. Lansing, and M. M. Mozell. 1975. Distribution of butanol molecules along bullfrog olfactory mucosa. *Nature (Lond.).* 254:617–618.

Hornung, D. E., and M. M. Mozell. 1977a. Factors influencing the differential sorption of odorant molecules across the olfactory mucosa. *J. Gen. Physiol.* 69:343–361.

———. 1977b. Odorant removal from the frog olfactory mucosa. *Brain Res.* 181:488–492.

———. 1980. Tritiated odorants to monitor retention in the olfactory and vomeronasal organs. *Brain Res.* 181:488–492.

———. 1981. Accessibility of odorant molecules to the receptors. In *Biochemistry of Taste and Olfaction.* R. H. Cagan and M. R. Kare, eds. New York: Academic Press, Inc., pp. 33–45.

Hornung, D. E., M. M. Mozell, and J. A. Serio. 1980. Olfactory mucosa/air partitioning of odorants. In *Olfaction and Taste VII* (Proceedings of the Seventh International Symposium). H. van der Starre, ed. London: IRL Press, pp. 167–170.

Kauer, J. S. 1980. Some spatial characteristics of central information processing in the vertebrate olfactory pathway. In *Olfaction and Taste VII* (Proceedings of the Seventh Internaitonal Symposium). H. van der Starre, ed. London: IRL Press, pp. 227–236.

Kauer, J. S., and D. G. Moulton. 1974. Responses of olfactory bulb neurones to odor stimulation of small nasal areas in the salamander. *J. Physiol. (Lond.)*. 243:717–737.

Kubie, J. L., A. Mackay-Sim, and D. G. Moulton. 1980. Inherent spatial patterning of response to odorants in the salamander olfactory epithelium. In *Olfaction and Taste VII* (Proceedings of the Seventh International Symposium). H. van der Starre, ed. London: IRL Press, pp. 163–166.

Mackay-Sim, A., and J. L. Kubie. 1981. The salamander nose: a model system for the study of spatial coding of olfactory quality. *Chem. Sens.* 6:249–257.

Mackay-Sim, A., P. Shaman, and D. G. Moulton. 1982. Topographic coding of olfactory quality: odorant specific patterns of epithelial responsivity in the salamander. *J. Neurophysiol. (Bethesda)*. 48:584–596.

Mathews, D. F. 1972. Response patterns of single neurones in the tortoise olfactory epithelium and olfactory bulb. *J. Gen. Physiol.* 60:166–180.

Moulton, D. G. 1981. Structure-activity relations in olfaction. In *Odor Quality and Chemical Structure*. H. R. Moskowitz and C. B. Warren, eds. Washington, DC: American Chemical Society, pp. 211–230.

Mozell, M. M. 1964a. Olfactory discrimination: electrophysiological spatiotemporal basis. *Science (Wash. DC)*. 143:1336–1337.

———. 1964b. Evidence for sorption as a mechanism of the olfactory analysis of vapors. *Nature (Lond.)*. 203:1181–1182.

———. 1966. The spatiotemporal analysis of odorants at the level of the olfactory receptor sheet. *J. Gen. Physiol.* 50:25–41.

———. 1970. Evidence for a chromatographic model of olfaction. *J. Gen. Physiol.* 56:46–63.

Mozell, M. M., and D. E. Hornung. 1981. Imposed and inherent olfactory mucosal activity patterns: an experimental design prompted by the work of David Moulton. *Chem. Sens.* 6:267–276.

Mozell, M. M., and M. Jagodowicz. 1973. ChSomatographic separation of odorants by the nose: retention times measured across in vivo olfactory mucosa. *Science (Wash. DC)*. 181:1247–1249.

Mozell, M. M., and M. Jagodowicz. 1974. Mechanisms underlying the analysis of odorant quality at the level of the olfactory mucosa. I. Spatiotemporal sorption patterns. *Ann. NY Acad. Sci.* 237:76–90.

O'Connell, R. J., and M. M. Mozell. 1969. Quantitative stimulation of frog olfactory receptors. *J. Neurophysiol. (Bethesda)*. 32:51–63.

Revial, M. F., A. Duchamp, and A. Holley. 1978. Odor discrimination by frog olfactory receptors: a second study. *Chem. Sens.* 3:7–21.

Revial, M. F., G. Sicard, A. Duchamp, and A. Holley. 1982. New studies on odor discrimination in the frog's olfactory receptor cells. I. Experimental results. *Chem. Sens.* 7:175–190.

Shibuya, T., and D. Tucker. 1967. Single unit responses of olfactory receptors in vultures. In *Olfaction and Taste II* (Proceedings of the Second International Symposium). T. Hayashi, ed. Oxford: Pergamon Press Ltd., pp. 219–234.

Takagi, S. F., and D. Omura. 1963. Responses of the olfactory receptor cells to odors. *Proc. Jpn. Acad.* 39:253–255.

12.

The Central Olfactory Connections and Their Cells of Origin

It is a pleasure to describe this work on the central olfactory projections in a symposium honoring Carl Pfaffmann. My own contribution to this area began in the Pfaffmann laboratory soon after he moved to The Rockefeller University. Like so many others who have worked with him, I was impressed with his article "The Pleasures of Sensation" (Pfaffmann, 1960) and encouraged by the skills I learned from both him and those he gathered into his laboratory.

The primary focus of this chapter is the pattern of organization of the central olfactory connections and how that pattern might relate to olfactory discrimination. I will consider the variety of olfactory bulb output cells and speculate whether they provide for separation of the central pathways into systems dedicated to different types of response to olfactory stimulation. This discussion of the olfactory projections will be selective, dealing only with the projections of the main olfactory bulb and only with those connections that bear on my thesis. I will suggest that the properties of the output cells vary with their type and with their projection. More comprehensive descriptions of the anatomy of these projections can be found in the review by Heimer (1976) and the research reports of Price (1973), Krettek and Price (1977), and Haberly and Price (1978a, b).

Outline of the Projection Paths

Figure 1 illustrates some of the secondary projection areas on the ventral surface of the brain. Output cell axons exit the olfactory bulb in a broad sheet, the lateral olfactory tract (LOT). The LOT passes over the anterior olfactory nucleus (AON) as a broad sheet from which many of these axons send collaterals to the AON. The LOT becomes a more compact bundle as it passes between the piriform cortex and olfactory tubercle. Axons innervating the secondary projection areas illustrated in Figure 1 form a thin outer layer (layer

---

John W. Scott, Department of Anatomy
Emory University, Atlanta, Georgia 30322

I α in the terminology used by Heimer [1976] and Price [1973]). These axons follow parallel paths without much divergence as they pass over the posterior piriform cortex and entorhinal cortex (Scott et al., 1980). The Golgi evidence of Stevens (1969) suggests that some of these axons have terminals along a considerable part of this trajectory. The width of the olfactory terminal region (layer IA) is not uniform over the cortex. It generally becomes thinner and

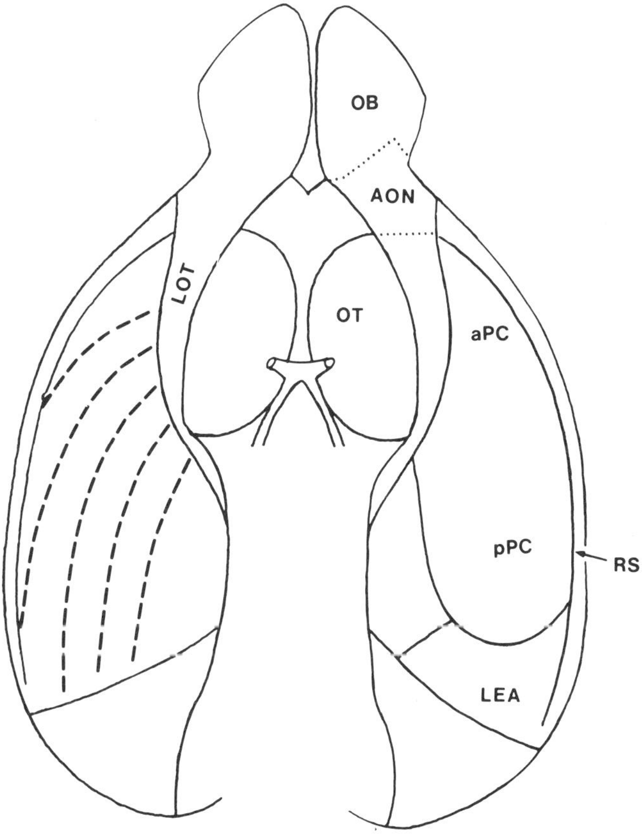

Figure 1. A schematic view of the rat brain from the ventral surface indicating the major secondary olfactory areas that will be discussed here. The dashed lines on the left indicate the paths of axons across the surface of the piriform and entorhinal cortices. Abbreviations: OB, olfactory bulb; AON, anterior olfactory nucleus; LOT, lateral olfactory tract; OT, olfactory tubercle; aPC, anterior portion of piriform cortex; LEA, lateral entorhinal cortex; RS, rhinal sulcus.

contains a lower density of terminals in regions of the cortex more distant from the LOT (Price, 1973; Schwob and Price, 1978).

The outputs of these secondary projection areas are projections to deeper structures and direct neocortical connections. I will discuss the projections to the frontal neocortex, which are both direct and through the mediodorsal nucleus of the thalamus, and the projections to the lateral hypothalamus and midbrain.

Neocortical Connections

In recent years, there have been several reports of a neocortical olfactory area. Cells of the orbitofrontal cortex of the Macaque have been shown to respond to odor stimuli (Tanabe et al., 1975b), as have cells of a nearby region of the rabbit cortex (Onoda and Iino, 1980). Anatomical studies indicate that this region corresponds to the ventral agranular and lateral orbital cortices (Krettek and Price, 1977; Price and Wiegand, 1982). This cortical area receives thalamic input from the central segment of the mediodorsal nucleus of the thalamus (Krettek and Price, 1977), which receives its olfactory input from cells of the deep layer of the olfactory tubercle and from the deep layer of the most medial part of the piriform cortex. These connections have been reviewed by Heimer (1976) and by Benjamin et al. (1982), who have also reviewed evidence that neurons of the central segment of the mediodorsal nucleus respond to electrical stimulation of the olfactory bulb. In addition to the transthalamic route to neocortex, there are apparently direct connections to this cortical region from piriform cortex, entorhinal cortex, and anterior olfactory nucleus. Several authors have indicated direct projections from the piriform cortex to ventral agranular insular cortex and lateral orbital cortex (Haberly and Price, 1978a; Krettek and Price, 1977; Motokizawa and Ino, 1981; Reep and Winans, 1982; Price and Wiegand, 1982). Motokizawa and Ino (1983) noted short latency (4 ms) responses in this region of cortex after stimulation of piriform cortex and noted that these responses disappeared after cuts along the rhinal sulcus, indicating that the transcortical route is functional.

Recordings of unit responses of neocortical cells to odor stimulation have suggested that these cells are more highly selective in their odor responses than are cells of the olfactory bulb or other parts of the olfactory projection (Tanabe et al., 1975b; Onoda and Iino, 1980). Behavioral experiments after lesions of the frontal neocortical region also indicate that it is very important in olfactory discrimination (Tanabe et al., 1975a; Eichenbaum et al., 1980). It is of interest here that Slotnick and Berman (1980) failed to demonstrate olfactory discrimination deficits after LOT lesions that denervated a large portion of the piriform cortex, entorhinal cortex, and amygdala. These lesions may have left input to the neocortical area through anterior piriform cortex, AON, and the mediodorsal nucleus pathways relatively intact.

Hypothalamic-Midbrain Connections

The olfactory projection through the lateral hypothalamus-midbrain pathway arises most heavily from the deep layers of the olfactory tubercle (Scott and Leonard, 1971; Scott and Chafin, 1975; Heimer, 1976). There is also a contribution from the ventral and posterior part of the AON (Haberly and Price, 1978a; Broadwell, 1975). Unit spike responses to odor stimulation and to electrical stimulation of the olfactory bulb can be recorded in the hypothalamus (Scott and Pfaffmann, 1972; Scott, 1977; Onoda and Kogure, 1981). In our studies, we found these responses and the associated axons restricted to a ventral lateral portion of the medial forebrain bundle. Although there is a strong terminal region in the nuclei gemini at the hypothalamic-midbrain border (Scott and Leonard, 1971), Heimer (1976) found little electron microscopic evidence of synaptic endings for this system in the anterior part of the hypothalamus. The recordings made from posterior hypothalamus, rostral to the nuclei gemini, showed some characteristics of soma spikes, i.e., large biphasic spikes of greater than 1–ms duration, whereas such spikes were not common from the anterior hypothalamus (Scott and Pfaffmann, 1972).

In contrast to the odor responses of frontal cortex units and the odor discrimination deficits after lesions of frontal cortex, which indicate that the frontal cortex functions in odor discrimination, the hypothalamic-midbrian pathway does not appear to have a discriminatory function. Instead, this pathway may serve a nonspecific arousal or orienting function. Motokizawa (1974) has reported that olfactory input reaches the midbrain reticular area via the medial forebrain bundle and that arousal of the cortical EEG by odor is dependent upon the integrity of the medial forebrain bundle (Motokizawa and Furuya, 1973). Hypothalamic odor responses poorly discriminate between odors (Scott and Pfaffmann, 1972; Onoda and Kogure, 1981). I have reported that hypothalamic odor responses habituate more rapidly than odor responses of neurons recorded in the olfactory bulb and have repeated the earlier observation that hypothalamic odor responses are dependent upon the state of the animal as measured by the cortical EEG (Scott, 1977).

Spatial Organization of Olfactory Projections

Three particular issues relate to the olfactory projections and their functions. The first is the question of spatial organization of these projections, the second is the question of morphological and functional differentiation between projection cell types, and the third is the interneuronal and centrifugal effects on these output cells.

Projection of the Receptors to the Olfactory Bulb

In this volume (Chapter 11), there is a description of the spatial and temporal factors in the stimulation process imposed by the physical and chemical properties of odors and of the nasal mucosa. These processes and selective distributions of cells with particular receptor properties were discussed by Moulton (1976) and appear to be the basis of a spatially organized receptor sheet. One question that has concerned some workers in the field of olfactory neuroanatomy is whether this spatial distribution of odor responses contributes to the process of discrimination. Furthermore, there has been interest in whether olfactory discrimination based on spatial distribution of sensitivity in the mucosa necessarily involves a spatial or "mucosotopic" organization in the central olfactory projection. There is general agreement that a spatial topography exists in the connections from the receptors to the olfactory bulb. This topography has been demonstrated in amphibians and in mammals by the use of electrical stimulation, degeneration after lesions, and axonal transport of labeled materials. Many of these experiments have been reviewed by Moulton (1976).

These reports have emphasized the diffuseness of this projection; however, there are factors in the makeup of the receptor surface that make the degree of precision of the topography difficult to evaluate. In the mammal, there are parts of the mucosa that cannot be easily reached. Lesions or electrical stimulation in posterior parts of the mucosa affect axons from more rostral parts, since the axons of the receptors lie just beneath the surface. It is also likely that amino acids or other labeled substances used to study transport by axons of the system will be spread in the mucus overlying the receptors. Nevertheless, while the exact details of this projection are still in doubt, it seems unlikely that the precision of this projection is nearly as high as that found in, for example, the retinogeniculate pathway.

Olfactory Bulb Output to Piriform and Entorhinal Cortices

The projection paths from the olfactory bulb to the olfactory cortices have been studied with many of the same techniques that were applied to the projections of the receptors. Many investigators have concluded that, while the LOT is organized as a topographic representation of the olfactory bulb, the projection from the bulb to the cortex and to the olfactory tubercle is very diffuse. Some have concluded that individual projection cells probably project via collaterals to all parts of the secondary olfactory area (Price and Sprich, 1975; Luskin and Price, 1982). The interpretation of these data is limited by the techniques and by the anatomical arrangement of the tissue. Among these difficulties are the length of dendrites (up to 1.8 mm for some mitral cells in the rat) and the fact that axons from rostral bulb pass near the projection cells of the posterior bulb. This means that, as in the mucosa, it is hard to make manipulations of a local region that do not affect cells at distal points. Nevertheless, it is clear that if there is topography in the overall projection of the bulb to the cortex, that topography is not very precise.

In my laboratory, we have investigated the organization of the bulbar output using retrograde transport of horseradish peroxidase (HRP) from the piriform cortex or olfactory tubercle (Scott, et al., 1980). The purpose was to test for a spatially organized distribution of projections by making small injections into layer I and statistically analyzing the distribution of labeled cells within the olfactory bulb. The rationale was that if the spatial distribution is meaningful and is constant from animal to animal, identical injections should lead to identical distributions of labeled cells. The similarity of these distributions of labeled cells should decrease systematically as injections are made far apart. To describe similarity of distribution of labeled cells, we divided the bulb into 30 standardized regions of approximately equal surface. We counted the cells in each region and compared the distributions in any two brains by a rank order correlation coefficient over the 30 regions. We called these correlation coefficients "similarity coefficients." Distance between injection sites was measured taking into account the parallel, curved paths of axons in layer Iα of the cortex. Each of 30 rats with cortical injections was compared with the other 29 by correlating the "similarity coefficients" and the distance between injection. Significant correlations ($P < 0.01$) were found for 20 of the 30 brains. An overall mean correlation was calculated from these values using the Z-transform. This gave a value of $R = -0.51$), which indicated a statistically significant ($P < 0.01$) tendency for the similarity of distributions of labeled cells to increase as the distance between injection sites decreased. This correlation accounts for 25% of the variance in the distribution of labeled cells. Considering the potential sources of error in this study (unreliability in assigning counts to regions, unreliability in assessing the injection site, etc.), it seems an impressive finding. It does not, however, indicate that there is a systematic topographic relationship, only that the cells of different parts of the bulb are not connected to all parts of the cortex with equal probability.

We drew composite maps of the projections to particular parts of the cortex, indicating the general form of the projection. These maps have distinct features such as a tendency for cells of the anterior dorsal medial portion of the olfactory bulb to project to the posterior part of the cortex, whereas cells of the anterior dorsal lateral bulb project to the anterior part of the cortex. On the other hand, these maps also show a diffuse quality in that all injections labeled cells from all parts of the bulb.

There are a number of possible models that could produce these broad distributions. One pair of alternatives would be that of cells with collaterals to many parts of the cortex vs. intermingling of axons of cells without collaterals. There are many cells that do send collateral branches to widespread parts of the piriform cortex. For example, in our HRP material, we were able to see the branch point of some of these axons as they joined the compact LOT from the fiber layer of the cortex (layer Iα), although we were not able to identify where the other branch went (Scott et al., 1980). With antidromic activation of mitral cells by electrical stimulation of the projection areas (Scott, 1981; Schneider and Scott, 1983), we found cells that could be antidromically activated from both the piriform cortex and olfactory tubercle or from two separated areas on the piriform cortex. Luskin and Price (1982) applied the double-label retrograde fluorescent tracer technique to this question and reported that a large portion (up to 67%) of mitral cells could be double-labeled by injection of two dyes into different parts of piriform cortex. They did not report that cells giving rise to these collaterals were located in any particular part of the olfactory bulb or showed any particular morphology.

Olfactory Bulb Projections to Olfactory Tubercle

One projection area that does appear to have some degree of spatially organized input from the olfactory bulb is the olfactory tubercle. Several orthograde transport and degeneration studies have indicated that the medial part of the olfactory tubercle receives its major input from the medial and posterior portion of the olfactory bulb. We reviewed some of these reports in a previous paper (Scott et al., 1980). Our results with retrograde HRP transport also supported the conclusion that the tubercle receives a preferential input from the ventral part of the olfactory bulb.

Mapping antidromically activated projection cells by electrical stimulation of the olfactory tubercle and piriform cortex also showed differential input to the tubercle from ventral olfactory bulb (Scott, 1981). As noted in the next section, many tufted cells could be antidromically activated from the olfactory tubercle but not from the posterior piriform cortex (Scott, 1981; Schneider and Scott, 1983).

It may be that this pattern of innervation of the olfactory tubercle is related to the types of odor responses seen from cells of the lateral hypothalamus. Hypothalamic odor-responsive neurons receive their input through the olfactory tubercle, and these cells appear to be very nonselective in their responses to odors. Determining whether this is a result of input through tufted cells, of input from the ventral portion of the olfactory bulb, or of circuitry within the olfactory tubercle and hypothalamus will require more information about the specific contacts made by mitral and tufted cells in the olfactory tubercle and their relationship to the output cells of the tubercle that ultimately project their axons into the lateral hypothalamus. The effects of the association connections to the olfactory tubercle from other secondary olfactory regions, such as the piriform cortex (Haberly and Price, 1978a), must also be taken into account in evaluating the role of the olfactory tubercle in reactions to olfactory information.

Projection Cell Subtypes

Morphology

Figure 2 is a schematic representation of the layers of the olfactory bulb and the major projection cells. This figure is based on material from our laboratory in which small iontophoretic injections of HRP have been used to fill and reconstruct individual output and interneuronal cells. The output cells have typically been divided into mitral and tufted cells. Usually the tufted cells have been divided into internal, middle, and external cells according to the depth of their somata within the external plexiform layer (EPL). This terminology, based on the work of Cajal, is followed in most modern work (e.g., Haberly and Price, 1977; Macrides and Schneider, 1982). Recently Macrides and Schneider, using Golgi-impregnated material from hamsters, subdivided the external tufted cell population into three classes: those with no secondary dendrites, those with short, densely branching secondary dendrites, and those with longer, sparsely branching secondary dendrites. In our material, we observe only the cells with longer, sparsely branching dendrites because the cells with shorter dendrites are obscured by the injection sites. We saw middle tufted cells and external tufted cells with long, sparsely branching dendrites that fit the description of Macrides and Schneider (1982). As they observed,

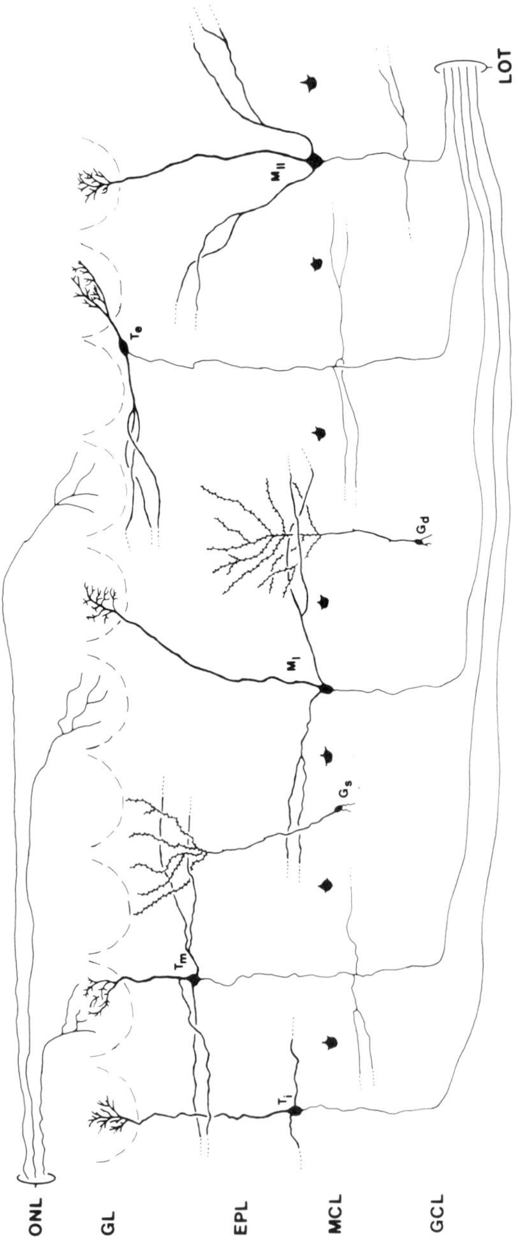
ONL
GL
EPL
MCL
GCL
LOT
M_{II}
T_e
G_d
M_I
G_s
T_m
T_i

we also note that the dendrites of these cells ramify in the superficial part of the external plexiform layer, with the dendrites of external tufted cells lying slightly superficial to those of middle tufted cells (Orona et al., 1983a). There are some other differences between these cells, particularly the fact that the dendritic ramifications of the external tufted cells are less symmetric.

Among the mitral cells, we see two classes: one with long basal dendrites that lie in the deep EPL near the mitral cell body layer as described by Scheibel and Scheibel (1975) and a second class with shorter basal dendrites (about the length of middle tufted cell dendrites) that mingle with middle tufted cell dendrites in the middle of the EPL. A few cells like those of this second class have recently been described by Kishi et al. (1982) after intracellular dye injections. We have called these cells type I and type II mitral cells, respectively. Type II mitral cells, as well as both types of tufted cells, emit collateral branches just beneath the mitral cell body layer or in the upper part of the granule cell layer near the coronal plane containing the cell body. When the type I mitral cells have collaterals in the olfactory bulb, they occur a few hundred micrometers caudal to the cell. None of these collaterals could be seen to reenter the EPL.

Figure 3 shows the distributions of labeled mitral and tufted cells of the various subtypes seen after small HRP injections into the EPL. The EPL could be divided into three sublaminae. Most of the labeled type I mitral cells were found after injections in the lower 40% of the EPL. Injections in the intermediate 40% of the EPL labeled approximately equal numbers of middle tufted cells and type II mitral cells. Injections in the upper 20% of the EPL labeled mostly external tufted cells. The few mitral cells and internal tufted cells labeled by superficial injections, all had apical dendrites that could be traced into the injection sites.

We have observed relatively few internal tufted cells with our HRP injections (Figure 3). The dendritic patterns of these cells were not exclusively of one type. We could not detect characteristics other than the position of the soma that reliably distinguish internal tufted cells from mitral cells.

Figure 2. A cross-sectional view of the lamination of output cells and granule cells in the olfactory bulb. Layers: ONL, olfactory nerve layer; GL, glomerular layer; EPL, external plexiform layer; MCL, mitral cell body layer; GRL, granule cell body layer. Cell types: T_e, external tufted cell (with sparsely branching basal dendrites); T_m, middle tufted cell; T_i, internal tufted cell; M_i, mitral cell with deep basal dendrites; M_{II}, mitral cell with superficial basal dendrites.

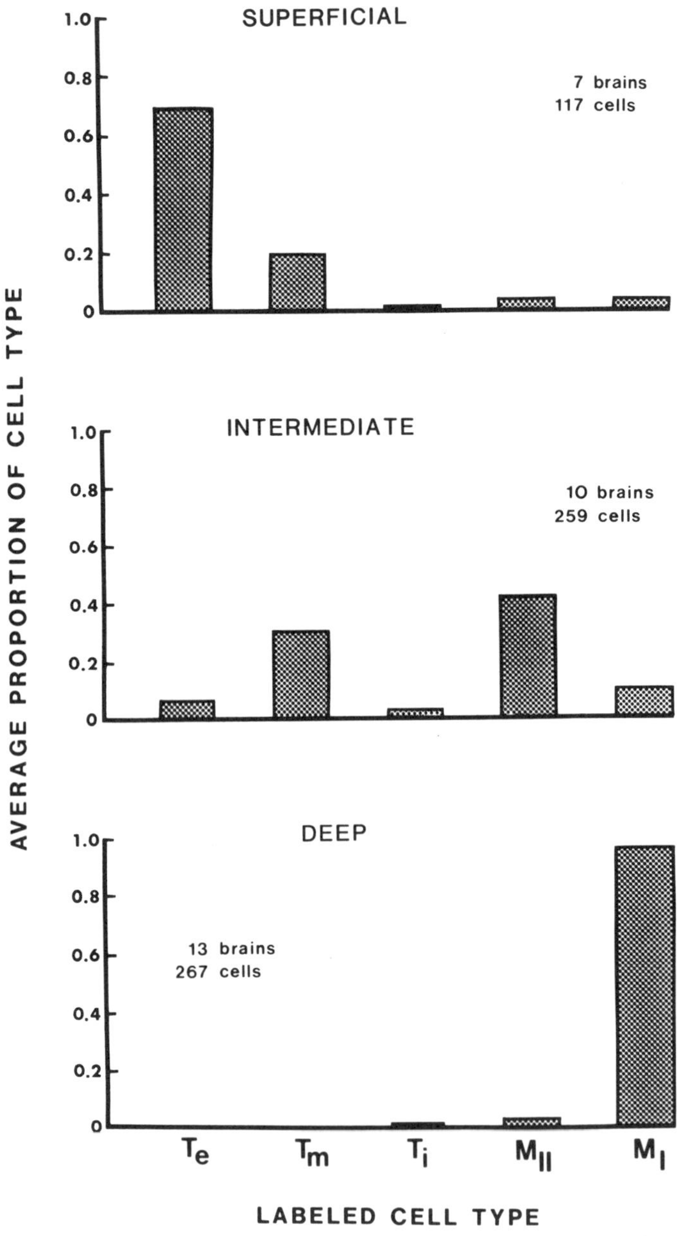
SUPERFICIAL
7 brains
117 cells
INTERMEDIATE
10 brains
259 cells
DEEP
13 brains
267 cells
1.0
0.8
0.6
0.4
0.2
0
AVERAGE PROPORTION OF CELL TYPE
Te
Tm
Ti
MII
MI
LABELED CELL TYPE

There is direct evidence that external and middle tufted cells differ in their central projections from mitral cells. The first applications of HRP retrograde tracing to this system were by Haberly and Price (1977) and by Skeen and Hall (1977), who demonstrated that while the axons of many tufted cells do project out of the olfactory bulb into the LOT, they do not project to the caudal parts of the secondary olfactory projection area. Haberly and Price (1977) reported that, with progressively more caudal injections, there were progressively fewer of the superficial tufted cells labeled. Scott et al. (1980) confirmed these findings on tufted cells and found no tufted cells labeled after injections into the posterior piriform cortex.

These observations provide some encouragement that functional cell types exist with different axonal projections. Since the type I and type II mitral cells differ in the collaterals they emit in the olfactory bulb, they may well differ in their terminations within the secondary olfactory areas. It is possible that by restricting study to one of those cell types one might see spatial patterns of distribution that we have not yet seen by studying the whole projection.

Electrophysiology

The evidence for differential distribution of mitral and tufted cell axons suggested that if these cell types are differentially sensitive to odors in terms of their response to intensity ranges or their response to quality ranges, they may selectively channel information to different central regions. We have done some electrophysiological studies of these cells, first demonstrating that we could reliably identify the cells during recording and, second, observing the responses to electrical stimulation of the olfactory nerve. We felt that the antidromic identification was crucial since, otherwise, we could not be sure that spikes recorded from the EPL were from tufted cell somata rather than from mitral cell dendrites. We approached the identification of projection cells by placing stimulating electrodes on the LOT and on the central projection areas to antidromically activate the mitral and tufted cells. From the HRP

Figure 3. Plots of numbers of projection cells of various types labeled by injections of HRP at different positions within the EPL. The injection positions were measured at the centers of the HRP injections sites. The sites were 50–100 μm in diameter. The positions are classified as superficial if the centers were in the upper 20% of the EPL (0.8–1.0), intermediate if in the middle 40% (0.4–0.79), and deep if in the lower 40% (0–0.39). The proportions of labeled cell types (represented on the ordinate) were calculated for each injection and averaged over all injections at each depth.

anatomical studies, we expected to be able to antidromically activate mitral cells from all of the secondary olfactory regions, including the posterior part of the piriform cortex (pPC). The tufted cells, however, we expected to be antidromically activated only by stimulation of the rostral regions including the LOT and the olfactory tubercle. Our first assay for the success of this strategy was the position of the recorded spikes in the bulb. Our results from two series of experiments (Scott, 1981; Schneider and Scott, 1983) show that spikes from presumed mitral cells (that is, cells antidromically activated from pPC), were observed from significantly deeper layers of the bulb than were spikes from presumed tufted cells (that is, cells that could be antidromically activated from the LOT or olfactory tubercle but not from the pPC). Some of these identifications of mitral and tufted cells were confirmed by intracellular marking with HRP.

In the second of the electrophysiological experiments (Schneider and Scott, 1983), we also observed the responses to olfactory nerve layer stimulation by antidromically identified mitral and tufted cells. We chose electrical stimulation of the olfactory nerve for the sake of rapidity and reliability of stimulation and because we also wished to look at the spatial organization of inputs to these cell types.

Our results, summarized in Table I indicate that, by a number of measures, antidromically identified tufted cells were more excited by olfactory nerve stimulation than were mitral cells. This effect was obvious in lower thresholds, shorter latencies, increased numbers of spikes in the response, and in a greater range of positions on the nerve layer from which the tufted cells could be excited. All values presented in this table are for the lowest threshold stimulus pole of a seven-electrode stimulus array, so that we are comparing cells for stimulation at their best site. Otherwise, with only a single stimulation site on the olfactory nerve, there would be substantial variability just from the variation in stimulation site relative to the nerve filiments innervating the cells being recorded.

In analyzing these data, we tested three possible criteria for distinguishing mitral and tufted cells to see which were the best predictors of orthodromic response. This was done by stepwise regression analysis using the antidromic activation pattern, antidromic latency, and recording depth as independent variables and using the threshold, numbers of spikes per response, and orthodromic latency as dependent variables. The antidromic activation pattern was the superior predictor of orthodromic response, as seen in Table II. We conclude that, next to intracellular filling of the recorded cell, the antidromic activation pattern is the best criterion of olfactory bulb projection cell type.

Table 1. *Summary of Orthodromic Responses of Olfactory Bulb Projection Cells Classified by Their Antidromic Patterns*

	Orthodromic responses to olfactory nerve stimulation				
Antidromic activation pattern	*Units excited*	*Mean threshold for excited units*	*Units showing multiple spikes*	*Mean minimum response latency*	*Mean number of effective stimulus poles*
	% (*n*)	μA (n)	% (*n*)	*ms* (*n*)	*no.* (*n*)
LOT cells	95 (21)	205 (11)	80 (10)	5.3 (12)	3.4 (14)
OT cells	96 (25)	256 (11)	38 (8)	5.9 (14)	3.3 (13)
pPC/pPC-OT cells	59 (68)	436 (10)	0 (14)	6.9 (14)	1.7 (17)

The number of units tested (*n*) in each entry varies because not all units were held long enough to establish threshold or other measures with confidence. Abbreviations: LOT cells were cells activated only from the LOT; OT cells were cells activated from olfactory tubercle (OT), but not posterior piriform cortex (pPC); pPC cells were activated from pPC but not OT; and pPC-OT cells were activated from both pPC and OT.

Table II. *Regression and Multiple Regression Analysis of Orthodromic Responses to Olfactory Nerve Stimulation*

	Depth	*Antidromic activation pattern*	*Antidromic latency*	*Combined*
ONL Threshold ($n = 27$)	0.019	0.228[1]	0.019	0.238
Number of effective ONL poles ($n = 36$)	0.398[2]	0.472[2]	0.052	0.601[3]
ONL response latency ($n = 36$)	0.002	0.237[2]	0.015	0.350[3]

Coefficients of Determination (R^2) from regression and multiple regression analysis of the orthodromic responses to olfactory nerve stimulation comparing three methods of classification of projection cells. The row headings refer to response measures and the column headings refer to ways of classifying the units. For the antidromic activation pattern, we calculated a multiple R^2 from the classification into *LOT*, *OT*, or *pPC/pPC-OT* groups using dummy variables. The combined values in the last column are for a regression equation using all three methods of classification.

1. An R^2 different from zero with $P < 0.05$.
2. An R^2 significantly different from zero, with $P < 0.01$.
3. An R^2 for the total regression greater than that for the antidromic activation pattern alone with $P < 0.05$.

In future experiments with odor stimulation of these cells, we will continue to use this antidromic identification procedure as well as intracellular filling. It will be important to determine whether tufted cells differ systematically from mitral cells in their thresholds to odor stimulation. Also, it is important to determine whether the differences in excitability of these cells are related to inhibitory mechanisms that may allow different abilities to selectively respond to odor quality. That is, it may be that mitral cells are less excitable because they are more involved in lateral inhibitory circuits that act in sharpening odor discrimination, as proposed by Holley and MacLeod (1977).

These observations on differential excitability of mitral and tufted cells are consistent with previous observations on the excitability of cells recorded from the mitral cell body layer and EPL of the olfactory bulb. Shepherd (1977) has summarized data showing that cells recorded superficially in the EPL were more excitable in response to olfactory nerve stimulation than cells recorded from the mitral cell body layer. However, it was not possible to unequivocally identify these cells until the distributions of their axons were determined.

As seen in Table II, the antidromic response latency was not a good predictor of the orthodromic response. This is somewhat surprising, since many authors seem to have expected that tufted cells would have slower conduction velocities. We did find that the antidromic latencies of tufted cells antidromically identified by response only to LOT stimulation (*LOT* cells) were significantly longer than those of mitral cells, but this was a mean difference of less than 1 ms (Schneider and Scott, 1983). This differs from the observations of Nicoll (1970), who reported antidromic activation of tufted cells with very long latencies on the order of 30 ms, whereas our longest antidromic latencies are on the order of 5 ms. This may be a difference in preparation or he may have activated very fine collaterals of some tufted cells.

It is also important to comment on the degree of certainty with which we can identify cells with these procedures. While the depth distributions of *LOT* cells or cells antidromically activated from olfactory tubercle but not pPC (*OT* cells) are statistically different from the distribution of those activated from the pPC, there is some overlap. It is not certain whether this overlap arises from dendritic spikes or some other source. We cannot clearly discriminate with the antidromic identification between external and middle tufted cells even though there was a tendency for *LOT* cells to be more superficial than *OT* cells. It is also possible that the variation in the response of identified mitral cells to olfactory nerve stimulation results from the variation in mitral cell types. At present, there is no evidence to support different distributions of type I and type II mitral cells outside of the olfactory bulb.

Interneuronal and Centrifugal Interactions with Projection Cells

Edward Orona, Betsy Rainer, and I have recently reported (Orona et al., 1983b) that there are subpopulations of granule cells that have the potential of selectively interacting with the basal dendrites of the various projection cells in the EPL (Figure 2). The granule cells are the most numerous interneuron type in the olfactory bulb. They have been shown to reciprocally interact with the basal dendrites of both mitral and tufted cells and to exert inhibitory influences on those cells (for a review, see Shepherd, 1977). We observed granule cells in rat brains after small iontophoretic injections of HRP and after Golgi impregnation. Superficial HRP injections filled granule cells with somata in the superficial part of the granule cell layer and with major distal dentritic ramification and greatest numbers of dendritic spines in the superficial half of the EPL. This finding was confirmed in Golgi impregnation (Figure 4). Other granule cells, with somata deep in the granule cell layer, could be demonstrated to have distal dendrites ramifying in the deep EPL but not reaching the

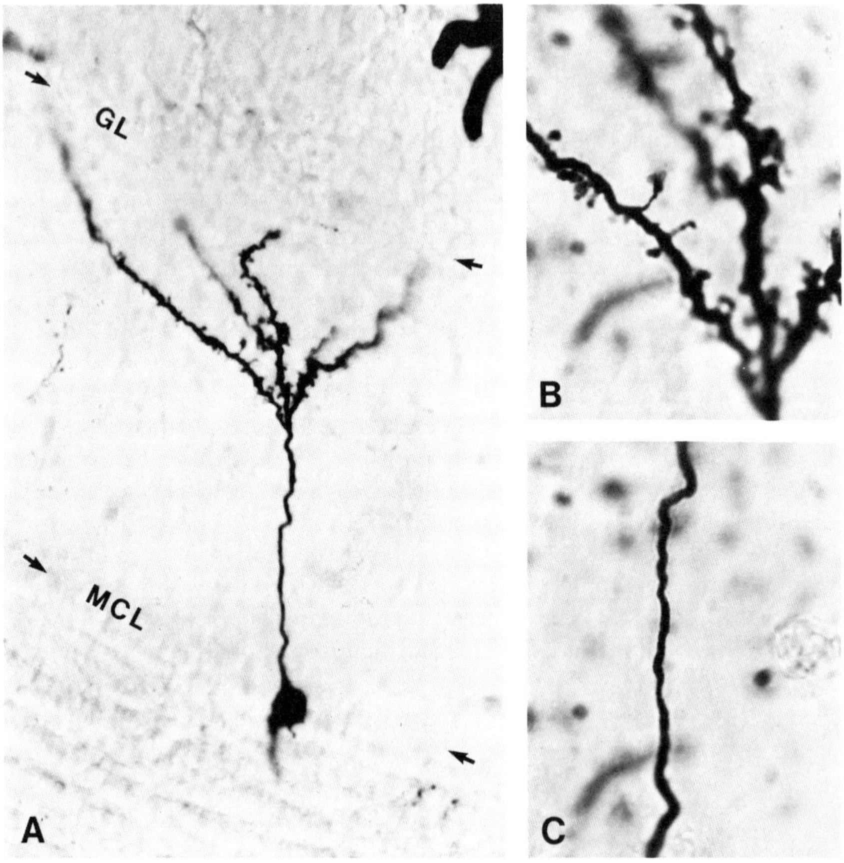

Figure 4. Photomicrograph of Golgi impregnated superficial granule cell. Insets B and C show the spine-studded distal dendrites and the spine-free shaft of the cell. Abbreviations as in Figure 2. The EPL in this section is 225 μm wide.

upper part of the EPL. While we found a few cases of granule cells with spines on dendrites in both superficial and deep EPL, these were a small minority in either (HRP) or Golgi material. We never found cells with somata in the superficial granule cell layer with dendrites confined to the deep EPL.

The granule cell spines are the point of reciprocal dendrodendritic interaction between the granule cells and the projection cells. These dendrodendritic contacts have been described morphologically with tufted cells as well as with mitral cells. Our intracellular recordings from HRP-marked tufted cells (Schneider and Scott, 1983) show the same type of EPSP-IPSP sequences that are ascribed to the reciprocal synapse in the case of the mitral cell (Shepherd, 1977; Mori, and Takagi, 1978). If there are separable granule cell populations that interact with the mitral and tufted cells, they might impose greater tonic inhibition on mitral cells and be the source of the physiological differences that Schneider and I observed in the responses of these cells. Bogan et al. (1982) have also provided some evidence that the granule cell populations in the rat differ in their properties. They noted enkephalinlike immunoreactivity of many granule cells, particularly those with somata deep in the granule cell layer. Furthermore, they saw processes of the enkephalinlike immunoreactive cells predominantly in the deep EPL. Whether enkephalin is a transmitter in these cells is not yet known, but the morphological and chemical differences between the deep and superficial granule cells suggest functional diversity.

The distribution of centrifugal imputs from the forebrain may also be critical in determining orthodromic responses such as those Schneider and I have reported. Davis and Macrides (1982) have reviewed investigations of the centrifugal inputs to the olfactory bulb indicating that these inputs contact interneurons in a laminarly organized fashion. Some of these inputs to the EPL are from the cholinergic horizontal nucleus of the diagonal band (Macrides et al., 1981). The distribution of these inputs and of bungarotoxin binding (Hunt and Schmidt, 1978) suggests that these inputs may be in an optimal position to selectively influence the type I mitral cell basal dendrites. Thus tonic activity by these centrifugal inputs may be the source of the depressed excitability of mitral cells. Other potential mechanisms must also be investigated, including different properties of the olfactory nerve afferents to mitral and tufted cells and differences in the input impedances of these cells.

One fact that does not appear to fit in with a granule cell origin for the differences between mitral and tufted cell responses is the observation that the type II mitral cells have dendrites that intermix with the dendrites of middle tufted cells. These mitral cells should have properties similar to those of middle tufted cells if the granule and centrifugal connections are determining their response. The fact that external tufted cell basal dendrites are distinctly different in position from mitral dendrites would contribute to our result. The result of Greer and Shepherd (1982), who saw narrowing of the inner portion of the EPL in mice with mitral cell degeneration, also supports a pattern of sublamination of the EPL in mice with mitral cell degeneration, also supports a pattern of sublimination of the EPL separating mitral cells from some tufted cells. In addition, we saw a range of mitral cell excitability and we cannot tell whether the more excitable mitral cells are type II cells until we have either recorded from several with HRP filling or until we can determine that they can be antidromically activated from some unique position.

Role of Spatial Organization and Specialized Cell Types in Olfactory Function

The anatomical and physiological investigations of mitral and tufted cell morphology, projections, and physiology suggest that there should be a serious study of the comparative neurobiology of sensory systems. The similarity of the synaptic arrangement of the olfactory bulb to that of the retina has already been pointed out (e.g., Shepherd, 1977). The issue of parallel projections by differing cell types has come up in the study of most sensory systems (e.g., Stone et al., 1979). In the visual system, the superior colliculus, which receives its input through one subset of ganglion cell types, appears to be involved in orienting the head and eyes toward visual stimuli, in contrast to area 17, which receives its input through a different subset of cell types and appears to be more involved in visual acuity. Similarly, the dorsal column–medial lemniscus system, necessary for the finest tactile discriminations, does not have collateral innervation of the reticular formation, whereas the anterior-lateral column system does have such inputs. Neurobiologists generally attribute to those reticular formation inputs the ability to alert the organism to significant stimuli. The observations of sublaminar distributions of projection cell basal dendrites and interneuron dendrites is also reminiscent of the interaction of retinal ganglion and amacrine cells (Nelson et al., 1978). In that system, on- and off-center ganglion cells have dendritic ramifications in different sublaminae of the internal plexiform layer of the retina where they interact with different bipolar cells and with different amacrine cells. Our present data do not allow us to say much about the functional significance of the sublaminar arrangement in the EPL of the olfactory bulb except that this arrangement appears to be associated with differences in the degree of excitability of output cells. These excitability differences may be associated with more complicated sensory processing.

The lack of strong evidence for a topographic organization of the projection from the olfactory bulb to piriform cortex and the evidence of extensive association connections that spread activation within the piriform cortex (Haberly and Shepherd, 1973; Haberly and Price, 1978a) have prompted several suggestions about mechanisms that code olfactory discrimination, without requiring spatial organization in the secondary projection areas. Freeman (1975), Haberly and Price (1978b), and Macrides et al. (1982) have all suggested mechanisms by which spatial aspects of stimulation of receptors may be coded temporally in the central olfactory system. Macrides has reported cells that fire in different phase relationships to the sniff or inhalation cycle and that these phase relations can change with the odor stimulus. He and his colleagues have also found olfactory bulb neurons that are entrained by the hippocampal theta rhythm. Macrides et al. (1982) have recently reviewed their work in this area and reported a temporal relationship between the theta rhythm and inhalations during active sniffing. They have suggested that the theta rhythm may act as a modulating signal for the spatiotemporal coding of olfactory information both in the olfactory bulb and the secondary projection areas. It would be of interest to know whether the activity of all cell types within the olfactory bulb exhibits these phase dependencies on the sniff cycle and on the hippocampal theta, particularly since the different output cells may be under different centrifugal control. My own position on this view of olfactory coding is that there is likely to be more spatial organization in the central projection than was originally thought, particularly when we take into account the possible differences in the projections of the subtypes of output cells. However, there is undoubtedly a broad distribution of collaterals and association connections in this system and it is important to develop some experimental models to test these suggestions for coding mechanisms.

The extent to which the different central projection areas receive their inputs from different cell types is not fully known although the differential projection of mitral cells to posterior parts of the cortex is well documented. I have speculated that the hypothalamus-midbrain pathway may receive its olfactory input through a population of cells which is, to some degree, different from those projecting to the posterior piriform cortex. I have suggested that these cells may impose different response properties on this path. The anatomical path of olfactory input to the frontal neocortex is still a matter of active investigation. It is thus premature to comment on whether this path has a selective input from one of the bulb subtypes. Considering the claims made for the odor specificity of neocortical cells, more information on the relative contribution of these cell types to this system will be important. The available data suggest that there are inputs to neocortex from a number of the secondary olfactory areas.

We may find that there is strong spatial organization in the projection of one of the types of mitral cells but not the other. They may have differing projection fields, as in the W-, X-, and Y-cell model. On the other hand, they may be more similar to the on- and off-center ganglion cells, which have different responses and participate in different circuits but appear to influence the same central projection regions.

The success of future experiments in functional analysis of the central olfactory projections depends upon more sophisticated analysis of the neural responses to odor. The observations of high odor-quality specificity for neocortical cells may be one of the most important recent observations made using odor stimulation. Another impressive series of experiments reviewed by Holley and MacLeod (1977) has indicated that there is a general progression from nonspecific to specific odor-quality responses as one goes from receptors to the olfactory bulb to piriform cortex. Difficulties arise in some of these experiments in definition of the response, in the choice of odor stimuli, and in choice of concentrations of odor stimuli. It is clear from the reports on the relationship of odor responses to the rhythmic sniff or inhalation cycle (Macrides et al., 1982; Scott, 1977) that quantitative measures of olfactory responses are going to be difficult to achieve. This situation will be even more difficult for those investigating these responses in behaving animals. However, ingenuity in this area combined with the developing information from anatomical and behavioral investigations hold great promise for understanding of central olfactory processing.

References

Benjamin, R. M., J. C. Jackson, G. T. Golden, and C. H. K. West. 1982. Sources of olfactory inputs to opossum mediodorsal nucleus identified by horseradish peroxidase and autoradiographic methods. *J. Comp. Neurol.* 207:358–368.

Bogan, N., N. Brecha, C. Gall, and H. J. Karten. 1982. Distribution of enkaphalin-like immunoreactivity in the rat main olfactory bulb. *Neuroscience.* 7:895–906.

Broadwell, R. D. 1975. Olfactory relationships of the telencephalon and diencephalon in the rabbit. II. An autoradiographic and horseradish peroxidase study of the efferent connections of the anterior olfactory nucleus. *J. Comp. Neurol.* 164:389–410.

Davis, B. J., and F. Macrides. 1981. The organization of centrifugal projections from the anterior olfactory nucleus, ventral hippocampal rudiment, and piriform cortex to the main olfactory bulb in the hamster: an autoradiographic study. *J. Comp. Neurol.* 203:475–493.

Eichenbaum, H., K. J. Shedlack, and K. W. Eckmann. 1980. Thalamocortical mechanisms in odor-guided behavior. I. Effect of lesions of the mediodorsal thalamic nucleus and frontal cortex on olfactory discrimination in the rat. *Brain Behav. Evol.* 17:255–275.

Freeman, W. J. 1975. *Mass Action in the Nervous System.* New York: Academic Press, Inc.

Greer, C. A., and G. M. Shepherd. 1982. Mitral cell degeneration and sensory function in the neurological mutant mouse Purkinje cell degeneration (PCD). *Brain Res.* 235:156–161.

Haberly, L. B., and J. L. Price. 1977. The axonal projection patterns of the mitral and tufted cells of the olfactory bulb in the rat. *Brain Res.* 129:152–157.

———. 1978a. Association and commissural fiber systems of the olfactory cortex of the rat. I.

System originating in the piriform cortex and adjacent areas. *J. Comp. Neurol.* 178:711–740.
———. 1978b. Association and commissural fiber systems of the olfactory cortex of the rat. II. Systems originating in the olfactory peduncle. *J. Comp. Neurol.* 181:781–808.
Haberly, L. B., and G. M. Shepherd. 1973. Current-density analysis of summed evoked potentials in opossum prepyriform cortex. *J. Neurophysiol. (Bethesda).* 36:789–802.
Heimer, L. 1976. The olfactory cortex and the ventral striatum. In *Limbic Mechanisms: The Continuing Evolution of the Limbic System Concept.* K. E. Livingstone and O. Hornydiewicz, eds. New York: Plenum Publishing Corp., pp. 95–189.
Holley, A., and P. MacLeod. 1977. Transduction et codage des informations olfactives chez les vertèbres. *J. Physiol. (Paris).* 73:725–828.
Hunt, S. P., and J. Schmidt. 1978. Some observations on the binding pattern of α-bungarotoxin in the central nervous system of the rat. *Brain Res.* 157:213–232.
Kishi, K., K. Mori, and Y. Tazawa. 1982. Three-dimensional analysis of dendritic trees of mitral cells in the rabbit olfactory bulb. *Neurosci. Lett.* 28:127–132.
Krettek, J. E., and J. L. Price. 1977. The cortical projections of the mediodorsal nucleus and adjacent thalamic nuclei in the rat. *J. Comp. Neurol.* 171:157–192.
Luskin, M. B., and J. L. Price. 1982. The distribution of axon collaterals from the olfactory bulb and the nucleus of the horizontal limb of the diagonal band to the olfactory cortex, demonstrated by double retrograde labeling techniques. *J. Comp. Neurol.* 209:249–263.
Macrides, F., B. J. Davis, W. M. Youngs, N. S. Nada, and F. L. Margolis. 1981. Cholinergic and catecholaminergic afferents to the olfactory bulb in the hamster: a nueroanatomical, biochemical and histochemical investigation. *J. Comp. Neurol.* 203:495–514.
Macrides, R., H. B. Eichenbaum, and W. B. Forbes. 1982. Temporal relationship between sniffing and the limbic (theta) rhythm during odor discrimination reversal learning. *J. Neurosci.* 2:1705–1717.
Macrides, F., and S. P. Schneider. 1982. Laminar organization of mitral and tufted cells in the main olfactory bulb of the adult hamster. *J. Comp. Neurol.* 208:419–430.
Mori, K., and S. F. Takagi. 1978. An intracellular study of dendrodendritic inhibitory synapses on mitral cells in the rabbit olfactory bulb. *J. Physiol. (Lond.).* 279:569–588.
Motokizawa, F. 1974. Electrophysiological studies of olfactory projection to the mesencephalic reticular formation. *Exp. Neurol.* 44:135–144.
Motokizawa, F., and N. Furuya. 1973. Neural pathway associated with the EEG arousal response by olfactory stimulation. *Electroencephalogr. Clin. Neurophysiol.* 35:83–91.
Motokizawa, F., and Y. Ino. 1981. A search for olfactory receiving areas in the cerebral cortex of cats. *Neuroscience.* 6:39–46.
———. 1983. A non-thalamic olfactory pathway to the orbital gyrus in the cat. *Brain Res. Bull.* 10:83–88.
Moulton, D. G. 1976. Spatial patterning of response to odors in the peripheral olfactory system. *Physiol. Rev.* 56:578–593.
Mozell, M. M., and D. Hornung. Peripheral mechanisms in the olfactory process. In *Taste, Olfaction, and the Central Nervous System.* D. Pfaff, ed. New York: The Rockefeller University Press, pp. 253–279.
Nelson, R., E. V. Famiglietti, and H. Kolb. 1978. Intracellular staining reveals different levels of stratification for on- and off-center ganglion cells in cat retina. *J. Neruophysiol. (Bethesda).* 41:472–483.
Nicoll, R. A. 1970. Identification of tufted cells in the olfactory bulb. *Nature (Lond.).* 227: 623–625.

Onoda, N., and M. Iino. 1980. Selective responses to odors of animal products in the neocortex neurons of rabbits. *Proc. Jpn. Acad.* 56(B):300–305.

Onoda, N., and S. Kogure. 1981. Odor-sensitive neurons in the lateral hypothalamic area of unanesthetized rabbits. In *Brain Mechanisms of Sensation.* Y. Katsuki, R. Norgren, and M. Sata, eds. New York: John Wiley & Sons, Inc., pp. 227–240.

Orona, E., J. W. Scott, and E. C. Rainer. 1983a. Dendritic and axonal organization of mitral and tufted cells in rat olfactory bulb. *Anat. Rec.* 205:148A.

———. 1983b. Different granule cell populations innervate superficial and deep regions of the external plexiform layer in rat olfactory bulb. *J. Comp. Neurol.* 217:227–237.

Pfaffmann, C. 1960. The pleasures of sensation. *Psychol. Rev.* 67:253–268.

Price, J. L. 1973. An autodariographic study of complementary laminar patterns of termination of afferent fibers to the olfactory cortex. *J. Comp. Neurol.* 150:87–108.

Price, J. L., and W. W. Sprich. 1975. Observations on the lateral olfactory tract of the rat. *J. Comp. Neurol.* 162:321–336.

Price, J. L., and S. Wiegand. 1982. Olfactory neocortical areas in the cat. *Neurosci. Abstr.* 8:201.

Reep, R. L., and S. S. Winans. 1982. Afferent connections of dorsal and ventral agranular insular cortex in the hamster *Mesocricetus auratus. Neuroscience.* 7:1265–1288.

Scheibel, M. E., and A. B. Scheibel. 1975. Dendrite bundles, central programs and the olfactory bulb. *Brain Res.* 95:407–421.

Schneider, S. P., and J. W. Scott. 1983. The orthodromic response properties of rat olfactory bulb mitral and tufted cells correlate with their projection patterns. *J. Neurophysiol.* (*Bethesda*). 50:358–378.

Schwob, J. E., and J. L. Price. 1978. The cortical projection of the olfactory bulb: development in fetal and neonatal rats correlated with quantitative variations in adult rats. *Brain Res.* 151:369–374.

Scott, J. W. 1977. A measure of extracellular unit responses to repeated stimulation applied to observations of the time course of olfactory responses. *Brain Res.* 132:247–258.

———. 1981. Electrophysiological identification of mitral and tufted cells and distributions of their axons in olfactory system of the rat. *J. Neurophysiol.* (*Bethesda*). 46:918–931.

Scott, J. W., and B. R. Chafin. 1975. Origin of the olfactory projections to the lateral hypothalamus and nuclei gemini of the rat. *Brain Res.* 88:64–68.

Scott, J. W., and C. M. Leonard. 1971. The olfactory connections of the lateral hypothalamus in the rat, mouse and hamster. *J. Comp. Neurol.* 141:331–344.

Scott, J. W., R. L. McBride, and S. P. Schneider. 1980. The organization of projections from the olfactory bulb to the piriform cortex and olfactory tubercle in the rat. *J. Comp. Neurol.* 194:519–534.

Scott, J. W., and C. Pfaffmann. 1972. Characteristics of responses of lateral hypothalamic neurons to stimulation of the olfactory system. *Brain Res.* 48:251–264.

Shepherd, G. M. 1977. The olfactory bulb: a simple system in the mammalian brain. *Handbk. Physiol.*, 1:945–968.

Skeen, L. C., and W. C. Hall. 1977. Efferent projections of the main and accessory olfactory bulb in the tree shrew (*Tupia glis*). *J. Comp. Neurol.* 172:1–36.

Slotnick, B. M., and E. J. Berman. 1980. Transection of the lateral olfactory tract does not produce anosmia. *Brain Res. Bull.* 5:141–145.

Stevens, C. F. 1969. Structure of cat frontal olfactory cortex. *J. Neurophysiol.* (*Bethesda*). 32:184–192.

Stone, J., B. Dreher, and A. Leventhal. 1979. Hierarchical and parallel mechanisms in the organization of the visual cortex. *Brain Res. Rev.* 1:345–394.

Tanabe, T., H. Yarita, M. Iino, Y. Ooshima, and S. F. Takagi. 1975a. An olfactory projection area in orbitofrontal cortex of the monkey. *J. Neurophysiol.* (*Bethesda*). 38:1269–1283.

Tanabe, T., M. Iino, and S. F. Takagi. 1975b. Discrimination of odors in olfactory bulb, pyriformamygdaloid areas, and orbitofrontal cortex of the monkey. *J. Neurophysiol.* (*Bethesda*). 38:1284–1296.

Olfactory Determination of Social Behaviors

13. Are There Labeled Lines in the Olfactory Pathway?

In the traditional view of the vertebrate olfactory system, receptors are broadly tuned to different odors, and the neural codes for different odors have been believed to be carried in complex "across-fiber" patterns of activity (Gesteland et al., 1965; Holley et al., 1974; Moulton, 1975). Our work over the past 10 years or so has provided evidence for a division of the olfactory system into several functional subsystems. This has suggested that, in addition to "across-fiber" patterns, one may begin to apply to some extent the concept of "labeled lines" to these subsystems.

It seems to me particularly appropriate on this occasion to discuss our results within the framework of these two competing concepts. The dialectic between them has animated much of Carl Pfaffmann's own work on the gustatory system. However, for the innocent Slawkenbergian, these terms carry risks, as I discovered several years ago when I first discussed the application of the labeled-line concept to the olfactory system, in a seminar at the Pierce Foundation, at the invitation of my colleague John Stitt. John's secretary telephoned to get the title of the seminar, and what I told her was identical to the title of the present lecture. The next day she called back to make quite sure she had it right, and I assured her that she had. When the time came for John to introduce me, he said that the title was an intriguing one, and he showed a slide of a sketch he had made of the scene conjured up by his secretary's interpretation of what I had told her; this is reproduced in Figure 1!

This has been a useful warning of the communications gap that may be inherent in discussing this subject. Being thus tutored by discretion, I shall try to summarize briefly the evidence that has suggested that some aspects of sensory processing in the olfactory system depend on specific subsystems. These results not only offer a new perspective on the uses of neural space in the processing of a nonspatial sensory modality, but they also carry, we believe, some potentially profound implications for our understanding of the molecular basis of olfactory reception, and for informing our research strategies in attacking these problems.

Gordon M. Shepherd, Section of Neuroanatomy
Yale University School of Medicine, New Haven, Connecticut 06510

Main Olfactory Bulb

Analysis of spatial organization has been central to the outstanding progress made in understanding other sensory systems over the past two or more decades. Since the olfactory system does not map external space (as does the visual system), nor the body surface (as does the somatosensory system), nor some obvious dimension of the stimulus energy (such as pitch in the auditory system), the olfactory system has been excluded from many of the advances in our understanding of sensory organization made by using traditional methods of electrophysiological recordings and controlled stimulation. Yet the three main levels of the peripheral olfactory system—olfactory receptor epithelium, olfactory bulb layers, and olfactory cortex—are all extensive sheets of neural components. The fundamental question thus arises: how is space used to process this nonspatial sensory modality? The hypothesis we wished to test was that space is available in this system to play a role in the encoding and abstraction of information about the stimulating molecules.

Our early anatomical studies, using Nauta degeneration techniques and amino acid transport autoradiography (Land et al., 1970; Land and Shepherd, 1974), provided evidence that there is a considerably more precise topographical organization of the olfactory receptor input to individual glomeruli in the olfactory bulb than had been realized. What was needed next was a

Figure 1. "Labeled lions in the old factory pathway." Cartoon by Dr. John Stitt.

method that would provide a means for monitoring the activity induced in this input by odor stimulation.

Such a method became available with the introduction of the 2-deoxyglucose (2DG) mapping technique by Louis Sokoloff and his co-workers (1977). With the support of Sokoloff, Frank Sharp, Ed Evarts, John Kauer, and I discovered very early in the development of the 2DG method that it is extremely effective in revealing spatial patterns of activity within the glomerular layer of the rat olfactory bulb in association with odor stimulation (Sharp et al., 1975, 1977). It is particularly striking that these patterns are revealed in the awake, naturally behaving animal. I do not know of any neural system in which the 2DG method is more effective in demonstrating patterns of activity under conditions of entirely natural behavior, down to threshold levels of weak stimuli.

Our studies with the 2DG method have shown that natural olfactory stimulation does not cause widespread, diffuse activity throughout the olfactory bulb; instead, there are limited regions within which groups of glomeruli show higher levels of 2DG uptake. These regions are reproducible from animal to animal, and form what we have termed "domains" (Stewart et al., 1979). Two characteristics of these domains are relevant to the encoding of olfactory information. First, the extent of activity within a domain varies with the concentration of the stimulus. Thus, odor concentration appears to be encoded, at least in part, by the numbers of active elements. At the glomerular level, where the foci are most intense, these elements include most prominently the terminals of the olfactory axons converging onto individual glomeruli. Thus, increasing stimulus concentration not only drives many receptors to higher impulse frequencies (this is known from unit recordings of receptors [e.g., Getchell and Shepherd, 1978]), but also activates increasing numbers of receptors with higher thresholds for a given odor molecule; such differing thresholds are also known from receptor recordings (see Getchell and Shepherd, 1978).

The other characteristic relevant to the encoding of olfactory information is that the domains for different odors are different. At this still early stage of testing various odors and odor conditions, the principle that seems to be emerging is that the domains are overlapping to some degree, but are distinguishably different from each other. Thus, the domains for amyl acetate (which has a fruity odor) and camphor are overlapping, but within the regions of overlap there are different patterns for the two, and there are regions active for one odor and not the other. Comparison of the map obtained by exposure of the animals only to the odor of their own cages (a heavy brew of urine, feces, and bedding material, it may be noted!) shows that the same principle applies; the distribution of activity mostly along the posterior aspect of the olfactory bulb overlaps the domains for amyl acetate and camphor, but is distinct from them (Stewart et al., 1979).

The extent to which these patterns are distinct suggested to us that this could contribute to the discrimination of different odors. It would be a way in which space would be used, not to map space as such, but some property of the stimulating molecule. It is a way, in other words, in which neural space can be used to map or analyze a nonspatial modality. By this we did not mean to imply an actual spatial "image" within the olfactory system for each odor, but rather a more general concept: that for each odor, limited and distinct ensembles of neural elements appear to be employed in the processing of that information. In view of the degree of topographical organization that had been discovered in the olfactory nerve input (see above), this has further implied that the limited and distinct ensembles of elements in the olfactory bulb might reflect to some extent correspondingly limited and distinct ensembles of olfactory receptors for each odor. These ensembles would not necessarily be thought of as constituting "labeled lines," but their degree of specificity is a step in that direction.

Modified Glomerular Complex

The patterns described above within the main olfactory bulb were not precisely reproducible from animal to animal, nor did we expect them to be, given the behavioral variability that was present during the experiments. The animals were merely oriented toward the current of odor-containing air, and they varied greatly in their levels of activity and orientation. Nor did we have any means of judging their actual interest in or attention to the odor itself.

The next step was obviously to test an odor that is important in mediating a naturally occurring behavioral response. The Sokoloff technique placed strict demands on the type of behavior that could be tested: the behavior must be well-defined, intense, and persistent throughout the 45 min from injection of the isotope to sacrifice of the animal. One possibility we investigated was the attraction of male hamsters by females, which is presumed to take place by the mediation of pheromones (Singer et al., 1976); however, our preliminary experiments indicated that this behavioral paradigm is too complex, and the subjects too fickle, for the method to work well.

A behavior that does meet all the criteria for optimal application of the 2DG technique is suckling. In the rat, suckling is the most crucial and well-defined behavior of early life. Rat pups suckle continuously from birth for the first 2 wk of postnatal life. Furthermore, in the rat, suckling is completely dependent on an odor cue on the mother's nipples. The work of Teicher and Blass (1976) and Hofer et al. (1976) had shown that suckling is eliminated by washing the nipples, and is reinstated by reapplying the rinse fluid. Thus, we had a well-defined, intense, and persistent behavior, and one that depended on an odor cue, the ideal conditions for our investigation.

In collaboration with Teicher, William Stewart, Kauer, and I injected suckling rat pups with 14C-2DG. Within the main olfactory bulb, the autoradiograms showed diffuse activity but no clear pattern. However, a striking finding was a small focus of activity high on the medial wall in the posterior part of the bulb. At first we thought this might be located within the accessory olfactory bulb (AOB), but careful histological correlation showed that the focus was never within the AOB, but instead was associated with one or several gomeruli that were immediately adjacent to it, distinct from both it and the main golmerular sheet. This group of glomeruli was reproducibly present from animal to animal. It had not been recognized previously. We termed it a "modified glomerular complex" (MGC) (Teicher et al., 1980).

In recent work, carried out in collaboration with Charles Greer, we have studied the ontogeny of this region (Greer et al., 1982). It is recognizable as a histological entity at least by the last few days of prenatal life. In an independent study, Beth Friedman and Joe Price (personal communication) have found that as early as E17 this complex can be identified using the Timm's stain, which is specific for heavy metals. Within the first 2 hr after birth, we have found that the MGC of the suckling pup is associated with focal 2DG uptake. The MGC is present as a histological entity throughout the first weeks of life, and our preliminary studies indicate that it persists into adult life. Focal 2DG activity is consistently present through the first 2 wk of life in the MGC; in the 3rd wk, as weaning begins and is complete, the activity becomes correspondingly less prominent.

These results suggested to us the hypothesis that the MGC represents a site where specific information related to the suckling odor cue is processed. The results imply that there is a set of receptors, activated by this odor cue, that make their synapses in these special glomeruli; from there, we hypothesize, the information is relayed to specific sites in the olfactory cortex which activate the motor pathways for control of suckling and related behaviors. We do not rule out the possibility that there may be two or more sites of receptors that converge on the MGC, and we also recognize the likelihood that some aspects of the suckling odor cue are processed in the main olfactory bulb, as indicated by the diffuse 2DG uptake present there. Taking these qualifications into account, it nonetheless remains established that suckling is associated with activity in an identifiable, histologically differentiated set of glomeruli, which constitutes evidence for a "labeled line" for at least some aspect of the suckling odor cue.

Accessory Olfactory Bulb (AOB)

These results provided strong evidence that olfactory function begins at least by the start of postnatal life in the rat pup. This raised a new and important question: does olfactory function actually begin in utero? To our knowlege, there has been no direct physiological evidence bearing on this point.

To test this question, Patricia Pederson, Stewart, Greer, and I performed the simple experiment of injecting a pregnant rat intraperitoneally with 2DG 2 d before term, and then delivering the pups by Ceasarean section the day before term, and immediately sacrificing them and preparing their olfactory bulbs for 14C-2DG autoradiography, without any chance for postnatal odor exposure. The unexpected and dramatic result seen on the autoradiographs was a dense focus of activity within the AOB (Pedersen et al., 1983). This contrasted with the lower levels of diffuse activity in the main olfactory bulb and MGC. This finding was all the more striking in that we have never seen 2DG foci within the AOB in all our experiments on postnatal and adult animals. These latter experiments included odor exposure (Sharp et al., 1975, 1977; Stewart et al., 1979), suckling (Teicher et al., 1980; Greer et al., 1982), electrical stimulation of the vomeronasal nerve (Meredith and Kana, unpublished observations), and exposure of male hamsters to female hamsters (Kauer, Meredith, Costanzo, O'Connell, and Shepherd, unpublished observations). I should stress that this does not prove that there was not physiological activity in the accessory bulb under those conditions; it means either that there was little or no significant activity, or that it was not of a form that was detected by the 2DG method.

The finding of differential 2DG uptake in the AOB was quite unexpected; there are no previous reports which implicate the AOB in any function in the pre- or early postnatal period. There is, of course, considerable evidence for a role of the vomeronasal organ (VN) and its central projection through the AOB in adult behaviors of several vertebrate species. These include especially prey searching by reptiles (see Kubie and Halpern, 1976), and sexual behavior of the male in rodents. In hamsters, there is evidence that sexual attraction of the male by the female depends in part on the VN-AOB pathway; section of the VN nerve reduces but does not eliminate male attraction and mating (Powers and Winans, 1975; Winans and Powers, 1977). Centrally, the AOB has a distinct projection to specific parts of the corticomedial nuclei of the amygdala, which in turn have distinct projections to central limbic structures, including the septum and hypothalamus (Scalia and Winans, 1975).

These previous studies have thus indicated that the AOB is part of a distinct olfactory pathway that lies in parallel with the main olfactory pathway. It therefore has been a good candidate for functioning as a kind of "labeled line" for mediation of specific information received in the vomeronasal organ. It is presumed that this organ is particularly sensitive to odorous substances in the liquid phase, placed at the inlet to the organ by physical contact, and possibly drawn inside by a pumping action of the organ (Meredith et al., 1980).

In light of these considerations, our results suggest several new hypotheses concerning VN-AOB function. The 2DG autoradiograms indicate that the AOB has a relatively high level of activity in the 1–2 d preceding birth. One possibility is that this reflects a high level of input from the vomeronasal receptors. This would imply that odorous substances circulating in the amniotic fluid are able to gain access to the vomeronasal organ. Another possibility is that the 2DG uptake in the AOB reflects increased activity of centrifugal fibers ending in the AOB. This would imply increased activity in the central olfactory and limbic regions from which these fibers arise (see Price and Powell, 1970; Broadwell, 1977; de Olmos et al., 1978). Experiments are in progress to attempt to test these alternative hypotheses.

Mitral and Tufted Cells

The three parts of the olfactory bulb I have discussed thus far are all anatomically distinct regions. Another kind of distinction is found in the differentiation between mitral and tufted cells within the main olfactory bulb. Historically, these were recognized as separate populations of neurons by the earliest work with the Golgi stain (Golgi, 1894; Ramon y Cajal, 1911). The tufted cell bodies lie in the exernal plexiform layer; their dendrites are shorter, their axon collaterals distribute differently (Ramon y Cajal, 1911; Shepherd, 1972; Macrides and Schneider, 1982), and the projections are overlapping but different (Haberly and Price, 1977). Nevertheless, at the fine structural level, the mitral and tufted cell dendrites are similar in their smooth contours, pale cytoplasm, and the fact that they are both presynaptic, by means of type I synapses, on the dendritic spines of granule cells and the dendrite of periglomerular cells (Pinching and Powell, 1971). For this reason, there has been a tendency to think of the tufted cells as simply smaller, more outwardly placed versions of the mitral cells.

Recently we have found that the mitral and tufted cell types are genetically distinct. The point of departure for our investigation was the report by Mullen et al. (1976) that in the mouse mutant Purkinje cell degeneration (pcd), there is degeneration of axons in the lateral olfactory tract. Charles Greer and I (Greer and Shepherd, 1982) studied the histology of the olfactory bulbs of mice homozygous (pcd/pcd) for the pcd gene. At 2 mo of age, the bulbs appear normal; at 4 mo, only ~ 20% of the normal number of mitral cell bodies are present, and at 8 mo only ~ 4% remain. In contrast, the tufted cell population appears unaffected; the cell bodies are present in their normal numbers and their normal distribution in the external plexiform layer. The numbers and distribution of other cell types (granule, PG) similarly appeared to be unaffected.

This selective degeneration of mitral cells induces an interesting reorganization of dendrodendritic synaptic connections within the olfactory bulb, a question we are presently pursuing (Greer et al., 1982). Here I would like to touch only on the implications for information processing. Regions of the nervous system vary in the degree of differentiation of their output neurons. On the one hand, in a region like the cerebellum, all the output neurons (the Purkinje cells) appear to be similar in their morphology, physiological properties, and neurotransmitters, and thus are presumed to be functionally equivalent. On the other hand, there are regions in which the output neurons have different morphologies and different functions. An example is the retina, in which the ganglion cells are subdivided into different morphological types (α, β, γ) (Boycott and Wassle, 1974), each associated with transmitting a different set of functional properties of the visual stimulus (Y, X, W) (Cleland and Levick, 1974).

Our results provide clear evidence that the genetic determinants of the mitral and tufted cell types are distinct. This gives strong support to the idea that they form two distinct morphological types. We hypothesize that these types are associated with different functions in analogy with the retina. Similarly, in visual cortex, simple and complex functional properties tend to be associated with cells with different morphologies, lying in different layers (cf. Gilbert and Wiesel, 1979). Furthermore, output cells in different cortical layers send their axons to different subcortical regions. All these features may have their counterparts in the division between mitral and tufted cells.

Discussion

In the traditional view of the vertebrate olfactory system, the olfactory epithelium is believed to contain a diffuse array of relatively broadly tuned receptors. The receptors are believed to project into the olfactory bulb in a diffuse fashion (e.g., receptors from a given site project widely in the bulb). The bulb in turn is believed to project diffusely into the olfactory cortex.

We have believed for some time that there is a much more precise structural and functional organization than is recognized in the traditional view. The work reviewed above contributes to a body of growing evidence supporting this view. In particular, three new types of organization of the main olfactory system have been revealed; these may be summarized in relation to the schematic diagram in Figure 2.

First, a given odorous substance does not elicit uniform activity throughout the olfactory bulb, as judged by the 2DG patterns. In addition, we have never observed the activity to be in the form of a gradient. The patterns instead tend to be regional, each regional "domain" being built up of punctate activation of individual glomeruli (cf. Leveteau and MacLeod, 1966). These domains could well give rise to apparent anterior-posterior gradients of activity, as seen by Adrian (1953) in his multiunit recordings of bulbar responses to odor stimulation. However, the 2DG results suggest that the underlying organization is in terms of circumscribed sets ("domains") of activated glomeruli, rather than gradients. In effect, each domain functions as a subsystem for a particular odor within the set of all of the glomeruli in the olfactory bulb.

We do not yet know whether the overlapping domains for different odors mean that the same glomerulus can be activated by different odors. However, we do know from the results of high-resolution 2DG studies that individual glomeruli tend to function as units (Lancet et al., 1982). Individual glomeruli may have high, medium, or low levels of 2DG uptake, but, whatever the level, it tends to be relatively uniform throughout the neuropil of a glomerulus. It is also striking that, even at threshold levels of odor concentration, when only one or two small groups of glomeruli are active, these groups may show very intense 2DG foci, suggesting not only that those glomeruli may function as units, but also that they may turn on or off in a quasi all-or-nothing manner.

A second type of organization, embedded in the first, consists of the subsystems of mitral and tufted cells. Recent studies have shown that mitral and tufted cells differ in their morphology (Macrides and Schneider, 1982; Orona and Scott, 1982), neurochemistry (Halasz et al., 1978), axon collateral patterns (Macrides and Schneider, 1982; Orona and Scott, 1982), functional properties (Orona and Scott, 1982), and cortical projection sites (Haberly and Price, 1977; Orona and Scott, 1982). Our results now provide decisive evidence that these two cell types are genetically distinct. They may thus provide a model system for further analysis of the molecular mechanisms underlying the differentiation of neuronal subtypes and the expression of gene products underlying their functional properties (cf. Lampson, 1983).

A third type of organization is seen in the AOB and the modified glomerular complex. This is the first anatomically distinct subdivision to be described in

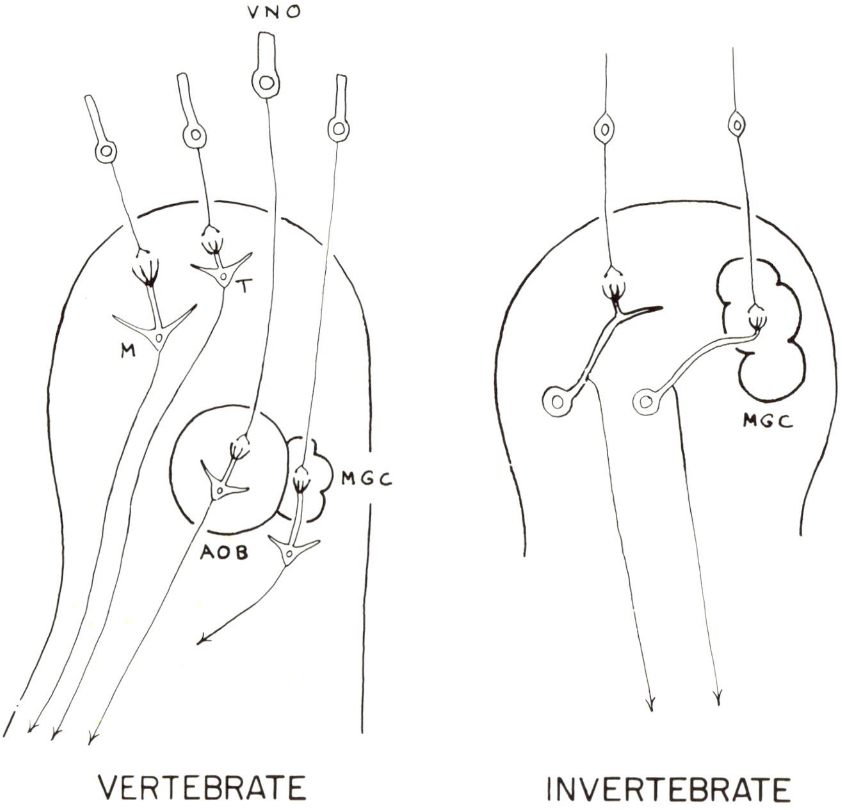

Figure 2. Schematic diagrams summarizing the subsystems that have been identified in the peripheral olfactory systems of vertebrates and higher invertebrates (insects). M, mitral cell; T, tufted cell; VNO, vomeronasal organ; AOB, accessory olfactory bulb; MGC, modified glomerular complex; MGC, macroglomerular complex (in males).

the olfactory system since the AOB was recognized and its relations to the vomeronasal organ, on the one hand, and the main olfactory bulb, on the other, were clarified (see McCotter, 1912; Tucker, 1971). That the MGC escaped notice is easy to understand; in a single frontal section at the level of the AOB, it is usually represented by only one or two glomeruli, which appear to be merely displaced from the main sheet of glomeruli to a position at the medial border of the AOB. The key to identifying it as an entity was the regular appearance there of 2DG foci in suckling rat pups; localization of function was thus crucial in determining the presence of a significant difference in structure.

The localized nature of the small number of glomeruli that make up the MGC means that the traditional view of olfactory processing as involving a complex "across-fiber pattern" of diffuse activity in a spatially widespread set of neuronal elements does not strictly apply to processing of the normal odor cue for suckling. This has been recognized by Erickson; in his valuable discussion of across-fiber and labeled-line concepts (see Chapter 6 of this volume), he notes that "pheromones are . . . discrete stimuli, and labeled lines would serve them well." Theories of olfactory coding now need to incorporate this new type of subsystem. This will become all the more important as new pheromones are identified in different mammalian species; we can speculate, for example, that the main olfactory system contains multiple, more or less discrete, "labeled-line" subsystems, one for each of the pheromones that elicit or control a specific behavior.

The identification of the MGC in the rat has had an interesting parallel in insects (see Figure 2). In the higher insects, work in several laboratories (Boeckh and Boeckh, 1979; Matsumoto and Hildebrand, 1981) has shown that the olfactory receptors in the antennae project their axons to glomeruli located in the antennal lobe (an analogue of the olfactory bulb). The antennal lobe is sexually dimorphic; in males, there is a large, multiglomerular group ("macroglomerulus"). Antennal lobe cells responsive to the sex-attractant pheromone released by a female are found only in males, and the dendritic processes of these cells invariably are connected to the macroglomerulus. Thus, it appears that this is a subsystem that subserves the function of relaying and processing the specific information about the sex-attractant pheromone. The significance of this information, of course, is that it is the necessary link in bringing about mating behavior; furthermore, it is information that does not require any "perceptual" (if that is a term appropriate for an insect!) or discriminating judgment, but rather must elicit an immediate and well-defined motor response.

It is attractive to hypothesize that, in vertebrates, similar factors may apply: olfactory-mediated behavior that is crucial to the survival of the species, and requires an immediate and preprogrammed sequence of motor actions, will involve transmission of the information about the pheromone or odor cue through a more or less specific subsystem within the olfactory pathway. This principle does appear to apply to the MGC and suckling. It suggests a useful strategy in analyzing the organization of the olfactory system, using, in effect, specific molecular probes (the pheromone molecules) to identify the neural subsets tuned to each probe.

Is the idea of discrete subsystems within a main sensory system peculiar only to the olfactory system? Even a brief consideration of other sensory systems indicates that, on the contrary, this is a very common and basic principle. A clear example is found in the auditory system. In the mustache bat, the regular tonotopic organization of the auditory cortex is interrupted by the presence of a relatively huge region for 61 kHz; this is precisely the frequency emitted by the bat itself, which is used in seeking prey and avoiding obstacles (Suga, 1978). In the visual system, it is clear that there are multiple submodalities, such as simple photosensitivity, night-day luminosity cycles, wave-length discrimination, movement detection, depth perception, and so forth (summarized in Shepherd, 1983). The specialization of the retinal fovea for high acuity and wave-length discrimination, the retinal periphery for movement detection, the corresponding ganglion cell types and their different central projections, and the separate retino-hypothalamic input for circadian luminosity rhythms, all would appear to represent the same principle of discrete subsystems for mediating discrete sensory submodalities. The fact that neural subsystems are widespread and are expressed in neural terms in such similar ways suggests the operation of relatively few common principles in the genetic mechanisms for specifying and assembling these circuits, though with many adaptations to the particular circumstances of each system.

We may thus return to the question: Are there labeled lines in the olfactory pathway? I have presented the evidence that there are specific identifiable subsystems for mediating specific types of olfactory information, and to that extent the answer would appear to be in the affirmative. However, the term "labeled line" has its history of connotation and controversies in other contexts, and it may or may not be useful to apply it to the anatomically distinct and functionally specific subsystems I have discussed here. What is clear is that the idea of the olfactory system as a diffusely organized single entity is now obsolete, as outdated and defunct as the "old factory" system pictured in the cartoon of Figure 1; what we need to do now is replace it with a new and modern facility!

Acknowledgments

The work upon which this review is based has been generously supported by grants NS-16993 from the National Institute for Neurological and Communicative Disorders and Stroke and BNS-78-16545 from the National Science Foundation. Many colleagues have collaborated in the experiments described here and contributed valuable discussions to the ideas expressed; among them most recently are Drs. John Kauer, William Stewart, Charles Greer, Martin Teicher, Doron Lancet, and Patricia Pedersen. I am also grateful to Drs. Linda Bartoshuk and Robert Erickson for attempting to educate me in some of the issues discussed here.

References

Adrian, E. D. 1953. Sensory messages and sensation: the response of the olfactory organ to different smells. *Acta Physiol. Scand.* 29:5–14.

Boeckh, J., and V. Boeckh. 1979. Threshold and odor specificity of pheromone-sensitive neurons in the dentrocerebrum of *Antheraea pernyi* and *A. polyphemus* (Saturnidae). *J. Comp. Physiol.* 132:235–242.

Boycott, B. B., and H. Wassle. 1974. The morphological types of ganglion cells of the domestic cat's retina. *J. Physiol. (Lond.).* 240:397–419.

Broadwell, P. D. 1977. Neurotransmitter pathways in the olfactory system. *Soc. Neurosci. Symp.* 3:131–166.

Cleland, B. G., and W. R. Levick. 1974. Brisk and sluggish concentrically organized ganglion cells in the cat's retina. *J. Physiol. (Lond.).* 240:421–456.

de Olmos, J., H. Hardy, and L. Heimer. 1978. The afferent connections of the main and the accessory olfactory bulb formations in the rat: an experimental HRP-study. *J. Comp. Neurol.* 181:213–244.

Gesteland, R. C., J. Y. Lettvin, and W. H. Pitts. 1965. Chemical transmission in the nose of the frog. *J. Physiol. (Lond.).* 181:525–559.

Getchell, T. V., and G. M. Shepherd. 1978. Responses of olfactory receptor cells to step pulses of odour at different concentrations in the salamander. *J. Physiol. (Lond.).* 282:512–540.

Gilbert, C. D., and T. N. Wiesel. 1979. Morphology and intracortical projections of functionally characterized neurones in the cat visual cortex. *Nature (Lond.).* 280:120–125.

Golgi, C. 1894. *Untersuchungen über den feineren Bau des centralen und peripherischen Nervensystems.* Jena: Gustav Fischer.

Greer, C. A., and G. M. Shepherd. 1982. Mitral cell degeneration and sensory function in the neurological mutant mouse Purkinje Cell Degeneration (pcd). *Brain Res.* 235:156–161.

Greer, C. A., W. B. Stewart, M. H. Teicher, and G. M. Shepherd. 1982. Functional development of the olfactory bulb and a unique glomerular complex in the neonatal rat. *J. Neurosci.* 2:1744–1759.

Haberly, L. B., and J. L. Price. 1977. The axonal projection patterns of the mitral and tufted cells of the olfactory bulb in the rat. *Brain Res.* 129:152–157.

Halasz, N., A. Llungdahl, and T. Hokfelt. 1978. Transmitter histochemistry of the rat olfactory bulb. II. Fluorescence histochemical, autoradiographic and electron microscopic localization of monoamines. *Brain Res.* 154:253–271.

Hofer, M. A., H. Shair, and P. Singh. 1976. Evidence that maternal ventral skin substances promote suckling in infant rats. *Physiol. Behav.* 17:131–136.

Holley, A., A. Duchamp, M. F. Revial, A. Juge, and P. MacLeod. 1974. Qualitative and quantitative discrimination in the frog olfactory receptors: analysis from electrophysiological data. Ann. NY Acad. Sci. 237:102–114.

Kubie, J. L., and M. Halpern. 1976. Chemical senses involved in garter snake prey trailing. *J. Comp. Physiol. Psychol.* 93:648–667.

Lampson, L. A. 1984. Molecular basis of neuronal individuality. In *Monoclonal Antibodies and Functional Cell Lines: Progress and Applications.* R. H. Kennett et al., eds. New York: Plenum Publishing Corp.

Lancet, D., C. A. Greer, J. S. Kauer, and G. M. Shepherd. 1982. Mapping of odor-related neuronal activity in the olfactory bulb by high-resolution 2-deoxyglucose autoradiography. *Proc. Natl. Acad. Sci. USA.* 79:670–674.

Land, L. J., and G. M. Shepherd. 1974. Autoradiographic analysis of olfactory receptor projections in the rabbit. *Brain Res.* 70:506–510.

Land, L. J., R. P. Eager, and G. M. Shepherd. 1970. Olfactory nerve projections to the olfactory bulb in rabbit: demonstration by means of a simplified ammoniacal silver degenerative method. *Brain Res.* 23:250–254.

Leveteau, J., and P. MacLeod. 1966. Olfactory discrimination in the rabbit olfactory glomerulus. *Science* (*Wash. DC*). 175:176–178.

Macrides, F., and S. P. Schneider. 1982. Laminar organization of mitral and tufted cells in the main olfactory bulb of the adult hamster. *J. Comp. Neurol.* 208:419–430.

Matsumoto, S. G., and J. G. Hildebrand. 1981. Olfactory mechanisms in the moth *Manduca sexta*: response characteristics and morphology of central neurons in the antennal lobes. *Proc. R. Soc. Lond. B. Biol. Sci.* 213: 249–277.

McCotter, R. E. 1912. The connections of the vomeronasal nerves with the accessory olfactory bulb in the opossum and other mammals. *Anat. Rec.* 6:299–318.

Meredith, M., D. M. Marques, R. V. O'Connell, and F. L. Stern. 1980. Vomeronasal pump: significance for male hamster sexual behavior. *Science* (*Wash. DC*). 207:1224–1226.

Moulton, D. G. 1975. Spatial patterning of response to odors in the peripheral nervous system. *Physiol. Rev.* 56:578–593.

Mullen, R. J., E. M. Eicher, and R. L. Sidman. 1976. Purkinje cell degeneration, a new neurological mutation in the mouse. *Proc. Natl. Acad. Sci. USA.* 73:208–212.

Orona, E., and J. W. Scott. 1982. Different populations of granule cells innervate superficial and deep sublaminae of the external plexiform layer of the rat olfactory bulb. *Soc. Neurosci. Abstrs.* 8:9.

Pedersen, P. E., W. B. Stewart, C. A. Greer, and G. M. Shepherd. 1983. Evidence for olfactory function *in utero*. *Science* (*Wash. DC*). 221:478–480.

Pinching, A. J., and T. P. S. Powell. 1971. The neuropil of the glomeruli of the olfactory bulb. *J. Cell Sci.* 9:347–377.

Powers, J. B., and S. S. Winans. 1975. Vomeronasal organ: critical role in mediating behavior of the male hamster. *Science* (*Wash. DC*). 187:961–963.

Price, J. L., and T. P. S. Powell. 1970. An experimental study of the site of origin and the course of the centrifugal fibres to the olfactory bulb in the rat. *J. Anat.* 107:215–237.

Ramon y Cajal, S. 1911. *Histologie du Système Nerveux de l'Homme et des Vertèbres*. Paris: Maloine.

Scalia, F., and S. S. Winans. 1975. The differential projections of the olfactory bulb and accessory olfactory bulb in mammals. *J. Comp. Neurol.* 161:31–56.

Sharp, F. R., J. S. Kauer, and G. M. Shepherd. 1975. Local sites of activity-related glucose metabolism in rat olfactory bulb during odor stimulation. *Brain Res.* 98:596–600.

———. 1977. Laminar analysis of 2-deoxyglucose uptake in olfactory bulb and olfactory cortex of rabbit and rat. *J. Neurophysiol. (Bethesda)*. 40:800–813.

Shepherd, G. M. 1972. Synaptic organization of the mammalian olfactory bulb. *Physiol. Rev.* 52:864–917.

———. 1983. *Neurobiology*. New York: Oxford University Press.

Singer, A. G., N. C. Agosta, R. J. O'Connell, C. Pfaffmann, D. V. Bowen, and F. Field. 1976. Dimethyl disulfide: an attractant pheromone in hamster vaginal secretion. *Science (Wash. DC)*. 191:948–950.

Sokoloff, L. 1977. Relation between physiological function and energy metabolism in the central nervous system. *J. Neurochem.* 27:13–26.

Stewart, W. B., J. S. Kauer, and G. M. Shepherd. 1979. Functional organization of the rat olfactory bulb analyzed by the 2-deoxyglucose method. *J. Comp. Neurol.* 185:715–734.

Suga, N. 1978. Specialization of the auditory system for reception and processing of species-specific sounds. *Fed. Proc.* 37:2342–2354.

Teicher, M. H., and E. M. Blass. 1976. Suckling in newborn rats: eliminated by nipple lavage, reinstated by pup saliva. *Science (Wash. DC)*. 193:422–425.

Teicher, M. H., W. B. Stewart, J. S. Kauer, and G. M. Shepherd. 1980. Suckling pheromone stimulation of a modified glomerular region in the developing rat olfactory bulb revealed by the 2-deoxyglucose method. *Brain Res.* 194:530–535.

Tucker, D. 1971. Nonolfactory responses from the nasal cavity: Jacobson's Organ and the trigeminal system. In *Handbook of Sensory Physiology*, vol. IV, *Chemical Senses*, pt. I, *Olfaction*. L. M. Beidler, ed. New York: Springer-Verlag, pp. 151–181.

Winans, S. S., and J. B. Powers. 1977. Olfactory and vomeronasal deafferentation of mole hamsters: histological and behavioral analyses. *Brain Res.* 126:325–344.

14. Olfactory and Vomeronasal Mechanisms of Communication

Although trained as a sensory physiologist, Carl Pfaffmann has not restricted his interests to that domain. Throughout his teaching and research career he has emphasized the links between the physiological and behavioral levels of analysis. In particular, Carl has often pointed out that even though we know less about the mechanisms of the chemical senses than we do about vision or audition, the chemical senses are often better model systems for investigations of behavior, since the functional significance of many stimuli is clearer than that of most visual or auditory stimuli. This view has had an important influence on many graduate students who have come into contact with Carl, including myself. Thus, despite the fact that most of my research has focused on the behavioral level of analysis, I thought that for this volume it would be most appropriate to attempt a synthesis of the sensory mechanisms underlying communication by chemical signals, in particular the functional attributes of the olfactory and vomeronasal systems. For reasons of space and consistency I will restrict this review to mammals; for more general reviews, see Wysocki (1979) and Meredith (1983a) and for related work on reptiles, see Halpern (1982).

In Table I, I suggest four ways in which the olfactory and vomeronasal systems may differ; in the remainder of the chapter, I review research from which these hypotheses were derived. First, it is likely that the two systems are specialized for receiving and responding to two different kinds of molecules: the olfactory system primarily handles relatively volatile chemicals, substances that are airborne and that have traditionally been thought of as odors, whereas the vomeronasal organ and accessory olfactory system primarily deal with large, relatively nonvolatile molecules that are contacted by the animal. Second, the two systems may be specialized for different broad classes of functions. The vomeronasal system is likely to be especially important in elicitation or facilitation of specific, functional responses, such as the initiation of mounting by a male or the release of luteinizing hormone from the

Robert E. Johnston, Department of Psychology
Cornell University, Ithaca, New York 14853

Table 1. Proposed Characteristics of the Vomeronasal and Olfactory Systems

	Vomeronasal	*Olfactory*
Type of molecules	Primarily responsive to large, nonvolatile molecules—physical contact with source often necessary	Primarily responsive to small, volatile molecules—i.e., airborne odorants
Kind of responses	Directly functional responses, such as release of hormones or facilitation of copulatory behavior	Provide information, response depends on context: e.g., individual sex and species discrimination
Role of experience	Learning not usually essential; specialized receptors and response mechanisms	Learning often important for meaning; mediates learned responses to odors
Nature of signals and sensory processing	Primarily responsive to specific chemical signals	Discrimination of complex odor mixtures

pituitary. In contrast, the main olfactory system should have more analytic functions with a less specific link to particular responses. For example, difficult discriminations may be the forte of this system, such as discriminations between individuals, kin, or species. Third, many of the functions of the olfactory system may depend on experience of individuals with the stimuli, whereas responses mediated by the vomeronasal system may be less dependent on such experience. Finally, as almost a corollary of the previous point, the two systems are likely to be involved in the detection and analysis of two different classes of signals, those with specific functions and those that just provide information. The vomeronasal system may be specialized to respond to specific chemical signals that have themselves been selected in the course of evolution for particular functions. In contrast, the olfactory system ought to be primarily involved in the analysis of complex scent signals that contain a large number of compounds and that, by virtue of slight changes in the ratios of such compounds, allow fine discriminations to be made, such as between individuals or between kin and non-kin. As suggested above, the olfactory system may be primarily designed to make discriminations between such complex signals. These predictions are offered in the spirit of a theory to stimulate research; I hope that they will prove to be correct in the majority of cases, but there will, no doubt, be exceptions. For example, a specific signal that releases a particular response but acts at a distance would necessarily have to be volatile, and responses to this signal would probably be mediated by the olfactory system. It should also be emphasized that I am not suggesting strict dichotomies; many responses may normally be influenced by both the olfactory and vomeronasal systems.

Peripheral Anatomy

The vomeronasal organ is a closed, tubular structure with an opening or pore in one end. The organ is situated at the base of the nasal cavity, usually at the front near the external nares; for all of the species I will be discussing, the pore is located at the anterior end of the organ (see Wysocki, 1979, for detailed review). In many mammals, a groove lined with epithelial tissue leads from the external nares to the vomeronasal organ, which lies in this groove.

Since the vomeronasal organ is a tube open only at one end, there are questions about how stimuli get inside to contact the receptors. In some species, the behavioral pattern known as *flehmen* may be involved. Although differing in detail across species, this behavior generally involves an immobile posture, a stereotyped position of the head, and a retraction of the upper lip. It is believed that these overt actions are manifestations of a response that opens the vomeronasal ducts or in some other way makes the interior of the organ more accessible to chemical stimuli (Estes, 1972). Physiological processes are implicated as well. For example, there may be active transport down the epithelial groove from the external nares (Wysocki, 1982), or there may be physical pumping of materials in and out, as has been described for the golden hamster (Meredith and O'Connell, 1979). In this species, firing of cells in the nasopalatine nerve causes the organ to contract, forcing out its contents, and then to expand, sucking new material inside.

Regardless of the exact mechanisms by which molecules gain access to the vomeronasal organ, it has been demonstrated that large, nonvolatile molecules do seem to be selectively taken up into the vomeronasal organ but not into the area of the olfactory sensory epithelium, at least in several species of rodents and in the domestic goat (Wysocki et al., 1980; Ladewig and Hart,

1980). Volatile molecules may also have access to vomeronasal receptors, although this has not been directly examined.

Thus, in mammals, the vomeronasal organ seems often to lie in close proximity to the outside world, specialized mechanisms have evolved to transport chemicals to the organ, and large, nonvolatile molecues do have access to the interior of the organ. These characteristics may be partly conservative characteristics in the evolution of the vomeronasal organ. In primitive vertebrates, apparently homologous tissues are contact chemical sense systems; in present-day snakes, for example, the vomeronasal organ opens into the buccal cavity and stimuli are transferred to it by the tongue (Halpern, 1982).

The anatomy of the peripheral olfactory system is much more widely known and I will not dwell on it, but it is interesting to contrast this system with the vomeronasal system in at least one way. The olfactory receptors are primarily situated near the back of the nasal cavity behind groups of turbinates and above the main airstream. Sniffing presumably aids the olfactory process by producing stronger eddy currents so that more air gets into contact with this convoluted receptor surface. Although it is likely that large molecules can gain access to these receptors, such molecules would have to be transported there in the mucus solution covering the epithelium, and such transport would take a considerable period of time. The experiments that demonstrated the access of large molecules to the vomeronasal organ indicated very little invasion of the olfactory mucosa (Wysocki et al., 1980; Ladewig and Hart, 1980).

Neuroanatomical Projections

Perhaps the major stimulus to the recent explosion of interest in the vomeronasal system was the discovery that the central projections of the two systems were distinct and almost entirely nonoverlapping, suggesting quite separate functions (Scalia and Winans, 1975). Anatomical work has continued to provide insight and to stimulate hypotheses about the two systems. In Figure 1, I have summarized the projections of the two systems; this scheme is based primarily on what is known about hamsters, but at this level of analysis it seems to be valid for the other mammalian species that have been investigated, such as rats and rabbits.

The receptor neurons in the olfactory epithelium send their axons in a roughly topographic way to the main olfactory bulb. The projections of the main olfactory bulb consist of connections to a diversity of areas, including the anterior olfactory nucleus, the anterior hippocampus, the olfactory tubercle, the olfactory cortex (piriform cortex), the entorhinal area, the nucleus of

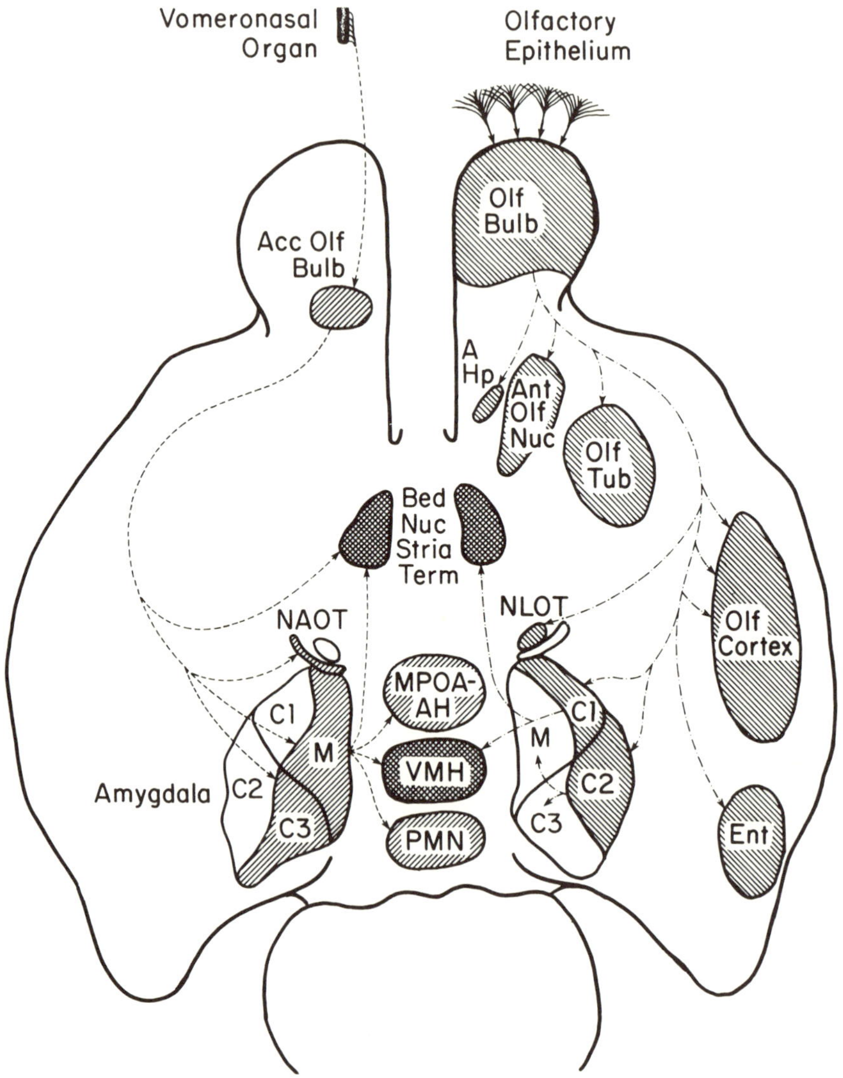

Figure 1. Projections of the olfactory and vomeronasal systems. Shown are all primary projections of the main olfactory bulb and accessory olfactory bulb, but only selected secondary projections. This generalized scheme is derived primarily from research on hamsters and rats. A Hp, anterior hippocampus; Acc Olf Bulb, accessory olfactory bulb; Ant Olf Nuc, anterior olfactory nucleus; Bed Nuc Stria Term, bed nucleus of the stria terminalis; Ent, entorhinal area; MPOA-AH, medial preoptic area–anterior hy-

the lateral olfactory tract, and two nuclei in the amygdala, the anterior cortical nucleus and the posterolateral cortical nucleus. It is interesting to note that within the amygdala these olfactory projection areas project in turn to the vomeronasal projection areas (M and C3) but not vice versa, suggesting that the olfactory system may modulate vomeronasal input or response systems but that the vomeronasal system may have little influence on olfactory processing (Kevetter and Winans, 1981b). There are two other relevant points to make about the third-order projections of the olfactory mucosa. First, one of the major projections of the olfactory cortex is to the mediodorsal nucleus of the thalamus, which in turn projects to prefrontal cortex. Lesions of the mediodorsal nucleus interfere with olfactory learning tasks, the only lesions of central structures that have so far been shown to have such effects (e.g., Sapolsky and Eichenbaum, 1980). Second, the olfactory system sends a variety of projections to hypothalamic regions, but, with the exception of the ventromedial nucleus, all of these third-order projections seem to be to the lateral hypothalamus (Scott and Leonard, 1971; Kevetter and Winans, 1981b).

In contrast, the vomeronasal system has a much more limited projection system (Figure 1; Scalia and Winans, 1975; Kevetter and Winans, 1981a). Cells originating in the accessory olfactory bulb send their axons primarily to the medial and posteromedial cortical nuclei of the amygdala and send lesser projections to the nucleus of the accessory olfactory tract and to the bed nucleus of the stria terminalis. The primary outputs from the vomeronasal amygdala arise in the medial cortical nucleus (M) and project to the bed nucleus of the stria terminalis and three medial hypothalamic areas, namely the medial preoptic-anterior hypothalamic region, the ventromedial nucleus, and the premammilary nucleus. The posteromedial nucleus of the amygdala apparently just projects back to the medial cortical nucleus (Kevetter and Winans, 1981a). This relatively specific set of projections suggests that the functions of the vomeronasal system may be more limited than those of the olfactory system. In addition, the exclusiveness of the projection of the medial nucleus of the amygdala to the medial hypothalamus and bed nucleus of the

---

pothalamus; NAOT, nucleus of the accessory olfactory tract; NLOT, nucleus of the lateral olfactory tract; Olf Cortex, olfactory cortex; Olf Tub, olfactory tubercle; PMN, premammilary nucleus; VMH, ventromedial nucleus of the hypothalamus. The nuclei of the amygdala are: C1, anterior cortical nucleus; C2, posterolateral cortical nucleus; C3, posteromedial cortical nucleus; M, medial nucleus.

stria terminalis suggests that the vomeronasal system has a more important role in regulation of neuroendocrine function than the olfactory system, or at least that a greater percentage of the vomeronasal system is taken up with such functions. A variety of experimental evidence indicates the importance of these medial areas for control of the anterior pituitary, such as brain stimulation eliciting pituitary hormone release (Beltramino and Taleisnik, 1978 and 1980), concentration of steroid hormone-binding cells in these regions (Pfaff and Kiner, 1973; Stumpf and Grant, 1975), and location of releasing hormones in these regions (Witkin et al., 1982).

Neuroendocrine Functions of the Vomeronasal System

Before 1975 there were no known functions for the vomeronasal system and consequently much recent research has involved attempting to discover such functions. Below I will summarize what is known about the neuroendocrine functions and then move on to the behavioral level of analysis.

It has been known for more than 20 years that odors are instrumental in causing a number of endocrinological changes among mammals, including alterations in estrous cycles, acceleration and retardation of puberty, and disruption of pregnancy. Whereas most of these effects require exposure to odors for several days, odors can also cause short latency release of hormones into the circulation. One example is the elicitation of LH surges in male mice in response to females or their odors (Maruniak and Bronson, 1976; Johnston and Bronson, 1982). In my own experiments, 40 sexually naive male mice were implanted with an indwelling cannula so that repeated blood samples could be withdrawn from awake, behaving animals with little disturbance. Our basic paradigm was to obtain five successive blood samples at 5-min intervals to establish baseline LH levels and then to spray 0.1 ml of urine as an aerosol into the male's home cage. We then obtained five more samples at 5-min intervals to determine whether there was a change in LH levels.

In Figure 2, I have shown some representative results. Urine from intact or ovariectomized females caused elevation of LH levels in the first sample after delivery of the stimulus, a delay of ~ 2.5 min. In contrast, urine from males or hypophysectomized females did not cause such LH surges. Thus chemical cues from females are sufficient to cause LH surges in males, and the factor in the urine that causes such responses is dependent on pituitary hormones but not ovarian hormones. An additional experiment indicated that exposure of males to hypophysectomized females themselves did not result in elevations in LH, suggesting that the pituitary may be necessary for such responses (Johnston and Bronson, 1982). It should be pointed out that the method of stimulus delivery constituted a contact situation, since the urine was sprayed into a very small area (29 × 14 × 14 cm), usually directly over the male.

It has subsequently been shown that the elevations of testosterone that are presumably caused by the prior LH surges are dependent on the presence of the vomeronasal organ (Wysocki et al., 1983). Surgical removal of the organ eliminated elevations in testosterone that occurred in sham-operated males upon exposure to anesthetized females and their bedding material. The role of the olfactory system was not investigated.

In Table II, I have summarized recent data on the primer pheromone effects that have been investigated to determine the possible role of the vomeronasal system. In addition to the example just discussed, all of the primer effects so far examined are dependent on a functional vomeronasal system, including the suppression of estrous cycles by cues from females (Reynolds and Keverne, 1979), acceleration of puberty onset in young female mice by cues from males (Kaneko et al., 1980; Lomas and Keverne, 1982), pregnancy block by exposure of the females to strange males (Bellringer et al., 1980; Lloyd-Thomas and Keverne, 1982), and reflex ovulation in rats maintained in constant estrus by exposure to continuous bright light (Johns et al., 1978). The primary weakness in this set of experiments is that in only one case, that of the pregnancy block effect, has it been shown that the olfactory system was *not* necessary. Nonetheless, it is striking that in every case that has so far been investigated, the presence of a functional vomeronasal organ and/or accessory olfactory system was necessary for the effects to occur.

In several of the examples in Table II, it is likely that nonvolatile signals are involved and the receiver must contact the chemical stimulus, for example, in the cases of puberty acceleration of female mice (Vandenbergh, 1975) and of reflex ovulation of constantly estrous rats (Johns et al., 1978). In the other cases, this question has not been adequately investigated.

Finally, it is valuable to ask whether experience is relevant in the mediation of these effects. In some cases, such as estrous cycle suppression, puberty acceleration, and pregnancy block, it is not clear what the relevant experience would be; the question doesn't seem to apply. In two cases, LH responses in male mice and reflex ovulation in constantly estrous female rats, sexual experience would seem to be a relevant variable, but sexually naive animals demonstrate the effects nonetheless.

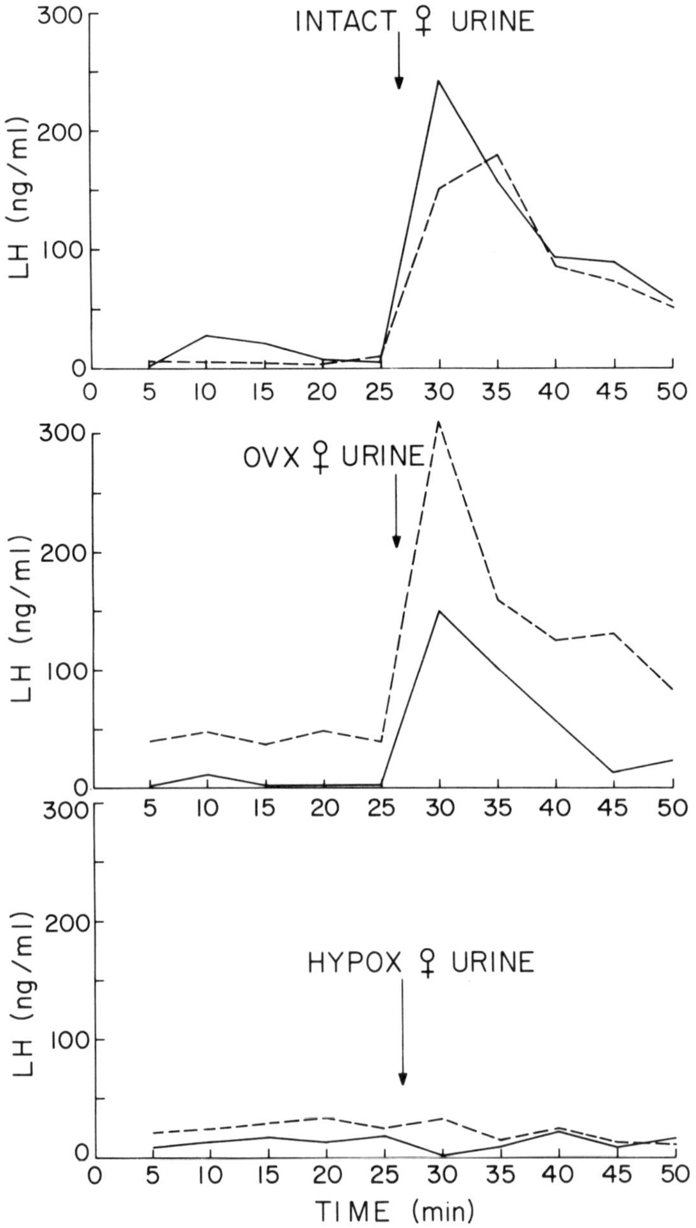
INTACT ♀ URINE
OVX ♀ URINE
HYPOX ♀ URINE
LH (ng/ml)
LH (ng/ml)
LH (ng/ml)
300
200
100
0
0 5 10 15 20 25 30 35 40 45 50
5 10 15 20 25 30 35 40 45 50
TIME (min)

Copulatory Behavior of Male Hamsters: Chemical Signals and Vomeronasal System Function

Murphy and Schneider (1970) demonstrated that removal of the olfactory bulb, and thus removal of both olfactory and vomeronasal input, eliminated mating in male hamsters. The first demonstration of a function for the vomeronasal system came when Powers and Winans (1975) showed that elimination of vomeronasal input in male hamsters caused severe mating deficits in ~ 40% of the operated animals. Elimination of the olfactory input alone had no effect, suggesting that the vomeronasal system has an important role in the elicitation of copulatory attempts of male hamsters. It is not entirely clear why

Table II. *Sensory Basis of the Investigated Primer Pheromone Effects*

Physiological effect	*Vomeronasal necessary?*	*Olfactory necessary?*
Suppression of estrous cycles by other females (mice)[1]	Yes	?
Acceleration of puberty in females by males (mice)[2]	Yes	?
Block of pregnancy by strange males (mice)[3]	Yes	No
Female elicited LH and testosterone surges in males (mice)[4]	Yes	?
LH release and reflex ovulation in constantly estrous female rats[5]	Yes	?

1. Reynolds and Keverne, 1979.
2. Kaneko et al., 1980; Lomas and Keverne, 1982.
3. Bellringer et al., 1980; Lloyd-Thomas and Keverne, 1982.
4. Wysocki et al., 1983.
5. Johns et al., 1978.

Figure 2. Representative records of luteinizing hormone (LH) levels in individual male mice before and after exposure to urine stimuli. Data from six individuals are shown, two in each odor condition (Johnston and Bronson, 1982).

the behavior of only 40% of the males was affected. Powers and Winans eliminated vomeronasal input by cutting the nerve from the organ to the accessory olfactory bulb, and in so doing they necessarily caused some damage to the main olfactory bulb. They did note that the extent of the lesion was correlated with the extent of the behavioral deficit but they concluded that this variable seemed relevant only in a limited percentage of cases (Winans and Powers, 1977). Nonetheless, elimination of vomeronasal input by means that did not damage olfactory structures might be valuable in clearing up this difficulty, and we have just begun a series of experiments of this type. We are also analyzing behavior in more detail so that we may be able to get a better idea of the nature of the processes subserved by the vomeronasal and olfactory systems.

Before presenting this new data, however, it is important to review what we know about the cues that influence male copulatory behavior in hamsters. During the course of a sexual encounter males spend a lot of time sniffing and licking females, especially the genital region, before mounting. Vaginal secretions produced by the female are known to have aphrodisiac effects on males, as illustrated in Figure 3. In this experiment males were paired with castrated males that were either untreated or had vaginal secretions added to their genital regions. Sexually naive males engaged in much more frequent mounting attempts toward castrated stimulus males if the latter were scented with vaginal secretions (Johnston, 1975). The low level of sexual behavior that was directed at the unscented castrates seemed to be due to experience with the scented castrates, since there were virtually no mounting attempts shown toward the unscented castrates before males encountered scented castrates. The components of vaginal secretions that have aphrodisiac effects are not yet known. If vaginal secretions are divided into volatile and nonvolatile components by vacuum distillation, both volatile and nonvolatile components have an effect on mounting but the nonvolatile fraction is more potent (Johnston, 1977; Frey, 1978; Singer et al., 1980).

We have recently discovered that other odor cues are also involved in sexual arousal. Four groups of males (n = 11 or 12 per group) were tested with naturally cycling, estrous females that differed in odor due to removal of one or more sources of scent, such as the vagina (VAGX females), the vagina plus flank glands and Harderian glands (3X females) or all of these and the ear glands (4X females). In the "normal" control condition, the females were vaginectomized but had been scented with vaginal secretions from other females (VAGX + V. SECR). Males mounted females that lacked vaginal secretions (VAGX) less often than such control females, as one would predict from the known aphrodisiac effects of the secretions (Figure 4). Additional reductions in mounting occurred when the females' flank and Harderian glands were removed in addition to the vagina (3X females). When the ear glands were also removed (4X females), no further deficits were observed. Thus there are several sources of odors that influence male copulatory behavior. Nothing is known about the chemistry of these other glands. It seems possible that they are effective because they contain the same substances that give vaginal secretions their potency; a much more interesting possibility is that these glands contain different chemicals that provide different kinds of information and influence sexual performance in different ways.

Returning to the role of the vomeronasal system in mediating the sexual behavior of male hamsters, graduate student John Kirn and I have completed a pilot experiment that suggests a specific function. We eliminated vomeronasal input using Wysocki's method of surgical removal of the organ itself (Wysocki et al., 1982). It is often possible to determine during surgery whether you have removed the entire organ, but unless you get it all in one piece, it is not always possible to know if you removed it all. We have not yet completed histological examination of all the animals, but in those that we have done there was either

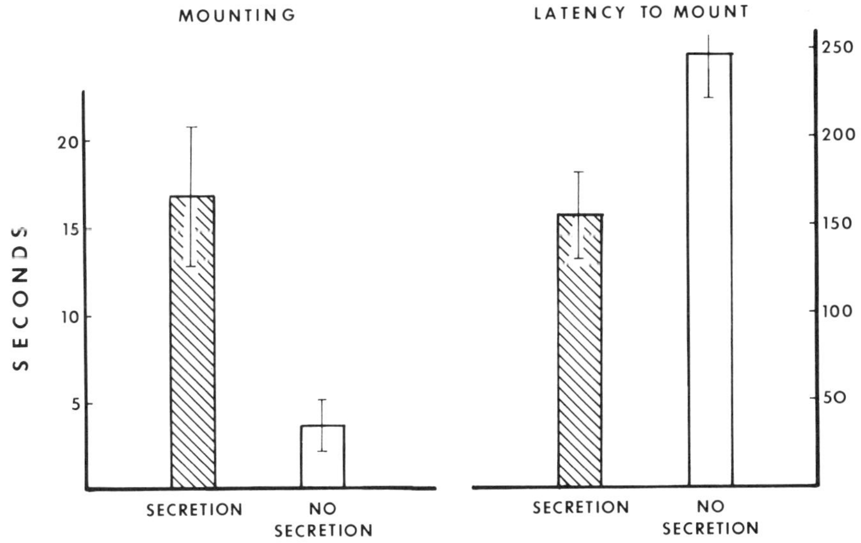

Figure 3. Sexual behavior of male hamsters toward castrated males that were either scented or not scented with vaginal secretions from females (Johnston, 1975).

extensive damage or total removal; in the few cases in which some of the organ remained, it was always the most anterior portion. Since the nerves from the anterior end run back along the organ it is unlikely that function remained in any of the experimental animals. There was no evidence of necrosis of the olfactory epithelium in animals with vomeronasal organ removals. Seven males with vomeronasal organ removal were compared with seven intact control animals; each male was tested once for 5 min with naturally estrous females.

Somewhat to our surprise we found that, for most measures of male behavior, there were no differences between the intact and operated males. As seen in Figure 5, there were no differences in the mean number of mounts, intromissions, or in intromission latency. The only difference we found in measures of sexual behavior was that males lacking the vomeronasal organ displayed a significantly greater latency to mount females than intact males did (Figure 5; $P = 0.007$, Mann-Whitney U test). Thus, initial sexual arousal, or, perhaps more specifically, the initiation of mounting behavior, seems to be impaired. Once these animals did mount, however, their copulatory performance was not distinguishable from that of intact males. One other difference

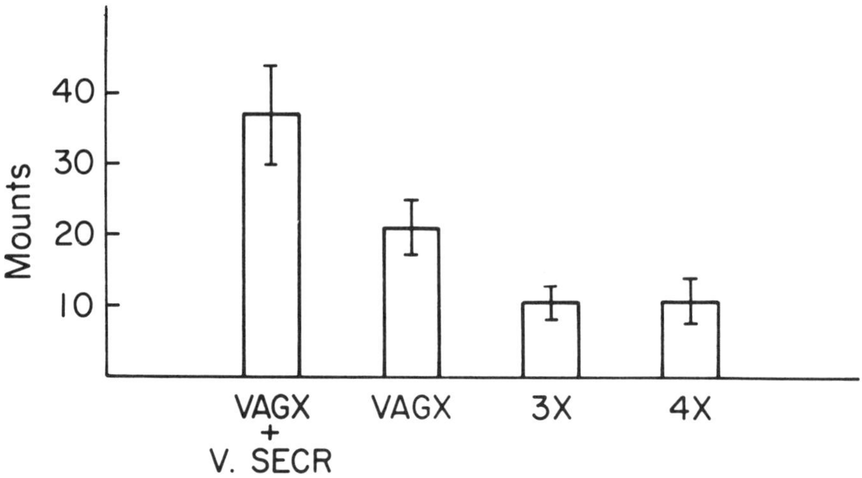

Figure 4. Number of mounts by male hamsters toward females that lacked various scent glands. Control females (VAGX + V. SECR) had been vaginectomized but were scented with vaginal secretions; VAGX females were vaginectomized; 3X females lacked vaginas and Harderian and flank glands; 4X females in addition lacked ear glands (Johnston, manuscript in preparation).

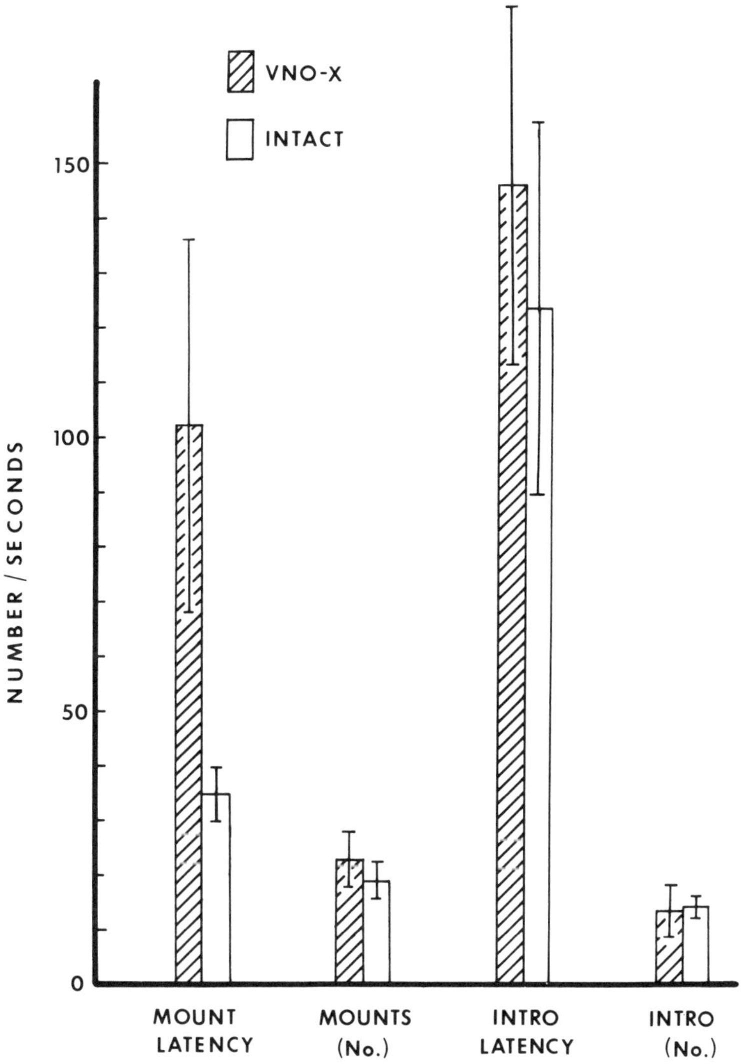

Figure 5. Sexual behavior of male hamsters that were either intact or had their vomeronasal organs removed (VNO-X) toward naturally cycling, estrous females.

between the intact and operated males may be important. Operated males spent more time sniffing the genital region of females (53.6 s) than intact males did (33.9 s, $P = 0.055$), whereas investigation of the head and body regions did not differ. It was as if the males that lacked the vomeronasal input persisted in sniffing the genital region to obtain the missing stimulation.

It should be noted that these findings contrast sharply with the original reports on the effects of vomeronasal deficiencies (Powers and Winans, 1975; Winans and Powers, 1977). We observed much less of an effect on male copulatory performance after removal of the organ than did Powers and Winans after sectioning the vomeronasal nerves; we have now observed 31 animals with vomeronasal lesions or removals and have observed that only two failed to mount naturally receptive females. As suggested above, one explanation for these differences may be that sectioning the vomeronasal nerves necessarily involved some damage to the main olfactory bulb and that this damage in combination with lack of vomeronasal input led to severe deficits (Winans and Powers, 1977). Recent data reported by Mike Meredith (1983b) indicate that sexual experience may be another important variable influencing the effects of such lesions. He found that sexually naive males showed much greater deficits than sexually experienced males. These results suggest that the vomeronasal system may mediate sexual arousal that has not been specifically conditioned by previous sexual experience, but that after a male has had some sexual experience with females the olfactory system may have a larger role in arousal via learned odor cues.

Olfactory System Functions

The unique functions of the olfactory system in reproductive activity have been to some extent ignored in recent years due to the excitement over discovering functions for the vomeronasal system. For example, it is recognized that both the olfactory and vomeronasal systems contribute to sexual arousal of male hamsters, but it is not known whether the two systems have redundant functions that influence arousal in the same way, or whether they have separate functions that influence different components of arousal. The olfactory system is necessary for attraction of male hamsters from a short distance to the odors of vaginal secretions (Powers et al., 1979), and thus may be important in the initial location of a mate. Since the volatiles in the secretion have some sexually arousing effects (Johnston, 1977; Frey, 1978; Singer et al., 1980), the olfactory system may also facilitate initial sexual arousal.

As suggested earlier, one type of function that may be uniquely a property of the olfactory system involves discriminations of complex odor stimuli, particularly those in which the significance of the two stimuli may depend on recent experience. We thought it would be interesting to investigate the role of the olfactory and vomeronasal systems in discrimination of individual females by males. One phenomenon that suggests the importance of individual discrimination is known as the Coolidge effect, in which a male is allowed to copulate to satiety with one female and then, when presented with another female, becomes rearoused and begins to copulate again. Male hamsters do show this effect (Bunnell et al., 1977), but when we started our experiments it was not clear whether it was due to discrimination of individual females or to discrimination between a female that had mated from one that had not. Also the relevance of the female's behavior was not known: a likely explanation for the effect could be that an unmated female would court the male much more strongly than one that he had already copulated with, and thus elicit more sexual behavior.

Our initial observations were designed to establish whether the Coolidge effect was due to discrimination of individual females by odor cues. Males were allowed to mate with a female until satiety, which we defined as no mounts within a period of 10 min; this criterion was reached after ~ 50–60 min of copulatory activity. At that point, males were removed from the mating arena and returned to their home cages for 8 min; then they were placed in a testing arena, a 2 × 3–ft, painted wooden box with a glass front to facilitate observation. In the test arena a male was confronted with two anesthetized females and allowed to interact with them for 10 min. This technique emphasized chemical cues over all other sources of information and eliminated the influence of the females' behavior. The simultaneous choice paradigm also provided an extremely sensitive measure of the male's motivation (Johnston, 1981). Three groups of 10 males were tested in three different choice situations. The first group was presented with a choice between the female they had just mated with (Same Female) vs. a new, unmated female (Fresh Female); this condition is thus a replication of the basic phenomenon using our restricted test situation. The second group was tested with the Same Female vs. one that had just mated with another male (Mated Female); this condition tests whether the reinvigoration of the males depended on discrimination of differences between individual females or between mated vs. unmated females. Finally, a third group was tested with a choice between a Fresh Female and a Mated Female; this condition tests the importance of the odors of mating in a situation in which both females are novel to the male.

A summary of the results of the experiment is shown in Figure 6 (Johnston,

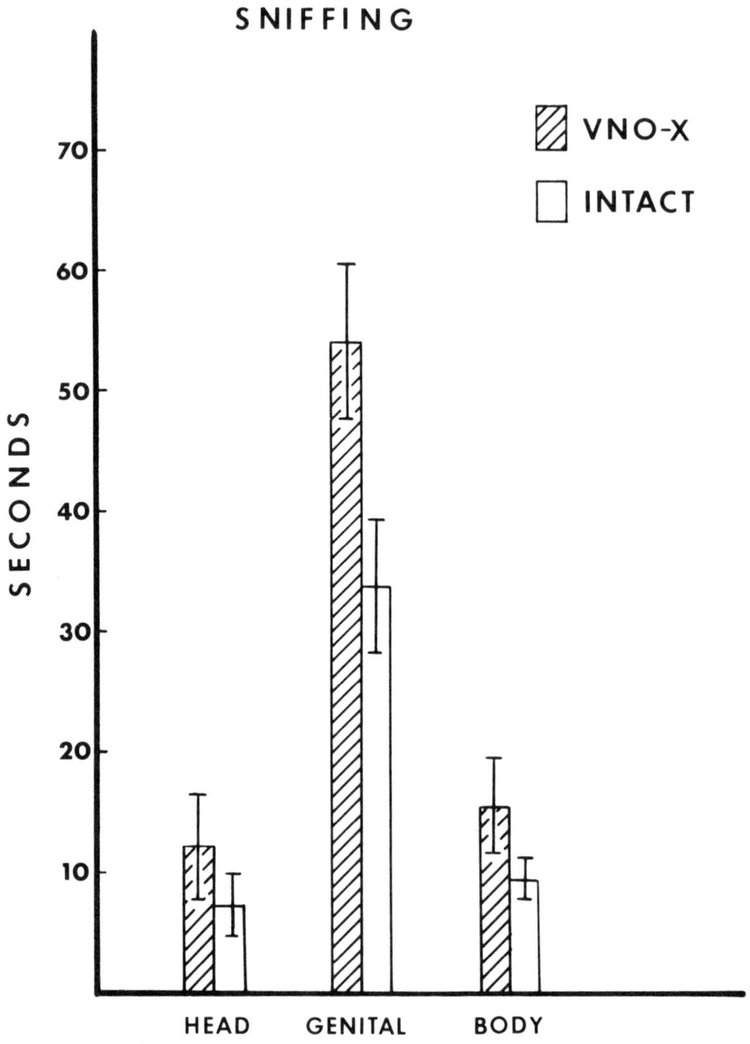

Figure 6. Sniffing investigation of females by males that were intact or had their vomeronasal organs removed.

1983; Johnston and Rasmussen, 1984). We recorded several measures of investigatory activity as well as mounting of the anesthetized females; since all of the individual measures show the same differences, I have combined them into a composite measure of male interest in females. It can be seen that males in the first group directed much more of their investigatory activity and mounting toward the Fresh Female than toward the female they had just mated with (Same Female). Thus, our test situation yields results similar to those observed when males are tested with awake females, indicating that chemical cues are sufficient for the Coolidge effect to be observed. The second group of males also demonstrated a strong preference for females that were new to them, even though such females had recently mated with other males, suggesting that the important property of a female in arousing sated males is individual identity and not recent mating history. This conclusion is strengthened by the results of the third group. These males were equally aroused by Fresh Females and novel, Mated Females. These experiments indicate that sexually satiated male hamsters are rearoused by chemical information from individuals with which they have not recently mated.

We investigated the basis of the olfactory discrimination in several additonal experiments (Johnston, 1983; Johnston and Rasmussen, 1984). An objection that could be raised to the experiments above is that males were not discriminating between two females, but rather were not attracted to a female that had the male's odor mixed with her own. We tested this idea by giving satiated males a choice between two Fresh Females, one of which had been rubbed with scent from the male. This manipulation had no effect: males were equally aroused by the scented and unscented females. Thus it seems that the most important variable is the difference in the smell of individual females. We also attempted to discover the source of odors that allowed individual discrimination by giving satiated males a choice between scents from different parts of the female's body. Instead of being presented with two anesthetized females, males were presented with two 3 × 5–in glass plates that had been scented with odors from Same Females or Fresh Females. Three groups of males were presented with odors from (a) the head region, presumably including secretions from the sexually dimorphic ear and Harderian glands, (b) the flank gland region, or (c) vaginal secretions. Males were not differentially attracted to the head or vaginal secretion scents, but spent significantly more time investigating the scent of the flank gland from Fresh Females than from Same Females. This intriguing result suggests that the flank gland is at least sufficient for individual discrimination and that it may be the most important source of such information.

Which sensory systems mediate sexual arousal when that arousal depends on discrimination of individuals? Previous research would suggest that the vomeronasal system is more important for sexual arousal than the olfactory system (Powers and Winans, 1975; Winans and Powers, 1977; Meredith, 1980), but my general theory of the functioning of these two systems suggests that individual discrimination should be an olfactory function.

We produced vomeronasal deficits in two ways. The main experiment was done by placing an electrode inside the nose directly over the organ and making a radio-frequency lesion (Johns et al., 1978); later we tested a second group of males in which we surgically removed the organ (Wysocki et al.,

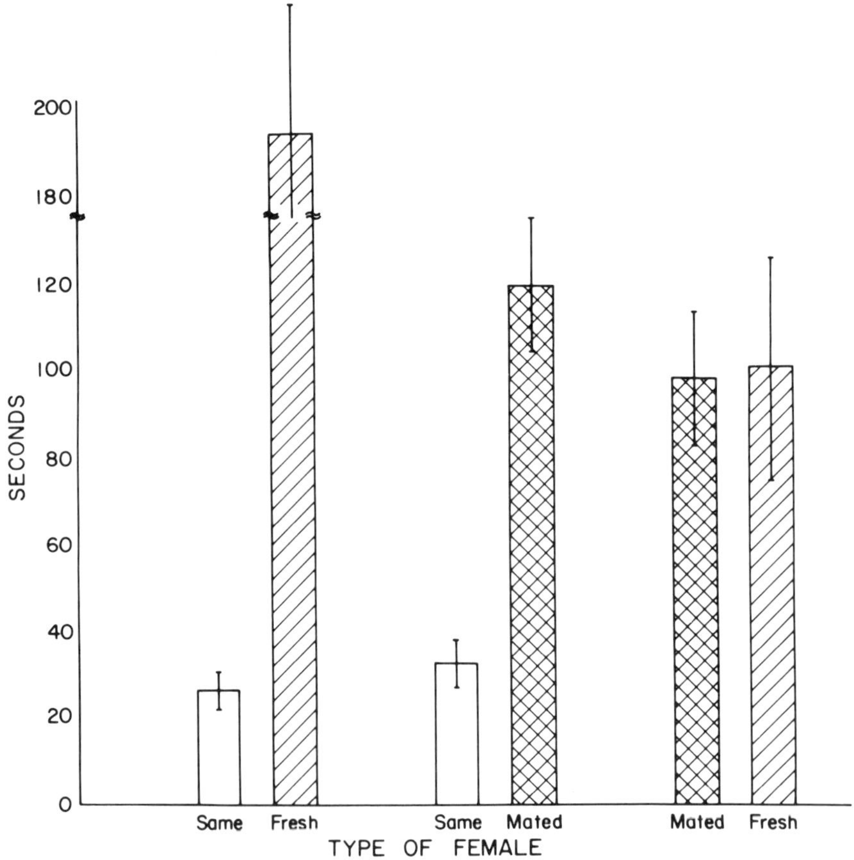

Figure 7. Total amount of investigation and sexual behavior shown by sexually satiated males toward two anesthetized females (see text for definitions of Same, Fresh, and Mated; Johnston, 1983; Johnston and Rasmussen, 1984).

1982). The results of vomeronasal lesions are shown in Figure 7 (Johnston, 1983; Johnston and Rasmussen, 1984). Both lesioned males and those with sham lesions continued to demonstrate a strong preference for Fresh Females and continued to be aroused by them but not by Same Females. Subsequent histology revealed that all males suffered severe damage to the vomeronasal organ; some lesions cauterized the whole organ while the least effective lesions destroyed the anterior half of the organ. Since these less effective lesions presumably interfered with access of substances to the organ we felt reasonably confident that these organs were nonfunctional, especially since there was no correlation between extent of damage and the behavior of the males. We ran a second experiment in which the vomeronasal organs of six males were surgically removed. Once again, sexually satiated males with vomeronasal organs surgically removed continued to prefer the Fresh Female (76.5-s investigating and mounting) to the Same Female (13.2s, $P < 0.01$, Wilcoxan Signed Rank test). Thus the vomeronasal organ and accessory olfactory system do not appear to influence sexual arousal in satiated males. In other words, when sexual arousal depends on discrimination or recognition of the presence of a new female, the vomeronasal system is not involved.

Next we examined the role of the olfactory system in mediating arousal in satiated males. The nasal cavities were irrigated with $ZnSO_4$, which temporarily destroys most of the peripheral olfactory receptors and may render the rest nonfunctional (Winans and Powers, 1977). Control males were treated with saline solution, which does no damage to the receptors. The results for 10 experimental males and 10 sham controls are shown in Figure 8. The experimental males treated with $ZnSO_4$ showed no preference for Fresh Females over Same Females, and indeed showed little evidence of interest in either type of partner. It is important to remember that these males all mated vigorously with their initial partner, but after satiation none of them engaged in any mounting behavior, although some did sniff the females. In contrast the controls treated with saline demonstrated the usual preference for Fresh Females. Thus elimination of main olfactory input also eliminates the sexual arousal of satiated males that is dependent on the discrimination of a novel female.

Several other studies suggest similar social discrimination functions for the olfactory system in hamsters. Murphy (1980) showed that the preference of male hamsters for anesthetized conspecific females over heterospecific females was not dependent on a functional vomeronasal-accessory olfactory system but was eliminated when the entire olfactory bulb was removed. Thus the preference for conspecific females was probably based on information obtained through the olfactory system. Males' preferences for the odors of females over those of other males also seem to be dependent on the olfactory system, since lesions of the mediodorsal nucleus of the thalamus eliminate such preferences. This nucleus receives its input from the main olfactory bulb via the olfactory cortex (Sapolsky and Eichenbaum, 1980).

In summary, I have presented evidence, primarily from work with several species of rodents, that supports the theory of olfactory and vomeronasal organ function presented in Table I. Although many of the available data fit these hypotheses, much more information is needed before they can be thoroughly assessed, even for a single species. The golden hamster is the species most thoroughly studied at the behavioral level of analysis, but many questions still remain about functions mediated by the two systems. An apparent exception to the hypothesis that learning should be important in behaviors influenced via the olfactory system is the case of attraction of males to vaginal

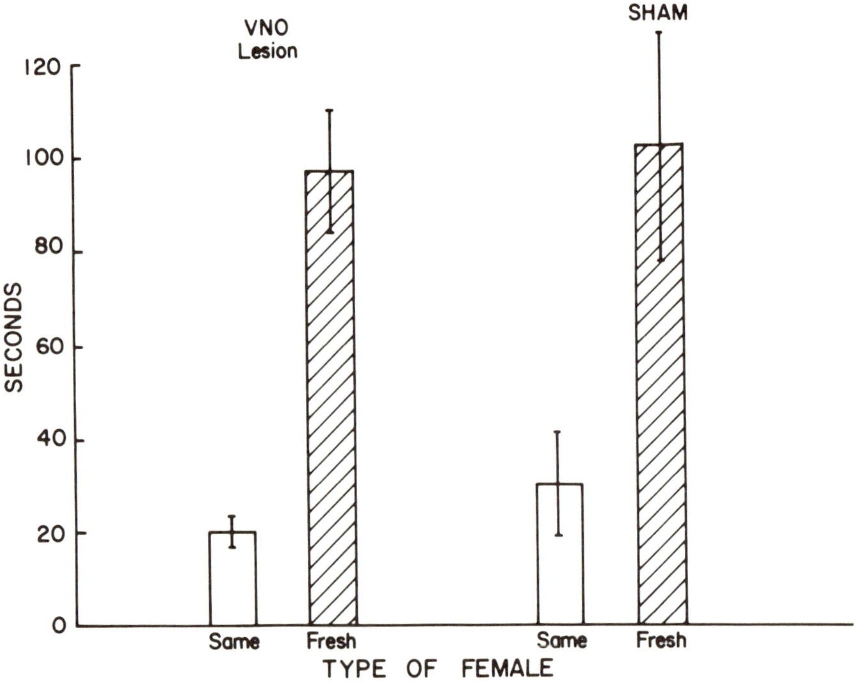

Figure 8. Total amount of investigation and sexual behavior shown by sexually satiated males with vomeronasal organ lesions or sham lesions toward two anesthetized females (Johnston, 1983; Johnston and Rasmussen, 1984).

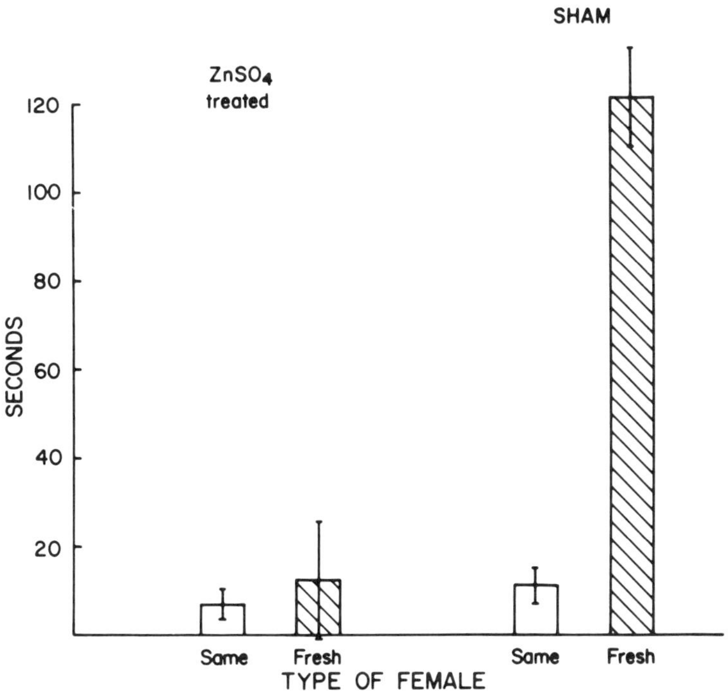

Figure 9. Total amount of investigation and sexual behavior shown by sexually satiated males that were intact (sham treatment) or had their olfactory system receptors destroyed by $ZnSO_4$ treatment. The males were tested with two simultaneously presented anesthetized females (Johnston, 1983; Johnston and Rasmussen, 1984).

secretion odors. This response is dependent on the olfactory system (Powers et al., 1979) but is not dependent on adult sexual experience (Johnston, 1974; Gregory et al., 1974). It is possible, however, that a different kind of experience is important. Pups do have experience with vaginal secretions of their mothers; in fact they pass through a stage at ~ 10–14 d of age when these odors are highly salient and attractive to them (Johnston and Coplin, 1979). It could be that this function of the olfactory system is, after all, dependent on learning, but the relevant experience occurs early in life. Other major gaps in our understanding of these two chemosensory systems in hamsters include the mechanisms underlying testosterone release in males to cues from females, the distinctive functions of the two systems in females, and the importance of the two systems for other behaviors such as scent marking, ultrasonic calling, and aggression. Although it is unclear how correct the hypotheses will be when applied to a wider variety of species, they should be valuable in organizing and stimulating research.

These ideas are, I hope, a fitting tribute to Carl Pfaffmann. He is and has been an innovative researcher and teacher, as well as an effective promoter of research in the chemical senses. He has had a major impact on research in this area, and, partly because of his influence, research in the chemical senses is beginning to have a major impact on our understanding of brain-behavior and brain-hormone relationships in general, rather than just those behaviors or hormones influenced by the chemical senses.

References

Bellringer, J. F., H. P. M. Pratt, and E. B. Keverne. 1980. Involvement of the vomeronasal organ and prolactin in pheromonal induction of delayed implantation in mice. *J. Reprod. Fertil.* 59:223–228.

Beltramino, C., and S. Taleisnik. 1978. Facilitatory and inhibitory effects of electrochemical stimulation of the amygdala on the release of luteinizing hormone. *Brain Res.* 144:95–107.

———. 1980. Dual action of electrochemical stimulation of the bed nucleus of the stria terminalis on the release of LH. *Neuroendocrinology.* 30:238–242.

Bunnell, B., B. D. Boland, and D. A. Dewsbury. 1977. Copulatory behavior of golden hamsters, *Mesocricetus auratus. Behaviour.* 61:180–206.

Estes, R. D 1972. The role of the vomeronasal organ in mammalian reproduction. *Mammalia.* 36:315–341.

Frey, K. F. 1978. Behavioral responses of male hamsters to relatively volatile and involatile fractions of the vaginal secretion. Master's dissertation, Cornell University.

Gregory, E., K. Engel, and D. Pfaff. 1974. Male hamster preference for odors of female hamster vaginal secretions: studies of experimental and hormonal determinants. *J. Comp. Physiol. Psychol.* 89:442–446.

Halpern, M. 1982. Nasal chemical senses in snakes. In *Advances in Vertebrate Neuroethology.* J. P. Ewert, R. R. Capranica, and D. J. Ingle, eds. New York: Plenum Publishing Corp., pp. 141–146.

Johns, M. A., H. H. Feder, B. R. Komisaruk, and A. D. Mayer. 1978. Urine-induced reflex ovulation in anovulatory rats may be a vomeronasal effect. *Nature* (*Lond.*). 272:446–448.

Johnston, R. E. 1974. Sexual attraction function of golden hamster vaginal secretion. *Behav. Biol.* 12:111–117.

———. 1975. Sexual excitation function of hamster vaginal secretion. *Anim. Learn. Behav.* 3:161–166.

———. 1977. Sex pheromones in golden hamsters. In *Chemical Signals in Vertebrates.* D. Muller-Schwarze and M. M. Mozell, eds. New York: Plenum Publishing Corp., pp. 225–249.

———. 1981. Attraction to odors in hamsters: an evaluation of methods. *J. Comp. Physiol. Psychol.* 95:591–960.

———. 1983. Mechanisms of individual discrimination in hamsters. In *Chemical Signals in*

Vertebrates III. D. Muller-Schwarze and R. M. Silverstein, eds. New York: Plenum Publishing Corp., pp. 245–258.

Johnston, R. E., and F. H. Bronson. 1982. Endocrine control of female mouse odors that elicit luteinizing hormone surges and attraction in males. *Biol. Reprod.* 27:1174–1180.

Johnston, R. E., and B. Coplin. 1979. Development of responses to vaginal secretion and other substances in golden hamsters. *Behav. Neural Biol.* 25:473–489.

Johnston, R. E., and K. Rasmussen. 1984. Individual recognition of female hamsters by males: role of chemical cues and the olfactory and vomeronasal systems. *Physiol. Behav.* 33:95–104.

Kaneko, N., E. A. Debski, M. C. Wilson, and W. K. Whitten. 1980. Puberty acceleration in mice. II. Evidence that the vomeronasal organ is a receptor for the primer pheromone in male mouse urine. *Biol. Reprod.* 22:873–878.

Kevetter, G. A., and S. S. Winans. 1981a. Connections of the corticomedial amygdala in the golden hamster. I. Efferents of the "vomeronasal amygdala." *J. Comp. Neurol.* 197:473–489.

———. 1981b. Connections of the corticomedial amygdala in the golden hamster. II. Efferents of the "olfactory amygdala." *J. Comp. Neurol.* 197:99–111.

Ladewig, J., and B. L. Hart. 1980. Flehmen and vomeronasal organ function in male goats. *Physiol. Behav.* 24:1067–1071.

Lloyd-Thomas, A., and E. B. Keverne. 1982. Role of the brain and accessory olfactory system in the block to pregnancy in mice. *Neuroscience.* 7:907–913.

Lomas, D. E., amd E. B. Keverne. 1982. Role of the vomeronasal organ and prolactin in the acceleration of puberty in female mice. *J. Reprod. Fertil.* 66:101–107.

Macrides, F., P. A. Johnston, and S. P. Schneider. 1977. Responses of the male golden hamster to vaginal secretion and dimethyl disulfide: attraction versus sexual behavior. *Behav. Biol.* 20:377–386.

Maruniak, J. A., and F. H. Bronson. 1976. Gonadotropic responses of male mice to female urine. *Endrocrinology.* 99:963–969.

Meredith, M. 1980. The vomeronasal organ and accessory olfactory system in the hamster. In *Chemical Signals in Vertebrates and Aquatic Invertebrates.* D. Muller-Schwarze and R. Silverstein, eds. New York: Plenum Publishing Corp., pp. 303–326.

———. 1983a. Sensory physiology of pheromone communication. In *Pheromones and Reproduction in Mammals.* J. G. Vandenbergh, editor. New York: Plenum Publishing Corp., pp. 199–252.

———. 1983b. Vomeronasal lesions before sexual experience impair male mating behavior in hamsters. Paper presented at the meeting of the Association for Chemoreception Sciences, Sarasota, Fl.

Meredith, M., and R. J. O'Connell. 1979. Efferent control of stimulus access to the hamster vomeronasal organ. *J. Physiol.* (*Lond.*). 286:301–316.

Murphy, M. R. 1980. Sexual preferences of male hamsters: importance of preweaning and adult experience, vaginal secretion, and olfactory or vomeronasal sensation. *Behav. Neural Biol.* 30:323–340.

Murphy, M. R., and G. E. Schneider. 1970. Olfactory bulb removal eliminates mating behavior in the male golden hamster. *Science* (*Wash. DC*). 167:302–304.

Pfaff, D., and M. Kiner. 1973. Atlas of estradiol-concentrating cells in the central nervous system of the female rat. *J. Comp. Neurol.* 151:121–158.

Powers, J. B., and S. S. Winans. 1975. Vomeronasal organ: critical role in mediating sexual behavior of the male hamster. *Science* (*Wash. DC*). 187:961–963.

Powers, J. B., R. B. Fields, and S. S. Winans. 1979. Olfactory and vomeronasal system participation in male hamsters' attraction to female vaginal secretions. *Physiol. Behav.* 22:77–84.

Reynolds, J. M., and E. B. Keverne. 1979. The accessory olfactory system and its role in pheromonally mediated suppression of oestrus. *J. Reprod. Fertil.* 57:31–35.

Sapolsky, R. M., and H. Eichenbaum. 1980. Thalamocortical mechanisms in odor-guided behavior. II. Effects of lesions of the mediodorsal thalamic nucleus and frontal cortex on odor preferences and sexual behavior in the hamster. *Brain Behav. Evol.* 17:276–290.

Scalia, F., and S. S. Winans. 1975. The differential projections of the olfactory bulb and accessory olfactory bulb in mammals. *J. Comp. Neurol.* 161:31–53.

Scott, J. W., and C. M. Leonard. 1971. The olfactory connections of the lateral hypothalamus in the rat, mouse and hamster. *J. Comp. Neurol.* 141:331–344.

Singer, A. G., F. Macrides, and W. C. Agosta. 1980. Chemical studies of hamster reproductive pheromones. In *Chemical Signals in Vertebrates and Aquatic Invertebrates.* D. Muller-Schwarze and R. Silverstein, eds. New York: Plenum Publishing Corp., pp. 365–375.

Stumpf, W. E., and L. D. Grant. 1975. *Anatomical Neuroendocrinology.* Basel: S. Karger.

Vandenbergh, J. G., J. M. Whitsett, and J. R. Lombardi. 1975. Partial isolation of a pheromone accelerating puberty in female mice. *J. Reprod. Fertil.* 43:515–523.

Winans, S. S., and J. B. Powers. 1977. Olfactory and vomeronasal deafferentation of male hamsters: histological and behavioral analyses. *Brain Res.* 126:325–344.

Witkin, J. W., C. M. Paden, and A.-J. Silverman. 1982. The luteinizing hormone-releasing hormone (LHRH) systems in the rat brain. *Neuroendocrinology.* 35:429–438.

Wysocki, C. J. 1979. Neurobehavioral evidence for the involvement of the vomeronasal system in mammalian reproduction. *Neurosci. Biobehav. Rev.* 3:301–341.

———. 1982. The vomeronasal organ: its influence upon the reproductive behavior and underlying endocrine systems. In *Olfaction and Endocrine Processes.* W. Breipohl, ed. London: IRL Press.

Wysocki, C. J., Y. Katz, and R. Bernhard. 1983. Male vomeronasal organ mediates female-induced testosterone surges in mice. *Biol. Reprod.* 28:917–922.

Wysocki, C. J., J. Nyby, G. Whitney, G. K. Beauchamp, and Y. Katz. 1982. The vomeronasal organ: primary role in mouse chemosensory gender recognition. *Physiol. Behav.* 29:315–327.

Wysocki, C. J., J. L. Wellington, and G. K. Beauchamp. 1980. Access of urinary nonvolatiles to the mammalian vomeronasal organ. *Science* (*Wash. DC*). 207:781–783.